Arturo Locatelli

Optimal Control

An Introduction

Springer Basel AG

Author

Arturo Locatelli
Dipartimento di Elettronica e Informazione
Politecnico di Milano
Piazza L. da Vinci 32
20133 Milano
Italy
e-mail: locatell@elet.polimi.it

2000 Mathematical Subject Classification 49-01

A CIP catalogue record for this book is available from the
Library of Congress, Washington D.C., USA

Deutsche Bibliothek Cataloging-in-Publication Data
Locatelli, Arturo:
Optimal control : an introduction / Arturo Locatelli. - Basel ; Boston ; Berlin :
Birkhäuser, 2001
ISBN 978-3-0348-9519-4 ISBN 978-3-0348-8328-3 (eBook)
DOI 10.1007/978-3-0348-8328-3

Originally published by Birkhäuser Verlag in 2001
Softcover reprint of the hardcover 1st edition 2001
Cover design: Micha Lotrovsky, CH-4106 Therwil, Switzerland
Printed on acid-free paper produced of chlorine-free pulp. TCF ∞

ISBN 978-3-0348-9519-4

9 8 7 6 5 4 3 2 1 www.birkhasuer-science.com

to Franca
and my parents

Contents

Preface ix

1 **Introduction** 1

I Global methods

2 The Hamilton-Jacobi theory

2.1 Introduction 9

2.2 Global sufficient conditions 10

2.3 Problems 18

3 The LQ problem

3.1 Introduction 21

3.2 Finite control horizon 22

3.3 Infinite control horizon 39

3.4 The optimal regulator 44

3.4.1 Stability properties 59

3.4.2 Robustness properties 71

3.4.3 The cheap control 76

3.4.4 The inverse problem 82

3.5 Problems 89

4 The LQG problem

4.1 Introduction 91

4.2 The Kalman filter 93

4.2.1 The normal case 93

4.2.2 The singular case 104

4.3 The LQG control problem 112

4.3.1 Finite control horizon 112

4.3.2 Infinite control horizon 115

4.4 Problems 122

5 The Riccati equations

5.1 Introduction 125

5.2 The differential equation 125
5.3 The algebraic equation 127
5.4 Problems 143

II Variational methods

6 The Maximum Principle
6.1 Introduction 147
6.2 Simple constraints 149
6.2.1 Integral performance index 151
6.2.2 Performance index function of the final event 164
6.3 Complex constraints 180
6.3.1 Nonregular final varieties 181
6.3.2 Integral constraints 184
6.3.3 Global instantaneous equality constraints 187
6.3.4 Isolated equality constraints 190
6.3.5 Global instantaneous inequality constraints 196
6.4 Singular arcs 200
6.5 Time optimal control 205
6.6 Problems 216

7 Second variation methods
7.1 Introduction 221
7.2 Local sufficient conditions 222
7.3 Neighbouring optimal control 235
7.4 Problems 246

A Basic background
A.1 Canonical decomposition 249
A.2 Transition matrix 252
A.3 Poles and zeros 253
A.4 Quadratic forms 254
A.5 Expected value and covariance 255

B Eigenvalues assignment
B.1 Introduction 259
B.2 Assignment with accessible state 260
B.3 Assignment with inaccessible state 267
B.4 Assignment with asymptotic errors zeroing 272

C Notation

Bibliography 283

List of Algorithms, Assumptions, Corollaries, ... 287

Index 291

Preface

From the very beginning in the late 1950s of the basic ideas of optimal control, attitudes toward the topic in the scientific and engineering community have ranged from an excessive enthusiasm for its reputed capability of solving almost any kind of problem to an (equally) unjustified rejection of it as a set of abstract mathematical concepts with no real utility. The truth, apparently, lies somewhere between these two extremes. Intense research activity in the field of optimization, in particular with reference to robust control issues, has caused it to be regarded as a source of numerous useful, powerful, and flexible tools for the control system designer. The new stream of research is deeply rooted in the well-established framework of linear quadratic gaussian control theory, knowledge of which is an essential requirement for a fruitful understanding of optimization. In addition, there appears to be a widely shared opinion that some results of variational techniques are particularly suited for an approach to nonlinear solutions for complex control problems. For these reasons, even though the first significant achievements in the field were published some forty years ago, a new presentation of the basic elements of classical optimal control theory from a tutorial point of view seems meaningful and contemporary.

This text draws heavily on the content of the Italian language textbook "Controllo ottimo" published by Pitagora and used in a number of courses at the Politecnico of Milan. It is, however, not a direct translation, since the material presented here has been organized in a quite different way, or modified and integrated into new sections and with new material that did not exist in the older work. In addition, it reflects the author's experience of teaching control theory courses at a variety of levels over a span of thirty years. The level of exposition, the choice of topics, the relative weight given to them, the degree of mathematical sophistication, and the nature of the numerous illustrative examples, reflect the author's commitment to effective teaching.

The book is suited for undergraduate/graduate students who have already been exposed to basic linear system and control theory and possess the calculus background usually found in any undergraduate curriculum in engineering.

The author gratefully acknowledges the financial support of MURST (Project Identification and Control of Industrial Systems) and ASI (ARS-98.200).

Milano, Spring 2000

Chapter 1

Introduction

Beginning in the late 1950s and continuing today, the issues concerning dynamic optimization have received a lot of attention within the framework of control theory. The impact of optimal control is witnessed by the magnitude of the work and the number of results that have been obtained, spanning theoretical aspects as well as applications. The need to make a selection (inside the usually large set of different alternatives which are available when facing a control problem) of a strategy both rational and effective is likely to be one of the most significant motivations for the interest devoted to optimal control. A further, and not negligable, reason originates from the simplicity and the conceptual clearness of the statement of a standard optimal control problem: indeed it usually (only) requires specifying the following three items:

(a) The equations which constitute the model of the controlled system;

(b) The criterion, referred to as the performance index, according to which the system behaviour has to be evaluated;

(c) The set of constraints active on the system state, output, control variables, not yet accounted for by the system model.

The difficulties inherent in points (a) and (c) above are not specific to the optimization context, while the selection of an adequate performance index may constitute a challenging issue. Indeed, the achievement of a certain goal (clearly identified on a qualitative basis only) can often be specified in a variety of forms or by means of an expression which is well defined only as far as its structure is concerned, while the values of a set of parameters are on the contrary to be (arbitrarily) selected. However, this feature of optimal control problems, which might appear as capable of raising serious difficulties, frequently proves to be expedient, whenever it is suitably exploited by the de-

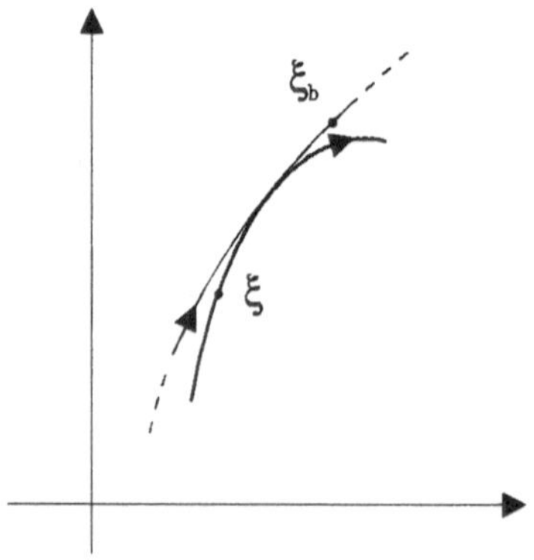

Figure 1.1: The *rendezvous* problem.

signer, in achieving a satisfactory trade-off among various, possibly conflicting, instances through a sequence of rationally performed trials.

The flexibility of optimal control theory together with the availability of suitable computing instruments has occasionally caused an excess of confidence in its capability to solve (almost all) problems, thus exposing it to severe criticisms and, as a reaction, giving rise to a similarly unjustified belief that it was a formally nice, but essentially useless, mathematical construction. The truth lying somewhere between these two extreme points, the contribution of optimal control theory can be evaluated in a correct way only if an adequate knowledge of its potentialities and limits has been acquired. In this perspective the motivation of the present book is twofold: from one side it aims to supply the basic knowledge of optimal control theory, while from the other side it provides the required background for the understanding of many recent and significant developments in the field (one for all, the control in Hardy spaces) which are undoubtedly and deeply rooted into such a theory.

A fairly precise idea of the kind of problems to be discussed here is given by the following simple examples.

Example 1.1 One of the less complicated statements of the so-called *rendezvous* problem makes reference to a pointwise object which is the target: its position is specified by a vector $\xi_b(t)$ of known functions of time. A second pointwise object must be driven, by suitably selecting a vector u of time functions (the control variables), so as to meet the target (see also Fig. 1.1). By denoting with $\xi(t)$ the position, at time t, of this second object, the *rendezvous* problem amounts to satisfying the relation $x(\tau) = x_b(\tau)$ at some time τ, where $x := \begin{bmatrix} \xi' & \dot{\xi}' \end{bmatrix}'$ and $x_b := \begin{bmatrix} \xi_b' & \dot{\xi}_b' \end{bmatrix}'$. Starting from this basic requirement a series of different scenarios can be set where the time τ is given or unspecified, equal or not equal to t_f, the (final) time when the control problem ends, t_f being given or unspecified. Furthermore, the control variables might be required to comply with a number of constraints either instantaneous (suited, for instance, to preventing the control from taking on arbitrarily large values) or integral (suited, for instance, to settling limitations on control resources, typically the available fuel). Also the performance index may be given various forms and hence substantially different meanings among which the duration of the control interval and global measures of the amount of involved control actions.

Three out of the many possible forms of *rendezvous* problems are now briefly presented in order to shed light on some typical aspects of optimal control problems: for this reason it is useless to mention explicitly the set of equations which describe the dynamic behaviour of the controlled system and constitute the main, always present, constraint.

Problem 1 The initial state x_0 and the initial time t_0 are given, while the final time t_f when the *rendezvous* takes place is free since it is the performance index J to be minimized. Thus

$$J = \int_{t_0}^{t_f} dt.$$

Besides the constraint $x(t_f) = x_b(t_f)$ (which is peculiar to the problem), other requirements can be set forth as, for instance, $u_m \leq u(t) \leq u_M$ which account for limits on the control actions.

Problem 2 The initial state x_0, the initial time t_0 and the final time t_f are given. The final state is only partially specified (for instance, the final position is given, while the final velocity is free inside a certain set of values) and the performance index aims at evaluating the global control effort (to be minimized) by means of an expression of the kind

$$J = \int_{t_0}^{t_f} \sum_{i=1}^{m} r_i u_i^2(t) dt, \; r_i > 0$$

where u_i is the i-th component of the control variable u. The peculiar constraint of the problem is $x(\tau) = x_b(\tau)$, where the time τ when the *rendezvous* takes place must satisfy the condition $t_0 < \tau < t_f$ and may or may not be specified.

Problem 3 This particular version of the *rendezvous* problem is sometimes referred to as the *interception* problem. The initial state x_0 may or may not be completely specified, while both the initial and final times t_0 and t_f are to be selected under the obvious constraint $t_0 < t_f \leq T$. The final state is free and the performance index is as in Problem 2. The peculiar constraint of the problem involves some of the state variables only (the positions), precisely, $\xi(\tau) = \xi_b(\tau)$, where the time τ when interception takes place may or may not be given and satisfies the condition $t_0 < \tau < t_f$.

Example 1.2 A typical *positioning* problem consists in transferring the state x of the controlled system from a given initial point P_0 to the neigbourhood of the point P_f with coordinates x_{fi}. The transfer has to accomplished in a short time by requiring control actions u of limited intensity. The problem is characterized by a given initial state x_0 and time t_0, while the performance index is given the form

$$J = \int_{t_0}^{t_f} \sum_{i=1}^{m} r_i u_i^2(t)dt + t_f + \sum_{i=1}^{n} s_i \left[x_i(t_f) - x_{fi}\right]^2, \ r_i > 0, \ s_i > 0.$$

In general the problem formulation may also include a set of constraints which set bounds on some of the state variables, typically, velocities and/or positions: they can be expressed by means of relations of the type $w(x(t), t) \leq 0$ to be satisfied along the whole control interval.

Example 1.3 A frequently encountered problem in space applications is the so-called *attitude* control problem which, in its simplest formulation, consists in orienting an object (satellite) in a specified direction, starting from an arbitrary wrong orientation. The problem can be stated in a way similar to the one in Example 1.2. Letting t_0 and t_f denote the extreme values (given or to be determined) of the control interval, a significant feature of this problem is the presence of integral constraints of the kind

$$\int_{t_0}^{t_f} w(u(t))dt \leq w^*$$

which emphasize the need that the required manoeuvre be accomplished by consuming a quantity of fuel not exceeding a given bound: the function w quantifies the instant consumption which is assumed to depend on the value of the control variable u applied at that time.

Example 1.4 The requirement of keeping the state x of a system as close as possible to x_d, after a disturbance has occurred, gives rise to an important class of optimal control problems. The perturbation δu to be given to the control u_d has to be determined in such a way as to make $\delta x := x - x_d$ small. Here u_d is the control which generates x_d in unperturbed conditions. In general, it is also desired that δu be small: hence these requirements can be taken into account by looking for the minimization of the performance index

$$J = \int_{t_0}^{t_f} \left[\sum_{i=1}^{n} q_i \delta x_i^2(t) + \sum_{i=1}^{m} r_i \delta u_i^2(t) \right] dt, \ q_i > 0, \ r_i > 0.$$

In this context, the initial time t_0 (when $\delta x \neq 0$ because of previous disturbances) and the final time t_f are given. The initial state $\delta x(t_0)$ is also given, even if generic, while the final state has to be considered as free.

A special yet particularly interesting case is the one where x_d is an equilibrium state. If the control interval is unbounded ($t_f \to \infty$), the problem can be seen as a *stabilization* problem since, at least from a practical viewpoint, the minimization of the performance index above amounts, whenever feasible, to asymptotically zeroing δx.

These examples suggest the structure of the optimal control problems to be dealt with in the following: actually, they are characterized by:

(a) The equations which describe the dynamic behaviour of the system. They are either differential or algebraic and their number is finite.

(b) The set of allowable initial states. It can entail the complete or partial specification of the initial state as well as its freedom.

(c) The set of allowable final states. It can entail the complete or partial specification of the final state as well as its freedom.

(d) The performance index. It is constituted by two terms which need not both be present: one of them is the integral of a function of the state, control and time, while the second one is a function of the final state and time.

(e) A set of constraints on the state and/or control variables. They can account for integral or instantaneous requirements and involve both equality or inequality relations.

(f) The control interval. It can be given or free, of finite or infinite duration.

The book has been structured into two parts which are devoted to *global methods* (sufficient conditions) and *variational methods* of the first order (necessary conditions) or second order (sufficient local conditions). The first part presents the so-called *Hamilton-Jacobi theory* which supplies sufficient conditions for global optimality together with its most significant achievements, namely the solution of the Linear Quadratic (LQ) and Linear Quadratic Gaussian (LQG) problems. Some emphasis is also given to the Riccati equations which play a particularly important role within the context of these problems. The second part of the book begins with the presentation of the so-called *Maximum Principle*, which exploits a first order variational approach and provides powerful necessary conditions suited to encompass a wide class of fairly complex problems. Successively, a second order variational approach is shown to be capable of guaranteeing, under suitable circumstances, the local optimality of solutions derived by requiring the fulfillment of the Maximum Principle conditions.

Part I

Global methods

Chapter 2

The Hamilton-Jacobi theory

2.1 Introduction

Results of a global nature will be presented relative to optimal control problems which are specified by the following six elements:

(a) The controlled system which is a continuous-time, finite dimensional dynamic system;

(b) The initial state x_0;

(c) The initial time t_0;

(d) The set of functions which can be selected as inputs to the system;

(e) The set of admissible final events (couples (x, t));

(f) The performance index.

The controlled system is described by the equation

$$\dot{x}(t) = f(x(t), u(t), t) \tag{2.1}$$

where $x \in R^n$, $u \in R^m$ and f is a continuously differentiable function. The boundary condition for eq. (2.1) is

$$x(t_0) = x_0 \tag{2.2}$$

and amounts to saying that the initial state is given, at a known initial time. The set $S_f \subseteq \{(x,t)|\ x \in R^n,\ t \geq t_0\}$ specifies the admissible final events, so that the condition (a constraint to be fulfilled)

$$(x(t_f), t_f) \in S_f \tag{2.3}$$

allows us to deal with a large variety of meaningful cases: *(i)* final time t_f given and final state free, ($S_f = \{(x,t)|\ x \in R^n, t = T = given\}$; *(ii)* final state given and final time free ($S_f = \{(x,t)|\ x = x_f = given, t \geq t_0\}$); and so on. The set of functions which can be selected as inputs to the system coincides with Ω^m, the space of m-vector functions which are piecewise continuous: thus the constraint $u(\cdot) \in \Omega^m$ must be satisfied. Finally, the performance index J (to be minimized) is the sum of an integral type term with a term which is a function of the final event, namely

$$J = \int_{t_0}^{t_f} l(x(t), u(t), t)dt + m(x(t_f), t_f) \tag{2.4}$$

where l and m are continuously differentiable functions.

In general, the performance index J depends on the initial state, the initial and final times and the whole time history of the state and control variables: thus a precise notation should be $J(x_0, t_0, t_f, x(\cdot), u(\cdot))$. However, for the sake of simplicity, only the items which are needed for a correct understanding will be explicitly displayed as arguments of J. For the same reason an analogously simplified notation will be adopted for other functions, whenever no misunderstanding can occur.

By denoting with $\varphi(t; t_0, x_0, u(\cdot))$ the solution at time t of eq. (2.1) with boundary condition (2.2) when the input is $u(\cdot)$, the optimal control problem to be discussed in the first part of the book can be stated as follows.

Problem 2.1 (Optimal control problem) *Determine a vector $u^o(\cdot) \in \Omega^m$ of functions defined on the interval $[t_0, t_f^o]$ in such a way that eq. (2.3) is satisfied and the performance index (2.4), evaluated for $x(\cdot) = \varphi((\cdot); t_0, x_0, u^o(\cdot))$ and $u(\cdot) = u^o(\cdot)$ over the interval $[t_0, t_f^o]$, takes on the least possible value.*

The next section will present the main results of the so-called *Hamilton-Jacobi theory*. In the given form they provide sufficient optimality conditions for only the above problem, even though they could have been stated so as to encompass a larger class of situations. Our choice has been made for the sake of keeping the discussion at the simplest possible level.

2.2 Global sufficient conditions

The presentation of the Hamilton-Jacobi theory relative to Problem 2.1 needs some preliminary, yet fairly intuitive, definitions.

Definition 2.1 (Admissible control relative to (x_0, t_0)) *The control $u(\cdot) \in \Omega^m$, defined on the interval $[t_0, t_f]$, $t_f \geq t_0$, is an admissible control relative to*

(x_0, t_0) for the system (2.1) and the set S_f if

$$(\varphi(t_f; t_0, x_0, u(\cdot)), t_f) \in S_f.$$

Definition 2.2 (Optimal control relative to (x_0, t_0)) *Let $u^o(\cdot) \in \Omega^m$ be a control defined on the interval $[t_0, t_f^o]$, $t_f^o \geq t_0$. It is an optimal control relative to (x_0, t_0) for the system (2.1), the performance index (2.4) and the set S_f if:*

(i) $(\varphi(t_f^o; t_0, x_0, u^o(\cdot)), t_f^o) \in S_f$;

(ii) $J(t_f^o, \varphi(\cdot; t_0, x_0, u^o(\cdot)), u^o(\cdot)) \leq J(t_f, \varphi(\cdot; t_0, x_0, u(\cdot)), u(\cdot))$.

Here $u(\cdot) \in \Omega^m$, defined on the interval $[t_0, t_f]$, is any admissible control relative to (x_0, t_0) for the system (2.1) and the set S_f.

Within the framework of optimal control theory a fundamental role is played by a function which is built up from two of the elements of the problem: the system to be controlled and the integral part of the performance index. The following definitions specify it together with some relevant characterizations.

Definition 2.3 (Hamiltonian function) *The function*

$$H(x, u, t, \lambda) := l(x, u, t) + \lambda' f(x, u, t) \tag{2.5}$$

where $\lambda \in R^n$, is the hamiltonian function (relative to the system (2.1) and the performance index (2.4)).

Definition 2.4 (Regularity of the hamiltonian function) *The hamiltonian function is said to be regular if, as a function of u, it admits, for each x, $t \geq t_0$, λ, a unique absolute minimum $u_h^o(x, t, \lambda)$, i.e., if*

$$H(x, u_h^o(x, t, \lambda), t, \lambda) < H(x, u, t, \lambda),\ \forall u \neq u_h^o(x, t, \lambda),$$
$$\forall x \in R^n,\ \forall t \geq t_0,\ \forall \lambda \in R^n. \tag{2.6}$$

Definition 2.5 (H-minimizing control) *Let the hamiltonian function be regular. The function u_h^o which verifies the inequality (2.6) is said to be the H-minimizing control.*

Example 2.1 With reference to eqs. (2.1), (2.4), let $f(x, u, t) = \sin(x) + tu$ and $l(x, u, t) = tx^2 + xu + u^2$, so that the hamiltonian function (2.5) is $H(x, u, t, \lambda) = tx^2 + xu + u^2 + \lambda(\sin(x) + tu)$. This function is regular and $u_h^o(x, t, \lambda) = -(x + t\lambda)/2$.

A partial differential equation (on which the sufficient conditions for optimality are based) can be defined, once the hamiltonian function has been introduced. It concerns a scalar function $V(z, t)$, where $z \in R^n$ and $t \in R$.

Definition 2.6 (Hamilton-Jacobi equation) *Let the hamiltonian function be regular. The partial differential equation*

$$\frac{\partial V(z,t)}{\partial t} + H(z, u_h^o(z,t,(\frac{\partial V(z,t)}{\partial z})'), t, (\frac{\partial V(z,t)}{\partial z})') = 0 \tag{2.7}$$

is the Hamilton-Jacobi equation (HJE).

A sufficient condition of optimality can now be stated.

Theorem 2.1 *Let the hamiltonian function (2.5) be regular and $u^o \in \Omega^m$, defined on the interval $[t_0, t_f^o]$, $t_f^o \geq t_0$, be an admissible control relative to (x_0, t_0), so that $(x^o(t_f^o), t_f^o) \in S_f$, where $x^o(\cdot) := \varphi(\cdot; t_0, x_0, u^o(\cdot))$. Let V be a solution of eq. (2.7) such that:*

(a1) *It is continuously differentiable;*

(a2) $V(z,t) = m(z,t), \ \forall (z,t) \in S_f$;

(a3)

$$u^o(t) = u_h^o(x^o(t), t, (\frac{\partial V(z,t)}{\partial z})'\Big|_{z=x^o(t)}), \ t_0 \leq t \leq t_f^o.$$

Then it follows that

(t1) $u^o(\cdot)$ *is an optimal control relative to* (x_0, t_0);

(t2) $J(x_0, t_0, t_f^o, x^o(\cdot), u^o(\cdot)) = V(x_0, t_0)$.

Proof. Define the function

$$\begin{aligned} l^*(z,u,t) &:= \frac{\partial V(z,t)}{\partial t} + H(z,u,t,(\frac{\partial V(z,t)}{\partial z})') \\ &= \frac{\partial V(z,t)}{\partial t} + l(z,u,t) + \frac{\partial V(z,t)}{\partial z} f(z,u,t) \end{aligned}$$

which is regular since it differs from the hamiltonian function because of a term which does not depend on u. Therefore it admits a unique absolute minimum in

$$u^*(z,t) = u_h^o(z,t,(\frac{\partial V(z,t)}{\partial z})') \tag{2.8}$$

and it follows that

$$l^*(z, u^*(z,t), t) = 0 \tag{2.9}$$

since V is a solution of eq. (2.7). If $u \in \Omega^m$, defined on the interval $[t_0, t_f]$, $t_f \geq t_0$, is any admissible control relative to (x_0, t_0) and x is the corresponding state motion, so that $(x(t_f), t_f) \in S_f$, then, in view of assumption *(a3)*, eqs.(2.8), (2.9) and recalling that l^* admits a unique absolute minimum in u^*, it follows that

$$\begin{aligned}\int_{t_0}^{t_f^o} l^*(x^o(t), u^o(t), t)dt &= \int_{t_0}^{t_f^o} l^*\ (x^o(t), u_h^o(x^o(t), t, (\frac{\partial V(z,t)}{\partial z})'\Big|_{z=x^o(t)}), t)dt \\ &= \int_{t_0}^{t_f^o} l^*(x^o(t), u^*(x^o(t), t), t)dt = 0 \\ &\leq \int_{t_0}^{t_f} l^*(x(t), u(t), t)dt. \end{aligned} \tag{2.10}$$

Assumption *(a1)* together with the definition of l^* imply that

$$\begin{aligned}\frac{dV(x^o(t), t)}{dt} &= \frac{\partial V(z,t)}{\partial t}\Big|_{z=x^o(t)} + \frac{\partial V(z,t)}{\partial z}\Big|_{z=x^o(t)} f(x^o(t), u^o(t), t) \\ &= l^*(x^o(t), u^o(t), t) - l(x^o(t), u^o(t), t), \end{aligned} \tag{2.11a}$$

$$\begin{aligned}\frac{dV(x(t), t)}{dt} &= \frac{\partial V(z,t)}{\partial t}\Big|_{z=x(t)} + \frac{\partial V(z,t)}{\partial z}\Big|_{z=x(t)} f(x(t), u(t), t) \\ &= l^*(x(t), u(t), t) - l(x(t), u(t), t) \end{aligned} \tag{2.11b}$$

Thanks to assumption *(a2)*, eqs. (2.10), (2.11) imply that

$$\begin{aligned} 0 &= J(x_0, t_0, t_f^o, x^o(\cdot), u^o(\cdot)) - V(x_0, t_0) \\ &\leq J(x_0, t_0, t_f, x(\cdot), u(\cdot)) - V(x_0, t_0) \end{aligned}$$

from which points *(t1)* and *(t2)* both follow.

Observe that $u^o(t),\ t_0 \leq \tau \leq t \leq t_f^o$ is obviously an admissible control relative to $(x^o(\tau), \tau)$ and that the initial event is absolutely arbitrary in the proof of Theorem 2.1. Therefore, $u^o(t),\ t_0 \leq \tau \leq t \leq t_f^o$ is optimal relative to $(x^o(\tau), \tau)$ and $J(x^o(\tau), \tau, t_f^o, x^o(\cdot), u^o(\cdot)) = V(x^o(\tau), \tau)$.

Example 2.2 Consider the electric circuit shown in Fig. 2.1. The optimal control problem computes the current u so as to waste the least possible energy in the resistor and bring the system, initially at rest ($x_1(0) = 0$ and $x_2(0) = 0$), to have, at a given time T, unitary voltage across the condenser and zero current flowing inside the inductor. Assuming that all components are ideal, the circuit equations are

$$\begin{aligned}\dot{x}_1 &= \frac{x_2}{L} \\ \dot{x}_2 &= \frac{u - x_1}{C}\end{aligned}$$

with $x(0) = 0$, while the performance index is

$$J = \int_0^T R\frac{u^2}{2}dt$$

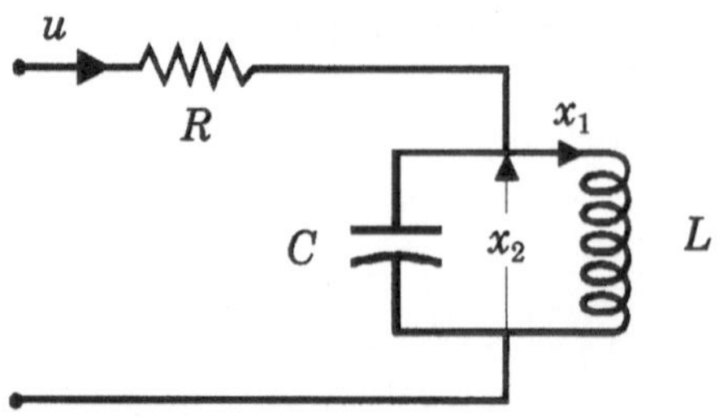

Figure 2.1: The electric circuit of Example 2.2.

where, for the sake of simplicity, $T := \pi/\omega$, $\omega := \sqrt{1/LC}$. Since both the final state and time are given, the set S_f is simply constituted by the single couple $([\,0 \quad 1\,]', T)$. It is easy to check that the control

$$u^o(t) = -2\frac{C}{T}\cos(\omega t),$$

which causes the state response

$$\begin{aligned} x_1^o(t) &= -\frac{1}{L\pi} t \sin(\omega t) \\ x_2^o(t) &= -\frac{1}{\pi}[\sin(\omega t) + \omega t \cos(\omega t)], \end{aligned}$$

is admissible. The hamiltonian function $H = R\frac{u^2}{2} + \lambda_1 \frac{x_2}{L} + \lambda_2 \frac{u - x_1}{C}$ is regular (the resistance R is positive) and the H-minimizing control is given by $u_h^o = -\frac{\lambda_2}{RC}$. Thus the HJE is

$$\frac{\partial V}{\partial t} + \frac{z_2}{L}\frac{\partial V}{\partial z_1} - \frac{z_1}{C}\frac{\partial V}{\partial z_2} - \frac{1}{2RC^2}\left(\frac{\partial V}{\partial z_2}\right)^2 = 0.$$

A solution of this equation and the relevant boundary condition

$$V(\begin{bmatrix} 0 \\ 1 \end{bmatrix}, T) = 0$$

is $V(z,t) = \alpha_1(t) z_1 + \alpha_2(t) z_2 + \beta(t)$, where

$$\begin{aligned} \alpha_1(t) &= \frac{2RC}{\pi}\sin(\omega t), \\ \alpha_2(t) &= \frac{2RC^2}{T}\cos(\omega t), \\ \beta(t) &= \frac{RC^2}{T}[1 + \frac{1}{T}(t + \frac{\sin(\omega t)\cos(\omega t)}{\omega})]. \end{aligned}$$

The optimality of u^o follows from Theorem 2.1.

It is worth emphasizing that Theorem 2.1 supplies an optimality condition which is sufficient only, so that if, for instance, point *(a3)* fails to hold corresponding to a certain solution of the HJE, we can not claim that the control at hand is not optimal. Indeed, it could well happen that there exist many solutions of such an equation and condition *(a3)* is satisfied corresponding to some of them and not satisfied corresponding to some others. This circumstance is made clear in the following example.

Example 2.3 Consider the first order system $\dot{x} = u$ with $x(0) = 0$ and the performance index

$$J = \int_0^1 2u^2 dt$$

to be minimized in complying with the requirement that $x(1) = 1$. The state response to the control $u^o(t) = 1,\ 0 \leq t \leq 1$ is $x^o(t) = t$, the hamiltonian function is $H = 2u^2 + \lambda u$ and regular with $u_h^o = -\lambda/4$ as H-minimizing control. Thus the HJE is

$$\frac{\partial V}{\partial t} - \frac{1}{8}(\frac{\partial V}{\partial z})^2 = 0.$$

The functions

$$\begin{aligned} V_1(z,t) &= -4z + 2(t+1), \\ V_2(z,t) &= \frac{2}{2-t}z^2 - \frac{8}{2-t}z + \frac{4+2t}{2-t}, \\ V_3(z,t) &= \frac{2}{2-t}z^2 - 2 \end{aligned}$$

solve the HJE and the relevant boundary condition $V(1,1) = 0$. However, we get

$$u_h^o(x^o(t), t, \left.\frac{\partial V_1(z,t)}{\partial z}\right|_{z=x^o(t)}) = 1 = u^o(t),\ 0 \leq t \leq 1,$$

$$u_h^o(x^o(t), t, \left.\frac{\partial V_2(z,t)}{\partial z}\right|_{z=x^o(t)}) = 1 = u^o(t),\ 0 \leq t \leq 1,$$

$$u_h^o(x^o(t), t, \left.\frac{\partial V_3(z,t)}{\partial z}\right|_{z=x^o(t)}) = -\frac{t}{2-t} \neq 1 = u^o(t),\ 0 \leq t \leq 1,$$

so that only the first two solutions allow us to conclude that u^o is optimal. Further, observe that V_1 and V_2 supply the correct optimal value of the performance index, namely, 2, while V_3 gives the wrong value -2.

Theorem 2.1 provides a mean of checking only whether a given control is optimal: however it is not difficult to restate it in such a way so as to allow us to determine an optimal control. This is shown in the forthcoming lemma the proof of which is not given since it requires only a little thought.

Corollary 2.1 *Let the hamiltonian function (2.5) be regular and* V *be a solution of the HJE (2.7) such that:*

(a1) *It is continuously differentiable;*

(a2) $V(z,t) = m(z,t),\ \forall (z,t) \in S_f$.

If the equation

$$\dot{x}(t) = f(x(t), u_h^o(x(t), t, (\frac{\partial V(z,t)}{\partial z})'\Big|_{z=x(t)}), t),$$
$$x(t_0) = x_0$$

admits a solution x_c *such that, for some* $\tau \geq t_0$,

$$(x_c(\tau), \tau) \in S_f,$$

then

$$u^o(t) := u_h^o(x_c(t), t, (\frac{\partial V(z,t)}{\partial z})'\Big|_{z=x_c(t)})$$

is an optimal control relative to (x_0, t_0).

Example 2.4 Consider a problem similar to the one discussed in Example 2.3: the final state is free and the term $(x(1)-1)^2$ is added to the performance index. The set of admissible final events is thus $S_f = \{(x,t) |\ x = free, t = 1\}$. The hamiltonian function and the relevant HJE remain unchanged, while the boundary condition becomes $V(z,1) = (z-1)^2$, $\forall z$. A solution of the HJE which satisfies this condition is

$$V(z,t) = \frac{2(z-1)^2}{3-t}.$$

Corresponding to this we get

$$u_h^o(x, t, \frac{\partial V(z,t)}{\partial z}\Big|_{z=x}) = \frac{1-x}{3-t}$$

and $x_c(t) = t/3$. Thus $u^o(t) = 1/3$ is an optimal control. The minimal value of the performance index can be computed by evaluating V at the initial event, yielding $J^o = \frac{2}{3}$.

Example 2.5 Consider the optimal control problem defined by the system $\dot{x} = u$ with $x(0) = 0$, $x(t_f) = 1$ and the performance index

$$J = \int_0^{t_f} (1 + \frac{u^2}{2}) dt.$$

The final time is free, so that the set S_f is constituted by all the couples $(1, \tau)$, $\tau \geq 0$. The hamiltonian function is regular with $u_h^o = -\lambda$. Thus the HJE is

$$\frac{\partial V}{\partial t} + 1 - \frac{1}{2}(\frac{\partial V}{\partial z})^2 = 0.$$

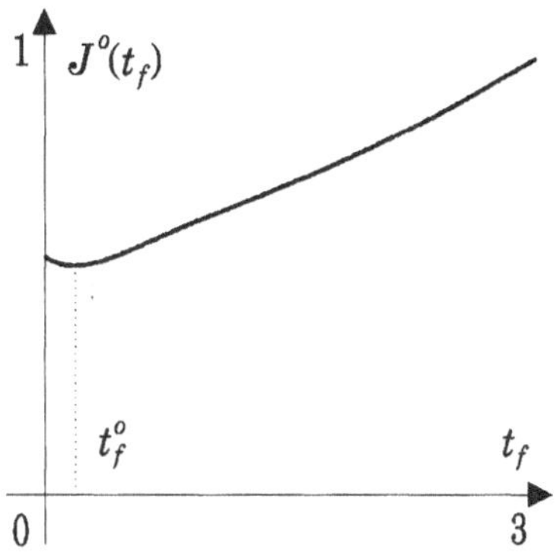

Figure 2.2: Example 2.6: the function $J^o(t_f)$.

A solution of this equation with the boundary condition $V(1,t) = 0,\ \forall t \geq 0$ is $V(z,t) = \sqrt{2}(1-z)$. Corresponding to this, we get $u^o = \sqrt{2}$, $x^o = \sqrt{2}t$ and $t_f^o = 1/\sqrt{2}$.

Example 2.6 Consider the optimal control problem defined by the system $\dot{x} = u$ with $x(0) = 1$ and the performance index

$$J = \frac{1}{2}\{\int_0^{t_f} (\frac{1}{2} + u^2)dt + f(t_f)x^2(t_f)\}$$

where $f(t_f) := 1 + t_f^2$. Thus the set S_f is constituted by all the couples (x,t) with $x \in R$ and $t \geq 0$. The problem is solved by resorting to point *(t2)* of Theorem 2.1 and Corollary 2.1, i.e., by first assuming that t_f is given, then computing the optimal value of the performance index corresponding to this final time, and finally performing a minimization with respect to t_f. Consistent with this, the term 1/2 in the performance index is first ignored, then it is easy to check that

$$V(z,t) = 0.5\frac{f(t_f)}{1 + f(t_f)(t_f - t)}z^2$$

is a solution of the HJE which is suited to determine an optimal control

$$u^o(t) = -\frac{f(t_f)}{1 + f(t_f)(t_f - t)}x_c(t)$$

where x_c satisfies the equation

$$\dot{x} = -\frac{f(t_f)}{1 + f(t_f)(t_f - t)}x,\ x_c(0) = 1.$$

The optimal value of the performance index turns out to be

$$J^o(t_f) = \frac{1}{2}(\frac{1 + t_f^2}{1 + t_f + t_f^3} + \frac{t_f}{2})$$

which is minimized by $t_f^o = 0.181$ (see also Fig. 2.2).

2.3 Problems

Problem 2.3.1 Find an optimal control law for the problem defined by the system $\dot{x}(t) = A(t)x(t) + B(t)u(t)$, $x(t_0) = 0$ and the performance index

$$J = \int_{t_0}^{t_f} \left[h'(t)x(t) + \frac{1}{2}u'(t)R(t)u(t) \right] dt$$

where $R'(t) = R(t) > 0$, t_0 and t_f are given, while the final state is free. Moreover, $A(t)$, $B(t)$, $R(t)$ and $h(t)$ are continuously differentiable functions.

Problem 2.3.2 Find an optimal control law for the problem defined by the system $\dot{x}(t) = Ax(t) + Bu(t)$, $x(0) = 0$, $x(t_f) = x_f$ and the performance index

$$J = \frac{1}{2} \int_0^{t_f} u'(t)Ru(t)dt$$

with $R' = R > 0$. In the statement above, t_f and x_f are given and the pair (A, B) is reachable.

Problem 2.3.3 Discuss the existence of a solution of the optimal control problem defined by the first order system $\dot{x}(t) = u(t)$, $x(0) = x_0$, $S_f = \{(x,t) | \; x \in R^n, \; t \geq 0\}$ and the performance index

$$J = \int_0^{t_f} \left[t^k + x(t) + \frac{1}{2}u^2(t) \right] dt$$

where k is a nonnegative integer.

Problem 2.3.4 Find an optimal control law for the problem defined by the system $\dot{x}(t) = A(t)x(t) + B(t)u(t)$, $x(t_0) = 0$ and the performance index

$$J = \frac{1}{2} \int_{t_0}^{t_f} u'(t)R(t)u(t)dt + h'x(t_f) + t_f^2$$

where $R'(t) = R(t) > 0$, t_0 is given, while the final state and time are free. Moreover, $A(t)$, $B(t)$ and $R(t)$ are continuously differentiable functions and h is a known n-dimensional vector.

Problem 2.3.5 Find a way of applying the Hamilton-Jacobi theory to the optimal control problem defined by the system $\dot{x}(t) = f(x(t), u(t), t)$ and the performance index

$$J = \int_{t_0}^{t_f} l(x(t), u(t), t)dt$$

where f and l are continuously differentiable functions. Here $x(t_f)$ and t_0 are given, while $x(t_0) \in R^n$ and $t_f \geq t_0$.

Problem 2.3.6 Find a solution of the optimal control problem defined by the first order system $\dot{x}(t) = u(t)$ and the performance index

$$J = \int_{t_0}^{t_f} \left[t^4 + u^2(t)\right] dt$$

where $x(t_f)$ and $x(t_0)$, $x(t_f) \neq x(t_0)$, while t_0 and $t_f \geq t_0$ are free.

Chapter 3

The LQ problem

3.1 Introduction

The *linear-quadratic* (LQ) problem, which is problably the most celebrated optimal control problem, is presented in this chapter. It refers to a *linear* system and a *quadratic* performance index according to the following statement.

Problem 3.1 (Linear-Quadratic problem) *For the system*

$$\dot{x}(t) = A(t)x(t) + B(t)u(t), \qquad (3.1)$$
$$x(t_0) = x_0$$

where x_0 and t_0 are given, find a control which minimizes the performance index

$$J = \frac{1}{2}\{\int_{t_0}^{t_f} [x'(t)Q(t)x(t) + u'(t)R(t)u(t)]dt + x'(t_f)Sx(t_f)\}. \qquad (3.2)$$

The final time t_f is given, while no constraints are imposed on the final state $x(t_f)$. In eqs. (3.1), (3.2) $A(\cdot)$, $B(\cdot)$, $Q(\cdot)$, $R(\cdot)$ are continuously differentiable functions and $Q(t) = Q'(t) \geq 0$, $R(t) = R'(t) > 0$, $\forall t \in [t_0, t_f]$, $S = S' \geq 0$.

Before presenting the solution of this problem, it is worth taking the time to shed some light on its significance, to justify the interest, both theoretical and practical, that the topics discussed in this book have received in recent years. Consider a dynamic system Σ and denote by $x_n(\cdot)$ its *nominal* state response, e.g. the response one wishes to obtain. Moreover, let $u_n(\cdot)$ be the corresponding input when $\Sigma = \Sigma_n$, that is when the system exhibits these nominal conditions. Unavoidable uncertainties in the system description and disturbances acting on the system apparently suggest that we not resort to an

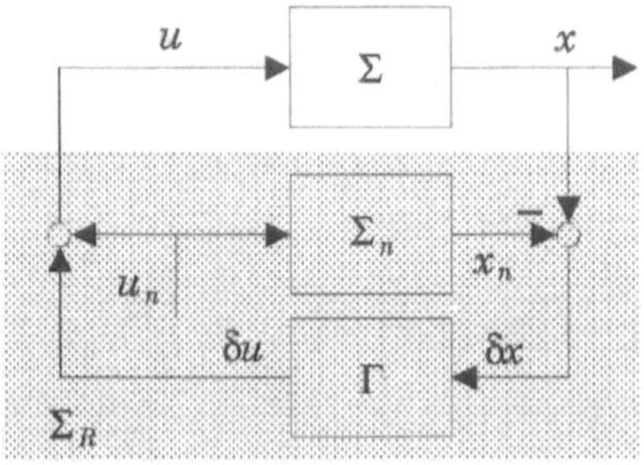

Figure 3.1: Closed loop control system

open loop control scheme but rather to a *closed loop* configuration as the one shown in Fig. 3.1. The controller Σ_R has to be such that system Γ supplies, on the basis of the deviation δx of the actual state from x_n, the correction δu to be given to u_n in order to make δx *small*. The convenience of not requiring large corrections δu suggests restating the objective above in terms of looking for the minimization of a (quadratic) performance index with a structure similar to the one given in eq. (3.2). Consistently, if the deviations δx and δu are actually small and Σ is described by $\dot{x} = f(x,u,t)$ with f sufficiently regular, the effect of δu on δx can be evaluated through the (linear) equation

$$\delta x = \left.\frac{\partial f(x, u_n(t), t)}{\partial x}\right|_{x=x_n(t)} \delta x + \left.\frac{\partial f(x_n(t), u, t)}{\partial u}\right|_{u=u_n(t)} \delta u.$$

Thus an LQ problem arises in a fairly spontaneous way and, according to circumstances, it might be convenient to state it on a *finite* or *infinite* time interval.

Problem 3.1, being apparently a particular case of Problem 2.1, can be approached via the Hamilton-Jacobi theory. This will be accomplished in the remaining part of the chapter, first with reference to a finite control interval, subsequently encompassing also the infinite horizon case.

3.2 Finite control horizon

The following result holds for the LQ problem over a finite horizon.

Theorem 3.1 *Problem 3.1 admits a solution for any initial state x_0 and for any finite control interval $[t_0, t_f]$. The solution is given by the control law*

$$u_c^o(x,t) = -R^{-1}(t)B'(t)P(t)x \tag{3.3}$$

where the matrix P solves the differential Riccati equation (DRE)

$$\dot{P}(t) = -P(t)A(t) - A'(t)P(t) + P(t)B(t)R^{-1}(t)B'(t)P(t) - Q(t) \qquad (3.4)$$

with boundary condition

$$P(t_f) = S. \qquad (3.5)$$

Further, the minimal value of the performance index is

$$J^o(x_0, t_0) = \frac{1}{2}x_0' P(t_0)x_0.$$

Proof. The proof will check that: 1) The hamiltonian function is regular; 2) The solution of the Riccati equation is symmetric whenever it exists and is unique; 3) The function $V(z,t) := z'P(t)z/2$ satisfies the HJE with the relevant boundary condition if P solves eqs. (3.4), (3.5); 4) The existence and uniqueness of the solution of eqs. (3.4), (3.5).

Point 1) It is easy to verify that the hamiltonian function is regular and that $u_h^o = -R^{-1}B'\lambda$. Indeed, by noticing that

$$\frac{1}{2}u'Ru + \lambda'Bu = \frac{1}{2}(u + R^{-1}B'\lambda)'R(u + R^{-1}B'\lambda) - \frac{1}{2}\lambda'BR^{-1}B'\lambda$$

and recalling that $R > 0$, there exists a unique absolute minimum in u_h^o.

Point 2) By letting $\Gamma := P'$ it is straightforward to check that Γ satisfies the same differential equation and boundary condition (S is symmetric).

Point 3) The symmetry of P implies that

$$\frac{\partial V}{\partial z} = z'P.$$

It is then easy to verify that the given function V solves the HJE with the boundary condition $V(z, t_f) = z'Sz/2$, $\forall z$, if P solves eqs. (3.4), (3.5). In doing this it is expedient to notice that

$$\frac{\partial V}{\partial z}Az = z'PAz = \frac{1}{2}(z'PAz + z'A'Pz).$$

Point 4) The continuity assumptions on data imply that: (*i*) a solution of eqs. (3.4), (3.5) exists and is unique in a neighbourhood of t_f; (*ii*) a solution of eqs. (3.4), (3.5) cannot be extended to t_0 only if there exists a time within the interval $[t_0, t_f]$ where at least one element of it becomes unbounded. Let $\hat{t} < t_f$ be the greatest $t \in [t_0, t_f)$ where the solution fails to exist. Then Problem 3.1 can be solved on the interval $(\hat{t}, t_f]$ and it follows that

$$J^o(x(\tau), \tau) = \frac{1}{2}x'(\tau)P(\tau)x(\tau) \geq 0, \ \tau \in (\hat{t}, t_f]$$

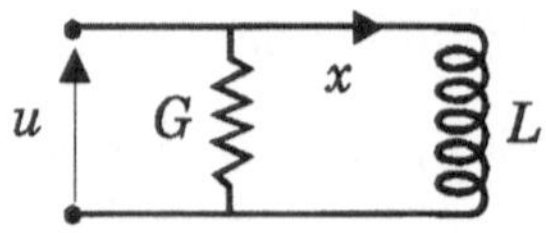

Figure 3.2: The electric circuit considered in Example 3.1.

where the inequality sign is due to the sign of matrices Q, R and S. Since $x(\tau)$ is arbitrary it follows that $P(\tau) \geq 0$, $\tau \in (\hat{t}, t_f]$, which in turn implies that if

$$\lim_{\tau \to \hat{t}+} |p_{ij}(\tau)| = \infty,$$

then

$$\lim_{\tau \to \hat{t}+} p_{hh}(\tau) = \infty, \quad h = i \text{ and/or } h = j$$

where p_{ij} is the (i, j) element of P. Indeed, should this not happen, it would then follow that

$$\lim_{\tau \to \hat{t}+} \frac{1}{2}(p_{ii}(\tau) + p_{jj}(\tau) \pm 2p_{ij}(\tau)) = -\infty$$

according to whether $p_{ij}(\tau)$ tends to $\mp\infty$, respectively. However, the previous equation supplies the limit value of $J^o([I]^i \pm [I]^j, \tau)$ which cannot be negative. Thus, if $p_{ii}(\tau) \to \infty$, we get

$$\infty = \lim_{\tau \to \hat{t}+} J^o([I]^i, \tau) > J([I]^i, \hat{t}, t_f, \varphi(\cdot; \hat{t}, [I]^i, 0), 0)$$

which is nonsense since, for τ sufficiently close to $\hat{t}$, the optimality of the adopted control law would be denied. Therefore the solution of the DRE must be extendable to t_0.

Theorem 3.1 supplies the solution to Problem 3.1 in terms of an optimal *control law*: if, on the other hand, the optimal *control* u^o is sought corresponding to a given initial state, it suffices to look for the solution x^o of the equation $\dot{x} = (A - BR^{-1}B'P)x$ with $x(t_0) = x_0$ (i.e., computing the response of the optimal closed loop system, namely of the system (3.1), (3.3)) and evaluating $u^o(t) = -R^{-1}(t)B'(t)P(t)x^o(t)$.

Example 3.1 Consider the electric circuit shown in Fig. 3.2 where x is the current flowing through the inductor ($L > 0$) and u is the applied voltage. At $t = 0$ the value of the current is x_0 and the goal is to determine u so as to lower this current without wasting too much energy in the resistor of conductance $G > 0$. Moreover, a nonzero final value of x should explicitly be penalized. According to these requirements an LQ problem can be stated on the system $\dot{x} = u/L$, $x(0) = x_0$ and the performance index

$$J = \int_0^1 (x^2 + Gu^2)dt + \sigma x^2(1)$$

where $\sigma \geq 0$. The solution of the DRE relevant to this problem is

$$P(t) = \tau \frac{\sigma + \tau + (\sigma - \tau)e^{2\frac{t-1}{\tau}}}{\sigma + \tau - (\sigma - \tau)e^{2\frac{t-1}{\tau}}}$$

where $\tau := L\sqrt{G}$. It is easy to verify that $P(t) \geq 0$ and

$$x^o(t) = \frac{\sigma + \tau - (\sigma - \tau)e^{2\frac{t-1}{\tau}}}{(\sigma + \tau)e^{\frac{t}{\tau}} - (\sigma - \tau)e^{\frac{t-2}{\tau}}} x_0.$$

When σ increases it should be expected that $x^o(1)$ tends to 0 and the control becomes more active. This is clearly shown in Fig. 3.3 where the responses of x^o and u^o are reported corresponding to some values of σ, $x_0 = 1$, $L = 1$ and $G = 1$. Analogously, it should be expected that increasing values of G would cause the control action to vanish. This outcome is clearly illustrated in Fig. 3.3 where the responses of x^o and u^o are plotted for various G, $\sigma = 1$, $L = 1$ and $x_0 = 1$.

Example 3.2 Consider an object with unitary mass moving without friction on a straight line (see Fig. 3.4). A force u acts on this object, while its initial position and velocity are $x_1(0) \neq 0$ and $x_2(0)$, respectively. The object has to be brought near to the reference position (the origin of the straight line) with a small velocity, by applying a force as small as possible. The task has to be accomplished in a unitary time interval. This control problem can be recast as an LQ problem defined on the system $\dot{x}_1 = x_2$, $\dot{x}_2 = u$ with initial state $x(0) = x_0 \neq 0$ and the performance index (to be minimized)

$$J = \int_0^1 u^2 dt + x_1^2(1) + x_2^2(1).$$

The solution of the relevant DRE can be computed in an easy way (see Problem 3.5.1) yielding

$$P(t) = \frac{1}{\alpha(t)} \begin{bmatrix} 12(2-t) & 6(t^2 - 4t + 3) \\ 6(t^2 - 4t + 3) & -4(t^3 - 6t^2 + 9t - 7) \end{bmatrix}$$

where $\alpha(t) := t^4 - 8t^3 + 18t^2 - 28t + 29$, from which all other items follow.

Remark 3.1 *(Coefficient $\frac{1}{2}$)* It should be clear that the particular (positive) value of the coefficient in front of the performance index is not important: indeed, it can be seen as a scale factor only and the selected value, namely $\frac{1}{2}$, is the one which simplifies the proof of Theorem 3.1, to some extent.

Remark 3.2 *(Uniqueness of the solution)* Theorem 3.1 implicitly states that the solution of the LQ problem is unique. This is true provided that two input functions which differ from each other on a set of zero measure are considered equal. In fact, assume that the pair (x^*, u^*) is also an optimal solution, so that $\dot{x}^* = Ax^* + Bu^*$, $x^*(t_0) = x_0$. By letting

$$v := u^* + R^{-1}B'Px^*$$

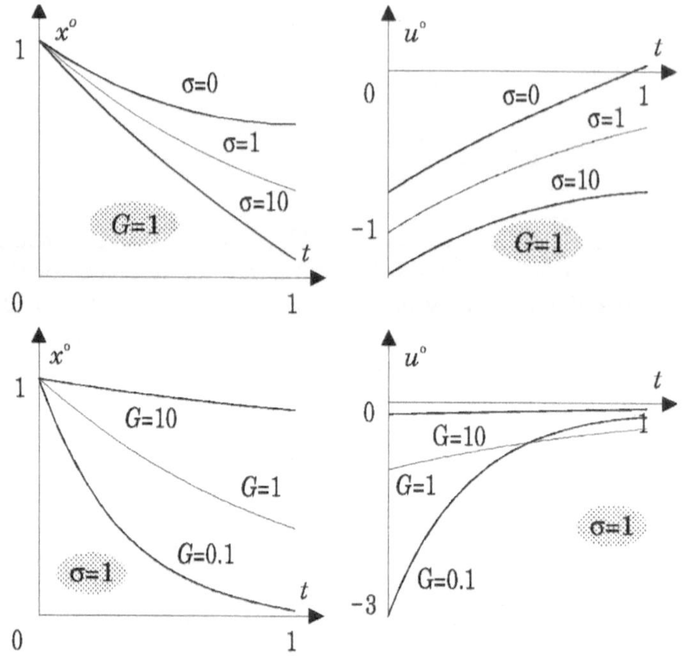

Figure 3.3: Example 3.1: responses of x^o and u^o for some values of σ and G.

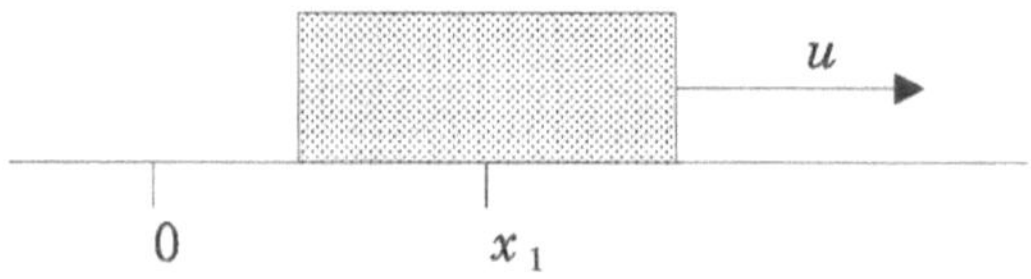

Figure 3.4: The system considered in Example 3.2.

where P is the solution of the DRE (3.4) with the boundary condition (3.5), it follows that
$$\dot{x}^* = (A - BR^{-1}B'P)x^* + Bv.$$

The expressions for u^* and Bv which can be deduced from these two relations can be exploited in computing the value of the performance index corresponding to the solution at hand. By recalling eqs. (3.4), (3.5), we obtain

$$\begin{aligned}
J(x^*(\cdot), u^*(\cdot)) &= \frac{1}{2}\{\int_{t_0}^{t_f} [v'Rv - x^{*\prime}\dot{P}x^* - 2x^{*\prime}P\dot{x}^*]dt + x^{*\prime}(t_f)Sx^*(t_f)\} \\
&= \frac{1}{2}\{\int_{t_0}^{t_f} [v'Rv - \frac{d}{dt}(x^{*\prime}Px^*)]dt + x^{*\prime}(t_f)Sx^*(t_f)\} \\
&= \frac{1}{2}\int_{t_0}^{t_f} v'Rvdt + \frac{1}{2}x^{*\prime}(t_0)P(t_0)x^*(t_0) \\
&= \frac{1}{2}\int_{t_0}^{t_f} v'Rvdt + J^o.
\end{aligned}$$

If the pair (x^*, u^*) is optimal, then

$$\int_{t_0}^{t_f} v'Rvdt = 0.$$

Since R is positive definite and $v \in \Omega^m$, this equation implies that $v(\cdot) = 0$ almost everywhere, so that $u^* = -R^{-1}B'Px^*$, and $x^* = x^o$.

Remark 3.3 *(Linear quadratic performance index)* A more general statement of the LQ problem can be obtained by adding *linear* functions of the control and/or state variables into the performance index, which thus becomes

$$\begin{aligned}
J = &\frac{1}{2}\int_{t_0}^{t_f} [x'(t)Q(t)x(t) + u'(t)R(t)u(t) + 2h'(t)x(t) + 2k'(t)u(t)]dt \\
&+\frac{1}{2}[x'(t_f)Sx(t_f) + 2m'x(t_f)]
\end{aligned}$$

where Q, R, S are as in Problem 3.1 and h, k are vectors of continuously differentiable functions. The resulting optimal control problem admits a solution for all initial states x_0 and for all finite control intervals $[t_0, t_f]$. The solution is given by the control law

$$u_c^o(x,t) = -R^{-1}(t)\{B'(t)[P(t)x + w(t)] + k(t)\}$$

where P solves eqs. (3.4), (3.5), while w is the solution of the linear differential equation
$$\dot{w} = -(A - BR^{-1}B'P)'w + PBR^{-1}k - h$$

satisfying the boundary condition

$$w(t_f) = m.$$

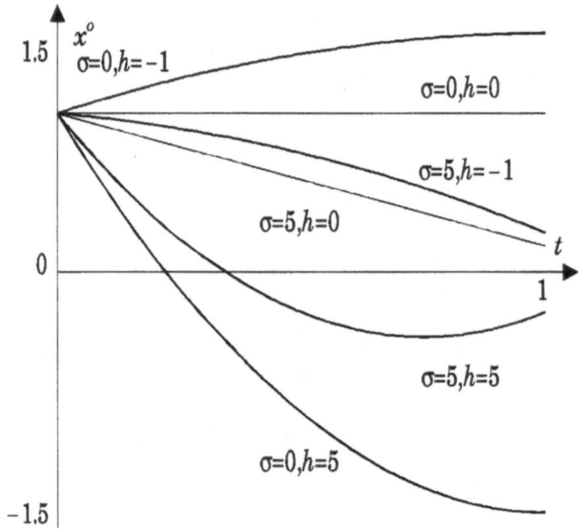

Figure 3.5: Example 3.3: responses of x^o for some values of h and σ.

The optimal value of the performance index is

$$J^o(x_0, t_0) = \frac{1}{2}x_0'P(t_0)x_0 + w'(t_0)x_0 + v(t_0)$$

where v is the solution of the linear differential equation

$$\dot{v} = \frac{1}{2}(B'w + k)'R^{-1}(B'w + k)$$

satisfying the boundary condition

$$v(t_f) = 0.$$

The proof of these statements can easily be done by checking that if P, w and v satisfy the differential equations above, then

$$V(z,t) := \frac{1}{2}z'P(t)z + w'(t)z + v(t)$$

is an appropriate solution of the relevant HJE.

Example 3.3 Consider the problem defined on the system $\dot{x} = u$ and the performance index

$$J = \int_0^1 [hx + \frac{u^2}{2}]dt + \sigma\frac{x^2(1)}{2}$$

where $h \in R$ and $\sigma \geq 0$ are given. By exploiting Remark 3.3 we find $u_c^o(x,t) = -P(t)x - w(t)$, with

$$P(t) = \frac{\sigma}{\sigma+1-\sigma t}, \; w(t) = \frac{h}{2\sigma}[\sigma + 1 - \sigma t - \frac{1}{\sigma+1-\sigma t}].$$

The responses of x^o corresponding to $x(0) = 1$ and some values of the parameters σ and h are shown in Fig. 3.5. As should be expected, $x^o(\cdot) = 1$ when $h = \sigma = 0$. Also the consequences of the choice $\sigma \neq 0$ can easily be forecast (the state remains closer to 0).

Remark 3.4 *(Performance index with a rectangular term)* A different extension of the LQ problem consists in adding the term $2x'(t)Z(t)u(t)$ to the integral part of the performance index, Z being a continuously differentiable function. Notice that the presence of this new term may substantially modify the nature of the problem, as the assumptions in Problem 3.1 are no longer *sufficient* to guarantee the existence of a solution. In fact, let

$$v := u + R^{-1}Z'x$$

so that the system description and the performance index become

$$\dot{x} = (A - BR^{-1}Z')x + Bv := A_c x + Bv$$

and

$$x'Qx + 2x'Zu + u'Ru = x'(Q - ZR^{-1}Z')x + v'Rv := x'Q_c x + v'Rv$$

respectively. The original problem has been transformed into a customary LQ problem where, however, matrix Q_c might no longer be positive semidefinite. Hence the existence of the solution of the DRE for arbitrary finite intervals is not ensured unless a further assumption of the kind

$$\begin{bmatrix} Q & Z \\ Z' & R \end{bmatrix} \geq 0, \; R > 0$$

(which is equivalent to the three conditions $Q \geq 0$, $Q_c \geq 0$ and $R > 0$) is added. Anyway, if the DRE

$$\dot{P} = -PA_c - A_c'P + PBR^{-1}B'P - Q_c$$

with the relevant boundary condition $P(t_f) = S$ admits a solution over the interval $[t_0, t_f]$ also when Q_c is not positive semidefinite, then the control law

$$u_c^o(x,t) = -R^{-1}(t)[B'(t)P(t) + Z'(t)]x$$

is optimal.

Example 3.4 Consider the LQ problem with a rectangular term defined by matrices

$$A = \begin{bmatrix} 0 & 1 \\ 0 & 0 \end{bmatrix}, \; B = \begin{bmatrix} 0 \\ 1 \end{bmatrix}, \; Q = \begin{bmatrix} 1 & 0 \\ 0 & 0 \end{bmatrix}, \; R = 1, \; Z = \begin{bmatrix} 1 \\ 0 \end{bmatrix}, \; S = I$$

and $t_0 = 0$, $t_f = \pi/2$. With reference to Remark 3.4 we find $Q_c = 0$, so that the DRE is solved by

$$P(t) = \frac{4}{\vartheta(t)} \begin{bmatrix} \alpha(t) & \beta(t) \\ \beta(t) & \gamma(t) \end{bmatrix}$$

where

$$\begin{aligned}
\alpha(t) &:= 4 + \pi - 2t + \sin(2t), \\
\beta(t) &:= 1 + \cos(2t), \\
\gamma(t) &:= 4 + \pi - 2t - \sin(2t), \\
\vartheta(t) &:= \alpha(t)\gamma(t) - \beta^2(t).
\end{aligned}$$

Remark 3.5 *(Sign of Q and S)* The assumptions on the sign of Q and S are no doubt *conservative* though *not unnecessarily conservative.* In other words, when these assumptions are not met with, there are cases where the DRE still admits a solution and cases where the solution fails to exist over the whole given finite interval. These facts are made clear in the following examples.

Example 3.5 Consider the interval $[0, t_f]$, $t_f > 0$ and the scalar equation

$$\dot{P} = -2P + P^2 + 1$$

with boundary condition $P(t_f) = S > 0$. Notice that $Q = -1$. In a neigbourhood of t_f the solution is

$$P(t) = \frac{1 + (1 - S)(t - t_f - 1)}{1 + (1 - S)(t - t_f)}$$

which cannot be extended up to $t = 0$ if $0 < S < 1$ and $t_f > (1 - S)^{-1}$. Indeed, letting $\tau := t_f - (1 - S)^{-1} > 0$, we get

$$\lim_{t \to \tau^+} |P(t)| = \infty.$$

On the other hand, the scalar equation

$$\dot{P} = 2P + P^2 + 1$$

with boundary condition $P(t_f) = S > 0$ admits a solution for each interval $[0, t_f]$ even if $Q = -1$. Such a solution is

$$P(t) = \frac{S + (1 + S)(t - t_f)}{1 - (1 + S)(t - t_f)}.$$

Observe that if

$$t_f > \frac{S}{1 + S}$$

then $P(0) < 0$ so that the optimal value of the performance index of the LQ problem which gives rise to the DRE at hand is negative unless $x(0) = 0$, consistent with the nonpositivity of Q. Finally, consider the scalar equation

$$\dot{P} = P^2 - 1$$

with boundary condition $P(t_f) = S$, $-1 \leq S < 0$. For each finite t_f it admits

$$P(t) = \frac{S + 1 + (S-1)e^{2(t-t_f)}}{S + 1 - (S-1)e^{2(t-t_f)}}$$

as a solution over the interval $[0, t_f]$.

Remark 3.6 *(Tracking problem)* A third extension of the LQ problem calls for adding to the controlled system an output

$$y(t) = C(t)x(t)$$

and considering the performance index

$$J = \frac{1}{2}\int_{t_0}^{t_f} \{[y'(t) - \mu'(t)]\hat{Q}(t)[y(t) - \mu(t)] + u'(t)R(t)u(t)\}dt$$
$$+\frac{1}{2}[y'(t_f) - \mu'(t_f)]\hat{S}[y(t_f) - \mu(t_f)].$$

Here C, $\hat{Q} = \hat{Q}' > 0$ and $R = R' > 0$ are continuously differentiable functions, $\hat{S} = \hat{S}' > 0$ and μ is a vector of given continuous functions. The aim is thus to make some linear combinations of the state variables (if $C = I$, actually the whole state) behave in the way specified by μ. This optimal control problem admits a solution for each finite interval $[t_0, t_f]$, initial state $x(t_0)$ and $\mu(\cdot)$. The solution is given by the control law

$$u_c^o(x,t) = -R^{-1}(t)B'(t)[P(t)x + w(t)] \tag{3.6}$$

where P solves the DRE

$$\dot{P} = -PA - A'P + PBR^{-1}B'P - C'\hat{Q}C \tag{3.7}$$

with boundary condition $P(t_f) = C'(t_f)\hat{S}C(t_f)$, while w is the solution of the (linear) differential equation

$$w = -(A - BR^{-1}B'P)'w + C'\hat{Q}\mu \tag{3.8}$$

with boundary condition $w(t_f) = -C'(t_f)\hat{S}\mu(t_f)$. Finally, the optimal value of the performance index is

$$J^o(x_0, t_0) = \frac{1}{2}x_0'P(t_0)x_0 + w'(t_0)x_0 + v(t_0) \tag{3.9}$$

where v is the solution of the (linear) differential equation

$$\dot{v} = \frac{1}{2}(w'BR^{-1}B'w - \mu'\hat{Q}\mu) \tag{3.10}$$

with boundary condition $v(t_f) = \mu'(t_f)\hat{S}\mu(t_f)/2$. These claims can easily be verified by exploiting Corollary 2.1 and checking that $V(z,t) := \frac{1}{2}z'P(t)z + w'(t)z + v(t)$ is an appropriate solution of the relevant HJE. Alternatively, one could resort to Remark 3.3.

Some comments are in order. First observe that since the boundary condition for w is set at the final time, the value of w at any time t depends on the *whole* history of μ. Therefore, the value of the optimal control $u^o(t)$ at a generic instant t depends on the *future* behaviour of the signal to be tracked. Second, if $\mu(\cdot) = 0$, then the solution above coincides with the one of the customary LQ problem, so that, in particular, it can be concluded that $P(t) \geq 0$, $\forall t$. Third, since the optimal value of the performance index is a linear-quadratic function of the initial state, the question whether there exists an optimal initial state is not trivial. The nonnegativity of the performance index together with $P \geq 0$ imply that the set of optimal initial states is given by

$$\{x_0| \ \frac{\partial J^o(x_0, t_0)}{\partial x_0} = 0\}$$

so that any vector of the form $x_0^o := -P^\dagger(t_0)w(t_0) + x_n$ is an optimal initial state. Here $P^\dagger(t_0)$ is the *pseudoinverse* of $P(t_0)$ and x_n is any element of $\ker(P)$: thus, the optimal initial state is not unique unless $P > 0$. Finally, particular attention is given to the case where the signal μ is the output of a linear system, namely

$$\begin{aligned}\dot{\vartheta}(t) &= F(t)\vartheta(t),\\ \mu(t) &= H(t)\vartheta(t),\\ \vartheta(t_0) &= \vartheta_0.\end{aligned}$$

Matrices F and H are continuous. In this particular case the optimal control law is

$$u_c^o(x,t) = -R^{-1}(t)B'(t)[P(t)x + \Gamma(t)\vartheta(t)]$$

where P solves the DRE (3.7) with the relevant boundary condition, while Γ is the solution of the (linear) differential equation

$$\dot{\Gamma} = -\Gamma F - (A - BR^{-1}B'P)'\Gamma + C'\hat{Q}H$$

with boundary condition $\Gamma(t_f) = -C'(t_f)\hat{S}H(t_f)$. The optimal value of the performance index is

$$J^o(x_0, t_0) = \frac{1}{2}x_0'P(t_0)x_0 + \vartheta_0'\Gamma'(t_0)x_0 + v(t_0)$$

where v is the solution of the (linear) differential equation

$$\dot{v} = \frac{1}{2}\vartheta'(\Gamma BR^{-1}B'\Gamma - H'\hat{Q}H)\vartheta$$

with boundary condition $v(t_f) = \frac{1}{2}\vartheta'(t_f)H'(t_f)\hat{S}H(t_f)\vartheta(t_f)$. The check of these claims can be performed by verifying that eqs. (3.8)–(3.10) are satisfied together with the relevant boundary conditions, if $w(t) := \Gamma(t)\vartheta(t)$.

Example 3.6 Consider the electric circuit shown in Fig. 3.6 and assume that the goal is to keep the value of the voltage x across the capacitor, initially 0, close to 1. The injected current u has to be determined so as to minimize the energy lost in the

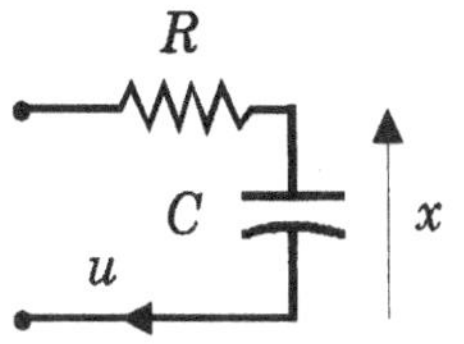

Figure 3.6: The electric circuit considered in Example 3.6.

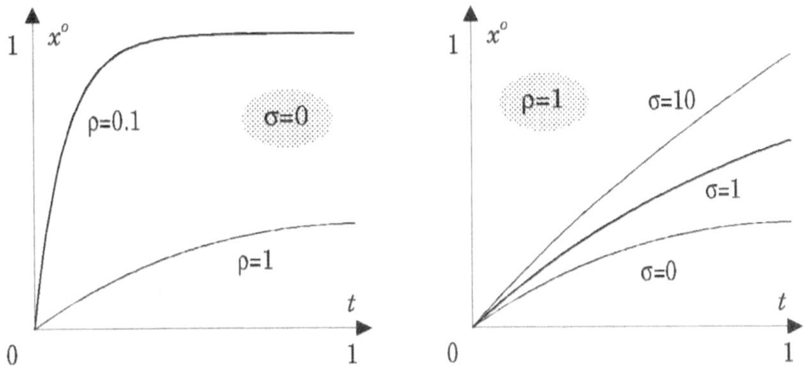

Figure 3.7: Example 3.6: responses of x^o for some values of σ and ρ.

resistor. Letting, for ease of notation, all electric parameters be unitary, the problem above can be cast into the optimal tracking problem defined on the system $\dot{x} = u$, $x(0) = x_0 = 0$ and the performance index

$$J(x_0) = \int_0^1 [(x-\mu)^2 + \rho^2 u^2]dt + \sigma[x(1) - \mu(1)]^2$$

where $\mu(\cdot) = 1$ while $\rho > 0$ and $\sigma \geq 0$ are two given parameters. Letting $\beta := (\sigma - \rho)/(\sigma + \rho)$, we obtain

$$P(t) = \rho \frac{1 + \beta e^{\frac{2}{\rho}(t-1)}}{1 - \beta e^{\frac{2}{\rho}(t-1)}}$$

$$w(t) = \frac{e^{\frac{t}{\rho}}}{1 - \beta e^{\frac{2}{\rho}(t-1)}}[(1 - \beta e^{-\frac{2}{\rho}})w(0) - \rho(e^{-\frac{t}{\rho}} - 1 + \beta e^{\frac{t-2}{\rho}} - \beta e^{-\frac{2}{\rho}})]$$

where

$$w(0) = \rho \frac{\beta + e^{\frac{2}{\rho}}}{\beta - e^{\frac{2}{\rho}}}.$$

The transients of x^o corresponding to some values of ρ and σ are shown in Fig. 3.7 when the optimal control law is implemented. The system response is closer to the desired one for large values of σ and/or small values of ρ. Finally, as could easily be forecast, the optimal value $x^o(0)$ for the initial state is 1.

Example 3.7 Consider again the system described in Example 3.2, namely an object with unitary mass moving without friction along a straight line with initial position $x_1(0)$ and velocity $x_2(0)$. It is desired that its position be close to $\mu(t) := \sin(t)$ for $0 \leq t \leq 1$ by applying a force as small as possible. This problem can be seen as an optimal tracking problem by adding to the system equations $\dot{x}_1 = x_2$, $\dot{x}_2 = u$, $y = x_1$ the performance index

$$J(x(0)) = \int_0^1 [(y-\mu)^2 + \rho u^2]dt + \sigma[y(1) - \mu(1)]^2$$

where $\rho > 0$ and $\sigma \geq 0$ are two given parameters. The system responses corresponding to some values of such parameters are shown in Fig. 3.8: the initial state is either $x(0) = 0$ or $x(0) = x_0^o$. The recorded plots agree with the forecast which is most naturally suggested by the selected values of the two parameters.

Remark 3.7 *(Not strictly proper system)* The tracking problem can be set also for systems which are not strictly proper, that is systems where the output variable is given by

$$y(t) = C(t)x(t) + D(t)u(t).$$

C and D being matrices of continuously differentiable functions. The performance index is

$$J = \frac{1}{2}\int_{t_0}^{t_f} \{[y'(t) - \mu'(t)]\hat{Q}(t)[y(t) - \mu(t)] + u'(t)\hat{R}(t)u(t)\}dt$$

where both matrices $\hat{Q}$ and $\hat{R}$ are symmetric, positive definite, continuously differentiable and μ is a given continuous function. Note that the adopted performance index is purely integral: this choice simplifies the subsequent discussion without substantially altering the nature of the problem. By exploiting Remarks 3.3 and 3.4 we easily find that the solution can be given in terms of the control law

$$u_c^o(x,t) = -R^{-1}(t)\{[D'(t)\hat{Q}(t)C(t) + B'(t)P(t)]x + B'(t)w(t) + k(t)\}$$

where $R := D'\hat{Q}D + \hat{R}$, $k := -D'\hat{Q}\mu$, P solves the DRE

$$\dot{P} = -PA_c - A_c'P + PBR^{-1}B'P - Q$$

with boundary condition $P(t_f) = 0$ and w is the solution of the (linear) equation

$$\dot{w} = -(A_c - BR^{-1}B'P)'w + PBR^{-1}k - h$$

satisfying the boundary condition $w(t_f) = 0$. In these two differential equations $Q := C'(\hat{Q} - \hat{Q}DR^{-1}D'\hat{Q})C$, $A_c := A - BR^{-1}D'\hat{Q}C$, $h := C'\hat{Q}(DR^{-1}D'\hat{Q} - I)\mu$. The optimal value of the performance index is

$$J^o(x_0) = \frac{1}{2}x_0'P(t_0)x_0 + w'(t_0)x_0 + v(t_0) + \frac{1}{2}\int_{t_0}^{t_f} \mu'(t)\hat{Q}(t)\mu(t)dt$$

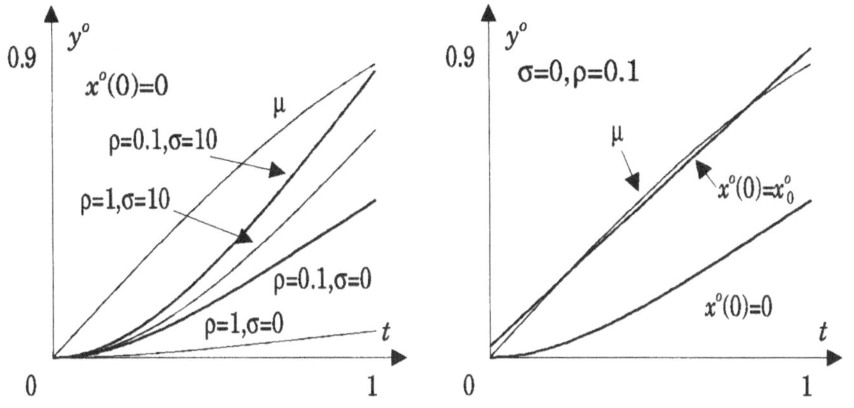

Figure 3.8: Example 3.7: responses of y^o for $x^o(0) = 0$ and some values of ρ and σ, or $\sigma = 0$, $\rho = 0.1$ and $x^o(0) = 0$, $x^o(0) = x_0^o$.

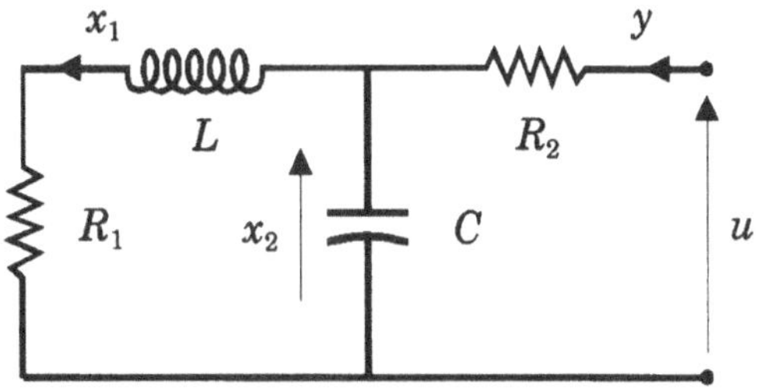

Figure 3.9: The electric circuit considered in Example 3.8.

where v solves the (linear) equation

$$\dot{v} = \frac{1}{2}(B'w + k)'R^{-1}(B'w + k)$$

with boundary condition $v(t_f) = 0$. Contrary to what was discussed in Remark 3.4, no further assumptions are needed to guarantee the existence of the solution of the DRE, in spite of the presence of a *rectangular* term. Indeed, the form of the integral part of J prevents the equivalent quadratic form for x from assuming negative values. Finally, it is straightforward to derive the solution of the LQ problem which results from setting $\mu(\cdot) = 0$, namely the problem where only the term $y'\hat{Q}y$ appears in the performance index.

Example 3.8 Consider the electric circuit shown in Fig. 3.9 where x_1 and x_2 are the current flowing through the inductor and the voltage across the capacitor, respectively. The external voltage u is the control variable, while the incoming current y

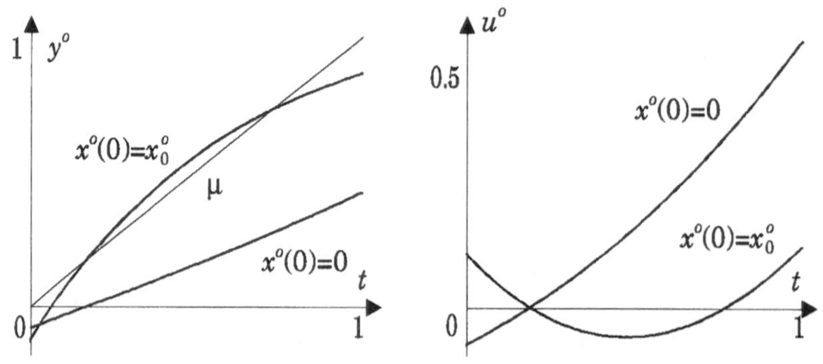

Figure 3.10: Example 3.8: responses of y^o and u^o for $x^o(0) = 0$ and $x^o(0) = x_0^o$.

is the output variable. Letting the electric parameters of the circuit be unitary for the sake of simplicity, the relevant equations are $\dot{x}_1 = -x_1 + x_2$, $\dot{x}_2 = -x_1 - x_2 + u$, $y = -x_2 + u$. In the performance index $\hat{Q} = \hat{R} = 1$, $\mu(\cdot) = 1$, $t_0 = 0$ and $t_f = 1$. The transient responses of y^o and u^o are shown in Fig. 3.10 corresponding to $x^o(0) = 0$ and $x^o(0) = x_0^o$, the latter value of the initial state being the optimal one. Finally, $J^o(0) = 0.11$ and $J^o(x_0^o) = 0.003$.

Remark 3.8 *(Penalties on the control derivative)* Frequently it is also convenient to prevent the first derivative of the control variable from taking on high values. This requirement can easily be cast into the problem formulation by adding to the integral part of the performance index the term $\dot{u}'(t)\hat{R}(t)\dot{u}(t)$. If the matrix $\hat{R}$ is positive definite and continuously differentiable, the problem can be brought back to a standard LQ problem (to which Theorem 3.1 can be applied) by viewing u as a further state variable satisfying the equation

$$\dot{u}(t) = v(t)$$

and letting v be the new control variable. Thus the given problem is equivalent to the LQ problem defined on the system

$$\dot{\hat{x}}(t) = \hat{A}(t)\hat{x}(t) + \hat{B}v(t)$$

and the performance index

$$J = \int_{t_0}^{t_f} [\hat{x}'(t)\hat{Q}(t)\hat{x}(t) + v'(t)\hat{R}v(t)]dt + \hat{x}'(t_f)\hat{S}\hat{x}(t_f)$$

where $\hat{x}(t) := \begin{bmatrix} x'(t) & u'(t) \end{bmatrix}'$ and

$$\hat{A}(t) := \begin{bmatrix} A(t) & B(t) \\ 0 & 0 \end{bmatrix}, \ \hat{B} := \begin{bmatrix} 0 \\ I \end{bmatrix}, \ \hat{Q}(t) := \begin{bmatrix} Q(t) & 0 \\ 0 & R(t) \end{bmatrix}, \ \hat{S} := \begin{bmatrix} S & 0 \\ 0 & 0 \end{bmatrix}.$$

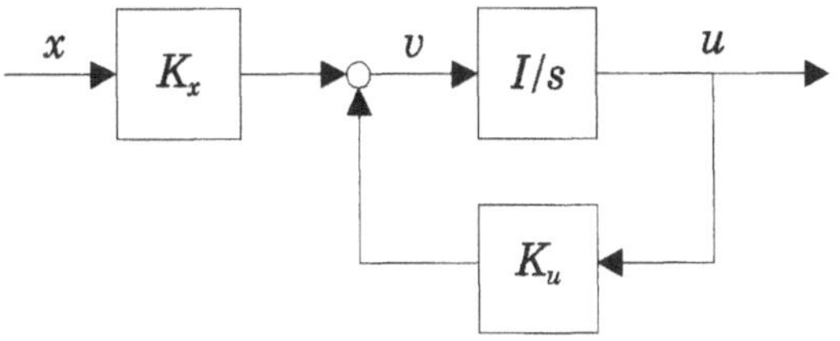

Figure 3.11: The regulator for the LQ problem with penalty on $\dot{u}$.

In view of Theorem 3.1 the solution is given by the control law

$$v_c^o(\hat{x}, t) = -\hat{R}^{-1}(t)\hat{B}'\hat{P}(t)\hat{x} := K_x(t)x + K_u(t)u,$$

$\hat{P}$ being the solution of the DRE

$$\dot{P} = -P\hat{A} - \hat{A}'P + P\hat{B}\hat{R}^{-1}\hat{B}'P - \hat{Q}$$

with boundary condition $P(t_f) = \hat{S}$. Note that the resulting controller, namely the device which computes the actual control variable u on the basis of the actual state variable x, is no longer a purely algebraic system as in the standard LQ context but rather a dynamic system (see Fig. 3.11) the order of which equals the number of the control variables. Finally, it is obvious how to comply with requirements concerning higher order derivatives of the control variable.

Example 3.9 Consider the LQ problem defined on a first order system with $A = 0$, $B = 1$ and the performance index characterized by $t_0 = 0$, $t_f = 5$, $Q = 1$, $R = 1$, $S = 0$. For $x(0) = 1$ the transient responses of x^o and u^o are shown in Fig. 3.12 and labelled with α. If high values for $\dot{u}$ have to be avoided and the design is carried on according to the discussion in Remark 3.8 by selecting $\hat{R} = 1$, the responses in Fig. 3.12 labelled with β are obtained corresponding to the same initial state and $u(0) = 0$. Observe that the smoother response of the control variable entails a less satisfactory response of the state.

Remark 3.9 *(Stochastic control problem)* The LQ problem can be stated also in a *stochastic* framework by allowing both the initial state and the input to the system to be uncertain. More precisely, assume that the controlled system is described by

$$\dot{x}(t) = A(t)x(t) + B(t)u(t) + v(t),$$
$$x(t_0) = x_0$$

where v is a zero mean gaussian white noise with intensity V and x_0 is a gaussian random variable with expected value $\bar{x}_0$ and variance matrix Π_0. Furthermore, it is assumed that x_0 is independent from v. The performance index to be minimized is

$$J_s = E\left[\int_{t_0}^{t_f} [x'(t)Q(t)x(t) + u'(t)R(t)u(t)]dt + x'(t_f)Sx(t_f)\right]$$

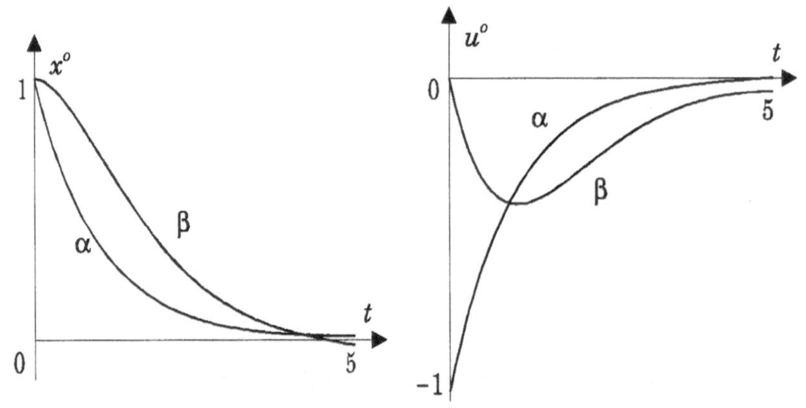

Figure 3.12: Example 3.9: responses of x^o and u^o when $\dot{u}$ is not weighted (α) and weighted (β).

where $Q \geq 0$, $S \geq 0$ and $R > 0$. It is not difficult to guess that if the state can be measured then the solution of the problem is constituted by the same control law which is optimal for its deterministic version, namely the law defined by eqs. (3.3)–(3.5). This claim can easily be proved if only the linear optimal control law has to be found, even if it holds also in the general case under the assumptions adopted here on v and x_0. In fact, corresponding to the control law $u(x,t) = K(t)x$ the value taken by the index J_s is

$$J_s = \mathrm{tr}[P_K(t_0)(\Pi_0 + \bar{x}_0\bar{x}_0') + \int_{t_0}^{t_f} VP_K(t)dt]$$

as in Appendix A, Section A.5. Matrix P_K is the solution of the Lyapunov differential equation

$$\dot{P} = -P(A+BK) - (A+BK)'P - (Q + K'RK)$$

with boundary condition $P(t_f) = S$. By letting $V = 0$ and $\Pi_0 = 0$ (that is operating in an essentially deterministic setting) $\bar{x}_0'P_K(t_0)\bar{x}_0$ (which is the value of J_s in such a setting) must be not smaller, due to an optimality argument, than $\bar{x}_0'P(t_0)\bar{x}_0$, where P is the solution of eqs. (3.4), (3.5). Since $\bar{x}_0$ and t_0 are both arbitrary, it follows that $P_K(t) \geq P(t)$, $t \in [t_0, t_f]$. Obviously $P_K(\cdot) = P(\cdot)$ when $K = -R^{-1}B'P$ (note that in such a case the Lyapunov and Riccati equations do coincide) so that J_s is minimized by this choice of K. Indeed, $\mathrm{tr}[P_K\Delta] \geq \mathrm{tr}[P\Delta]$, for all $\Delta = \Delta' \geq 0$, as this inequality is equivalent to $\mathrm{tr}[(P_K - P)\Delta] \geq 0$ which is no doubt verified since the eigenvalues of the product of two positive semidefinite matrices are real and nonnegative.

3.3 Infinite control horizon

By no means can the linear-quadratic optimal control problem over an *infinite* horizon be viewed as a trivial extension of the problem over a *finite* horizon, which has been considered to some extent in the previous section. As a matter of fact, the assumptions which have proved to be sufficient in the latter case are no longer such in the former one, as shown in the following simple example.

Example 3.10 Consider the system in Fig. 3.13: it is constituted by three objects with mass m which can move without friction along a straight trajectory. The first and second of them are linked to the third one through a spring with stiffness k. Thus, if x_i, $i = 1, 2, 3$ and x_4, x_5, x_6 denote their positions and velocities, respectively, the model for this system is

$$\begin{aligned}
\dot{x}_1 &= x_4, \\
\dot{x}_2 &= x_5, \\
\dot{x}_3 &= x_6, \\
\dot{x}_4 &= -x_1 + x_3, \\
\dot{x}_5 &= -x_2 + x_3, \\
\dot{x}_6 &= u + x_1 + x_2 - 2x_3
\end{aligned}$$

where, for the sake of simplicity, $m = 1$, $k = 1$ while u is the force which can be applied to the third object. The goal is to make the first two objects move as close as possible to each other without resorting to large control actions. If the control interval is infinite, these requirements can adequately be expressed by the criterion

$$J = \int_0^\infty [(x_1 - x_2)^2 + (x_4 - x_5)^2 + u^2]dt.$$

By letting $\varepsilon_1 := x_1 - x_2$ and $\varepsilon_2 := x_4 - x_5$ it is easy to check that

$$\begin{aligned}
\dot{\varepsilon}_1 &= \varepsilon_2, \\
\dot{\varepsilon}_2 &= -\varepsilon_1
\end{aligned}$$

so that

$$\begin{aligned}
\varepsilon_1(t) &= \varepsilon_2(0)\sin(t) + \varepsilon_1(0)\cos(t), \\
\varepsilon_2(t) &= \varepsilon_2(0)\cos(t) - \varepsilon_1(0)\sin(t).
\end{aligned}$$

Thus the performance index is

$$J = 2\int_0^\infty (\varepsilon_1^2(0) + \varepsilon_2^2(0))dt + \int_0^\infty u^2 dt$$

and is obviously unbounded whenever $\varepsilon_1^2(0) + \varepsilon_2^2(0) \neq 0$.

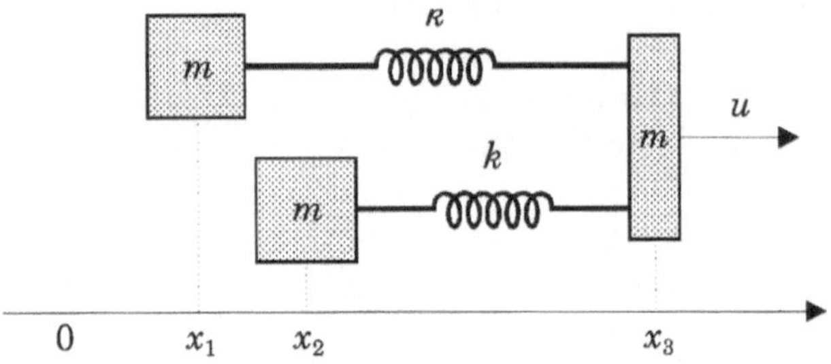

Figure 3.13: The system considered in Example 3.10.

This example suggests that controllability is the further assumption to be added, if the existence of the solution has to be guaranteed. Indeed, the solution fails to exist because the performance index (explicitly) depends upon state variables which are not initially zero but belong to the uncontrollable part of the system, this part being not asymptotically stable. As shown in the sequel, the new assumption is sufficient, together with the previous ones (continuity and sign definiteness), to make the solution exist for each initial state.

Problem 3.1 will now be discussed for $t_f = \infty$ and $S = 0$. This particular choice for S is justified mainly by the fact that in the most significant class of LQ problems over an infinite horizon, the state asymptotically tends to zero and a nonintegral term in the performance index would be useless. The LQ problem over an infinite horizon is therefore stated in the following way.

Problem 3.2 (Linear-quadratic problem over an infinite horizon) *Given the system*

$$\begin{aligned} \dot{x}(t) &= A(t)x(t) + B(t)u(t), \\ x(t_0) &= x_0 \end{aligned} \tag{3.11}$$

where x_0 and t_0 are specified, find a control which minimizes the performance index

$$J = \frac{1}{2}\int_{t_0}^{\infty} [x'(t)Q(t)x(t) + u'(t)R(t)u(t)]dt.$$

The final state is free and $A(\cdot)$, $B(\cdot)$, $Q(\cdot)$, $R(\cdot)$ are continuously differentiable functions; further $Q(t) = Q'(t) \geq 0$, $R(t) = R'(t) > 0$, $\forall t \geq t_0$.

A solution of this problem is provided by the following theorem.

Theorem 3.2 *Let system (3.11) be controllable for each $t \geq t_0$. Then Problem 3.2 admits a solution for each initial state x_0 which is specified by the control* *law*

$$u_c^o(x,t) = -R^{-1}(t)B'(t)\bar{P}(t)x \tag{3.12}$$

where

$$\bar{P}(t) := \lim_{t_f \to \infty} P(t, t_f), \tag{3.13}$$

$P(\cdot, t_f)$ *being the solution of the differential Riccati equation*

$$\dot{P} = -PA - A'P + PBR^{-1}B'P - Q \tag{3.14}$$

satisfying the boundary condition

$$P(t_f, t_f) = 0, \ \ t_0 < t_f < \infty. \tag{3.15}$$

Further, the optimal value of the performance index is

$$J^o(x_0, t_0) = \frac{1}{2} x_0' \bar{P}(t_0) x_0. \tag{3.16}$$

Proof. The proof of this theorem consists of four steps: 1) Existence of $\bar{P}$; 2) Check that $\bar{P}$ solves eq. (3.14); 3) Evaluation of J when the control law (3.12) is implemented; 4) Optimality of the control law (3.12).

Point 1) Controllability for $t \geq t_0$ implies that for each $x(t)$ there exists a bounded control $\hat{u}(\cdot) \in \Omega^m$ defined over the finite interval $[t, T]$ (in general, T depends on $x(t)$ and t) such that $\hat{x}(T) = 0$, where $\hat{x}(\cdot) := \varphi(\cdot; t, x(t), \hat{u}(\cdot))$. Letting $\hat{u}(\tau) = 0, \ \tau > T$ it follows, for $t_0 \leq t \leq t_f < \infty$,

$$\begin{aligned} \frac{1}{2} x'(t) P(t, t_f) x(t) &= J^o(x(t), t, t_f) \\ &\leq J(x(t), t, t_f, \hat{x}(\cdot), \hat{u}(\cdot)) \\ &\leq J(x(t), t, \infty, \hat{x}(\cdot), \hat{u}(\cdot)) \\ &= J(x(t), t, T, \hat{x}(\cdot), \hat{u}(\cdot)) < \infty. \end{aligned}$$

These relations, which imply the boundedness of $x'(t)P(t, t_f)x(t)$ for each finite t_f, follow from (i) the existence of the solution of the LQ problem over any finite horizon; (ii) the nonnegativity of the quadratic function in the performance index; (iii) $\hat{x}(\tau) = 0$, $\hat{u}(\tau) = 0$, $\tau > T$; (iv) the boundedness of $\hat{u}(\cdot)$ and $\hat{x}(\cdot)$. Let now $t_0 \leq t \leq t_{f1} \leq t_{f2}$, $t \leq \xi \leq t_{f2}$, $K^o(\xi, \eta) := -R^{-1}(\xi)B'(\xi)P(\xi, \eta)$, $A_c(\xi, \eta) := A(\xi) + B(\xi)K^o(\xi, \eta)$, $Q_c(\xi, \eta) := Q(\xi) + K^{o\prime}(\xi, \eta)R(\xi)K^o(\xi, \eta)$ and $x_c(\xi, \eta)$ be the solution of

$$\frac{dx(\xi)}{d\xi} = A_c(\xi, \eta) x(\xi)$$

with boundary condition, at the initial time t, $x(t)$. We obtain

$$
\begin{aligned}
x'(t)P(t,t_{f2})x(t) &= \int_t^{t_{f1}} x_c'(\xi,t_{f2})Q_c(\xi,t_{f2})x_c(\xi,t_{f2})d\xi \\
&+ \int_{t_{f1}}^{t_{f2}} x_c'(\xi,t_{f2})Q_c(\xi,t_{f2})x_c(\xi,t_{f2})d\xi \\
&\geq x'(t)P(t,t_{f1})x(t)
\end{aligned}
$$

since the second integral is nonnegative ($Q_c \geq 0$) and $x_c(\xi, t_{f2})$, which is optimal over $[t, t_{f2}]$, is not, in general, optimal over $[t, t_{f1}]$. Thus $x'(t)P(t,\cdot)x(t)$ is a nondecreasing function which, being also bounded, admits a limit as $t_f \to \infty$. Therefore eq. (3.13) is well defined because $x(t)$ is arbitrary.

Point 2) Let $P(\cdot, t_f, S)$ be the solution of eq. (3.14) with $P(t_f, t_f, S) = S$ and $\tau \leq t_f$. Then, $P(t, t_f, 0) = P(t, \tau, P(\tau, t_f, 0))$ so that

$$
\begin{aligned}
\bar{P}(t) &= \lim_{t_f\to\infty} P(t,t_f,0) = \lim_{t_f\to\infty} P(t,\tau,P(\tau,t_f,0)) \\
&= P(t,\tau, \lim_{t_f\to\infty} P(\tau,t_f,0)) = P(t,\tau,\bar{P}(\tau)).
\end{aligned}
$$

Indeed the solution of the DRE depends continuously on the boundary condition and it is therefore possible to evaluate the limit inside the function. Thus $\bar{P}(t)$ equals, for each t, the solution of eq. (3.14) with boundary condition, at *any* instant τ, given by $\bar{P}(\tau)$: in other words, it satisfies such an equation.

Point 3) Letting $\bar{x}$ be the solution of the equation

$$
\frac{dx}{d\xi} = (A - BR^{-1}B'\bar{P})x
$$

with initial condition, at time t, $x(t)$ and setting $\bar{u}(\cdot) := -R^{-1}(\cdot)B'(\cdot)\bar{P}(\cdot)\bar{x}(\cdot)$, we obtain

$$
\begin{aligned}
2J(x(t),t,t_f,\bar{x}(\cdot),\bar{u}(\cdot)) &= \int_t^{t_f} \bar{x}'(\xi)[Q(\xi) + \bar{P}(\xi)B(\xi)R^{-1}(\xi)B'(\xi)\bar{P}(\xi)]\bar{x}(\xi)d\xi \\
&= \int_t^{t_f} \bar{x}'(\xi)[-\dot{\bar{P}}(\xi) - \bar{P}(\xi)A(\xi) - A'(\xi)\bar{P}(\xi) \\
&\quad +2\bar{P}(\xi)B(\xi)R^{-1}(\xi)B'(\xi)\bar{P}(\xi)]\bar{x}(\xi)d\xi \\
&= \int_t^{t_f} -\frac{d}{d\xi}[\bar{x}'(\xi)\bar{P}(\xi)\bar{x}(\xi)]d\xi \\
&= x'(t)\bar{P}(t)x(t) - \bar{x}'(t_f)\bar{P}(t_f)\bar{x}(t_f) \\
&\leq x(t)\bar{P}(t)x(t)
\end{aligned}
$$

having exploited Point 2. For the inequality sign, note that for $t_0 \leq t_f \leq \tau_f$ and for each $z \in R^n$, $z'P(t_f, \tau_f)z$ is nonnegative since it equals twice the optimal value of the performance index for the LQ problem over the interval $[t_f, \tau_f]$ and initial state

z. It has already been checked that such a quantity is a nondecreasing function of τ_f, so we can conclude that

$$\lim_{\tau_f \to \infty} \bar{x}'(t_f)P(t_f, \tau_f)\bar{x}(t_f) = \bar{x}'(t_f)\bar{P}(t_f)\bar{x}(t_f) \geq 0.$$

Therefore, $J(x(t), t, t_f, \bar{x}(\cdot), \bar{u}(\cdot))$, which is apparently a nondecreasing function of t_f, is also bounded, so that there exists its limit as $t_f \to \infty$ and

$$\lim_{t_f \to \infty} J(x(t), t, t_f, \bar{x}(\cdot), \bar{u}(\cdot)) \leq \frac{1}{2}x'(t)\bar{P}(t)x(t).$$

On the other hand, an optimality argument implies that

$$J(x(t), t, t_f, \bar{x}(\cdot), \bar{u}(\cdot)) \geq \frac{1}{2}x'(t)P(t, t_f)x(t) = J^o(x(t), t, t_f).$$

Since the limits of both functions exist, it follows that

$$\lim_{t_f \to \infty} J(x(t), t, t_f, \bar{x}(\cdot), \bar{u}(\cdot)) \geq \frac{1}{2}x'(t)\bar{P}(t)x(t).$$

This relation together with what has previously been found allows us to conclude that

$$J(x(t), t, \infty, \bar{x}(\cdot), \bar{u}(\cdot)) = \frac{1}{2}x'(t)\bar{P}(t)x(t).$$

Point 4) By contradiction assume $u^\star(\cdot) \neq \bar{u}(\cdot)$ to be such that

$$\lim_{t_f \to \infty} J(x(t), t, t_f, x^\star(\cdot), u^\star(\cdot)) \leq \lim_{t_f \to \infty} J(x(t), t, t_f, \bar{x}(\cdot), \bar{u}(\cdot)),$$

$x^\star$ being the state motion corresponding to the input $u^\star$. Then

$$\begin{aligned}\lim_{t_f \to \infty} J^o(x(t), t, t_f) &= \lim_{t_f \to \infty} \frac{1}{2}x'(t)P(t, t_f)x(t) = \frac{1}{2}x'(t)\bar{P}(t)x(t) \\ &\geq \lim_{t_f \to \infty} J(x(t), t, t_f, x^\star(\cdot), u^\star(\cdot)).\end{aligned}$$

This inequality is nonsense as it would entail $J(x(t), t, t_f, x^\star(\cdot), u^\star(\cdot)) \leq J^o(x(t), t, t_f)$ for t_f sufficiently large.

Finally, observe that the value of the performance index corresponding to $\bar{u}$ and $\bar{x}$ proves eq. (3.16)

Example 3.11 Consider the LQ problem defined by the first order system

$$\dot{x} = \frac{x}{t^2} + u$$

and the performance index

$$J = \int_1^\infty (\frac{x^2}{t^2} + t^2u^2)dt.$$

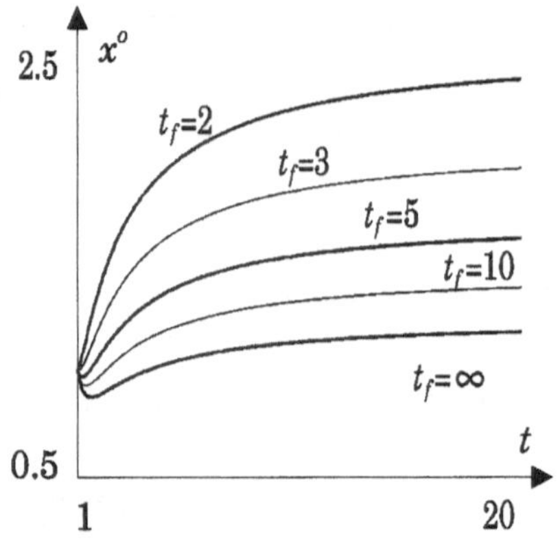

Figure 3.14: Example 3.11: responses of x^o for some values of t_f.

We find

$$P(t,t_f) = (1+\sqrt{2})\frac{1-e^{2\sqrt{2}\frac{t-t_f}{tt_f}}}{1+(1+\sqrt{2})^2 e^{2\sqrt{2}\frac{t-t_f}{tt_f}}}$$

as solution of eqs. (3.14), (3.15) so that

$$\bar{P}(t) = (1+\sqrt{2})\frac{1-e^{-\frac{2\sqrt{2}}{t}}}{1+(1+\sqrt{2})^2 e^{-\frac{2\sqrt{2}}{t}}}.$$

It is simple to check that $\bar{P}$ solves the DRE relevant to the problem at hand and that it is positive definite. In Fig. 3.14 the responses of x^o are shown corresponding to $x(1) = 1$ and control laws which are optimal for some values of the parameter t_f. Note the way such time responses (which are indeed optimal for control intervals ending at t_f) tend to the response resulting from adopting the control law defined by $\bar{P}$.

Remark 3.10 *(Uniqueness of the solution)* The discussion on the uniqueness of the solution (which has been presented in Remark 3.2 with reference to a finite control interval) applies to the present case, provided a suitable limit operation is performed.

3.4 The optimal regulator

Due to the importance of the results and the number of applications, the LQ problem over an infinite horizon when both the system and the performance index are time-invariant, that is when A, B, Q, R are constant matrices is particularly meaningful. The resulting problem is usually referred to as the *optimal regulator* problem and apparently is a special case of the previously considered LQ problem over an infinite horizon. However, it is worth discussing it in detail since the independence of data from time implies a substantial simplification of the relevant results, making their use extremely simple. Thus the problem at hand is

Problem 3.3 (Optimal regulator problem) *For the time-invariant system*

$$\dot{x}(t) = Ax(t) + Bu(t), \tag{3.17}$$
$$x(0) = x_0$$

where x_0 is given, find a control that minimizes the performance index

$$J = \frac{1}{2}\int_0^\infty [x'(t)Qx(t) + u'(t)Ru(t)]dt. \tag{3.18}$$

The final state is unconstrained and $Q = Q' \geq 0,\ R = R' > 0$.

Observe that, thanks to time-invariance, the initial time has been set to 0 without loss of generality. The following result holds for the problem above.

Theorem 3.3 *Let the pair (A, B) be reachable. Then Problem (3.3) admits a solution for each x_0. The solution is specified by the control law*

$$u_{cs}^o(x) = -R^{-1}B'\bar{P}x \tag{3.19}$$

where $\bar{P} = \bar{P}' \geq 0$ solves the algebraic Riccati equation (ARE)

$$0 = PA + A'P - PBR^{-1}B'P + Q \tag{3.20}$$

and is such that $\bar{P} = \lim_{t_f \to \infty} P(t, t_f)$, $P(\cdot, t_f)$ being the solution of the differential Riccati equation $\dot{P} = -PA - A'P + PBR^{-1}B'P - Q$ with boundary condition $P(t_f, t_f) = 0$. Further, the optimal value of the performance index is

$$J^o(x_0) = \frac{1}{2}x_0'\bar{P}x_0. \tag{3.21}$$

Proof. Obviously, it suffices to show that the limit of the solution of the DRE is constant, since the limit itself is a solution of the DRE (see the proof of Theorem 3.2), then it must solve the ARE as well. From the time-invariance of the problem it follows that the optimal values of the performance index (3.18), when the control intervals are $[t_1, \infty)$ or $[t_2, \infty)$, must coincide if the initial state is the same. Thus

$$x_0'\bar{P}(t_1)x_0 = x_0'\bar{P}(t_2)x_0,\ \forall t_1, t_2, x_0$$

which implies that $\bar{P}(\cdot)$ =cost.

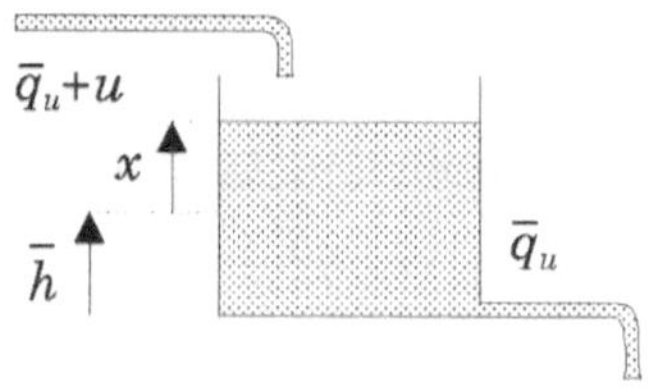

Figure 3.15: The system considered in Example 3.12.

Remark 3.11 *(Control in the neighbourhood of an equilibrium point)* From a practical point of view the importance of the optimal regulator problem is considerably enhanced by the discussion at the beginning of this chapter. Indeed equation (3.17) can be seen as resulting from the linearization of the controlled system about an equilibrium state, say ξ_n. For this system the state ξ is desired to be close to such a point, without requiring, however, large deviations of the control variables η from the value η_n which, in nominal conditions, produces ξ_n. In this perspective, x and u are, with reference to the quoted equation, the state and control deviations, respectively, and the meaning of the performance index is obvious. Further, should the control law (3.19) force the state of system (3.17) to tend to 0 corresponding to any initial state, then it would be possible to conclude that the system has been *stabilized* in the neighbourhood of the considered equilibrium.

Example 3.12 Consider the system shown in Fig. 3.15 where x denotes the difference between the actual and the reference value $\bar{h}$ of the liquid level, while u is the difference between the values of the incoming flow and $\bar{q}_u$, the outgoing flow. Assuming constant and unitary the area of the tank section, the system is described by the equation $\dot{x} = u$, while a significant performance index is

$$J = \int_0^\infty (x^2 + \rho u^2)dt$$

where $\rho > 0$ is a given parameter. The system is reachable and

$$P(t, t_f) = \sqrt{\rho}\frac{1 - e^{\frac{2}{\sqrt{\rho}}(t-t_f)}}{1 + e^{\frac{2}{\sqrt{\rho}}(t-t_f)}}$$

so that

$$\bar{P} = \sqrt{\rho}.$$

It is easy to check that $\bar{P}$, apparently positive, satisfies eq. (3.20) which, however, admits also the (negative) solution $-\sqrt{\rho}$. Thus $\bar{P}$ could have been determined by resorting to the ARE only. The optimal system motion can easily be evaluated after the control law $u^o_{cs}(x) = -x/\sqrt{\rho}$ has been computed. It results that $x^o(t) = e^{-t/\sqrt{\rho}}x_0$. In Fig. 3.16 the responses of x^o and the related optimal control u^o are plotted for $x_0 = 1$. Note that the system response becomes more rapid for low values of ρ at the price of a more demanding control action.

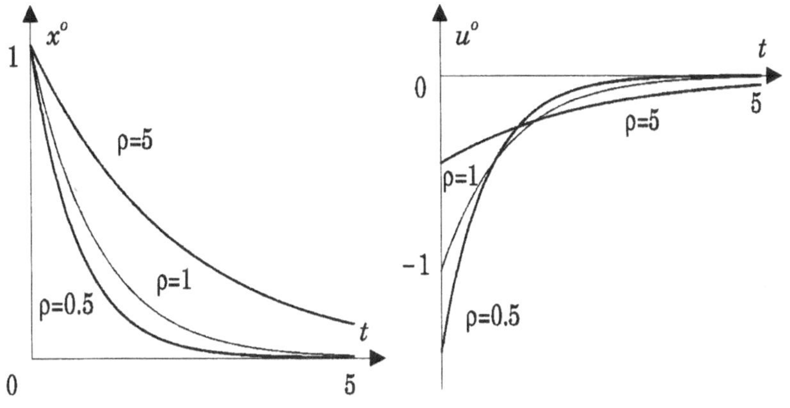

Figure 3.16: Example 3.12: responses of x^o and u^o for some values of ρ.

Example 3.13 Consider an object with unitary mass which can move on a straight line subject to an external force u and a viscous friction. Letting the coefficient relevant to the latter force be unitary, the system description is $\dot{x}_1 = x_2$, $\dot{x}_2 = -x_2 + u$, where x_1 and x_2 are the object position and velocity, respectively. The adopted performance index is

$$J = \int_0^\infty (x_1^2 + u^2)dt.$$

The system is reachable and the ARE admits a unique symmetric, positive semidefinite solution, namely

$$\bar{P} = \begin{bmatrix} \sqrt{3} & 1 \\ 1 & \sqrt{3}-1 \end{bmatrix},$$

so that the optimal control law is $u^o_{cs}(x) = -x_1 - (\sqrt{3}-1)x_2$. If the term $33x_2^2(t)$ is introduced into the performance index with the aim of setting a significant penalty on the deviations of the velocity from 0, an ARE possessing a unique symmetric positive semidefinite solution results and the control law $u^o_{cs}(x) = -x_1 - 5x_2$ is found. In Fig. 3.17 the state responses are reported for $x_1(0) = 1$, $x_2(0) = 1$ and both choices of Q, the first one corresponding to the subscript 1: the consequences of the changes in matrix Q are self explanatory.

Example 3.14 Consider Problem 3.3 with

$$A = \begin{bmatrix} 1 & 1 \\ 0 & 0 \end{bmatrix}, \; B = \begin{bmatrix} 0 \\ 1 \end{bmatrix}, \; Q = \begin{bmatrix} 0 & 0 \\ 0 & 2 \end{bmatrix}, \; R = 1.$$

The pair (A, B) is reachable and the ARE admits two positive semidefinite solutions

$$P_1 = \begin{bmatrix} 0 & 0 \\ 0 & \sqrt{2} \end{bmatrix}, \; P_2 = \begin{bmatrix} 2(\sqrt{2}+1)^2 & 2(\sqrt{2}+1) \\ 2(\sqrt{2}+1) & 2+\sqrt{2} \end{bmatrix}$$

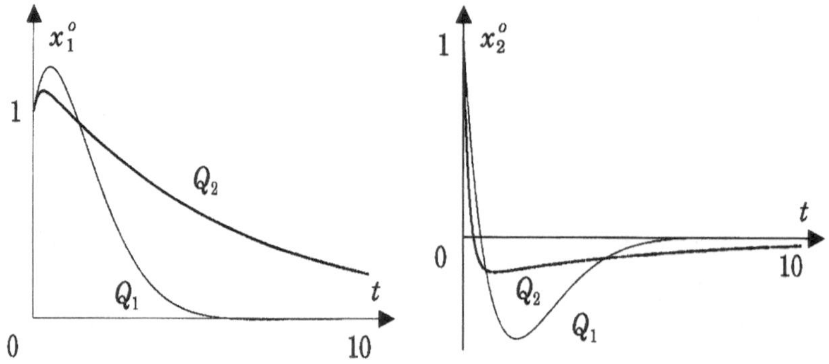

Figure 3.17: Example 3.13: responses of x_1^o and x_2^o for different Q's.

so that it is not possible to determine the optimal control low (at least by exploiting the up to now acquired results) without resorting to integration of the DRE. We find

$$P(t,t_f) = \begin{bmatrix} 0 & 0 \\ 0 & \sqrt{2}\frac{1-e^{2\sqrt{2}(t-t_f)}}{1+e^{2\sqrt{2}(t-t_f)}} \end{bmatrix} \quad \text{and} \quad \bar{P} = \lim_{t_f \to \infty} P(t,t_f) = P_1.$$

Example 3.14 has shown that more than one symmetric positive semidefinite solutions of the ARE may exist: one of them is $\bar{P}$, the matrix which defines the optimal control law (if it exists). The following result allows us to characterize, at least in principle, such a particular solution, provided that the set

$$\mathcal{P} := \{P |\ PA + A'P - PBR^{-1}B'P + Q = 0,\ P = P' \geq 0\}$$

is known.

Theorem 3.4 *Assume that the pair (A, B) is reachable and let P_a be any element of the set $\mathcal{P}$. Then $P_a - \bar{P} \geq 0$.*

Proof. Let $J_a(x_0)$ be the value of the performance index when the control law $u_a(x) := -R^{-1}B'P_a x$ is enforced. Then, by proceeding as in Point 3) of the proof of Theorem 3.2, it follows that

$$J_a(x_0) = \frac{1}{2}x_0' P_a x_0 - \frac{1}{2}\lim_{t_f \to \infty} x_a'(t_f) P_a x_a(t_f) \leq \frac{1}{2}x_0' P_a x_0$$

where x_a is the system motion under the control law u_a. By an optimality argument we can conclude that

$$J^o(x_0) = \frac{1}{2}x_0' \bar{P} x_0 \leq J_a(x_0) \leq \frac{1}{2}x_0' P_a x_0$$

which implies $P_a - \bar{P} \geq 0$, since x_0 is arbitrary.

Example 3.15 Consider the optimal regulator problem defined in Example 3.14: the set $\mathcal{P}$ is made out of two elements only and it is easy to check that

$$P_2 - P_1 = \begin{bmatrix} 2(\sqrt{2}+1)^2 & 2(\sqrt{2}+1) \\ 2(\sqrt{2}+1) & 2 \end{bmatrix} \geq 0.$$

Theorem 3.3 supplies the solution to Problem 3.3 under a *reachability* assumption for the pair (A, B). Apparently, this assumption is unnecessary in general. Indeed assume that the controlled system is constituted by two subsystems $\mathcal{S}_1$ and $\mathcal{S}_2$, independent of each other. One of them, $\mathcal{S}_1$, is reachable while the control has no influence on the second one, so that the whole system is *not* reachable. However, if only the control and the state pertaining to subsystem $\mathcal{S}_1$ appear in the performance index, then the problem can obviously be solved. Thus it is worth seeking the least restrictive assumption which ensures the existence of the solution. The question is made clear by Theorem 3.5 which relies on the following lemma where reference is made to a *factorization* of matrix Q (see Appendix A, Section A.4).

Lemma 3.1 *Let Q be a symmetric positive semidefinite matrix and C_1 and C_2 two distinct factorizations of it. Let A be a square matrix with the same dimension as Q. Then the unobservable subspace of the pair (A, C_1) coincides with the unobservable subspace of the pair (A, C_2).*

Proof. If x is such that $C_1 A^i x = 0$, $i \geq 0$, then $0 = x' A^{i'} C_1' C_1 A^i x = x' A^{i'} C_2' C_2 A^i x$, from which $C_2 A^i x = 0$. Therefore, if x belongs to the unobservable subspace of the pair (A, C_1), that is if $C_1 A^i x = 0$, $i \geq 0$, then it is also $C_2 A^i x = 0$, $i \geq 0$ and x is an element of the unobservable subspace of the pair (A, C_2). By interchanging the subscripts 1 and 2 it is possible to conclude that each element of the unobservable subspace of the pair (A, C_2) also belongs to the unobservable subspace of the pair (A, C_1). Thus the two subspaces coincide.

In view of this lemma the notion of observability for the pair (A, Q) can be introduced in a sharp way, simultaneously allowing us to state the following theorem which gives a necessary and sufficient condition for existence of the solution of the optimal regulator problem. Such a condition is fairly evident if reference is made to the canonical decomposition (see Appendix A, Section A.1) of system (A, B, C), where C is any factorization of Q: for this reason the proof is not given here.

Theorem 3.5 *Problem 3.3 admits a solution for each initial state x_0 if and only if the observable but unreachable part of the triple (A, B, Q) is asymptotically stable.*

Example 3.16 Consider the problem presented in Example 3.10: the system is not reachable and a factorization of Q is

$$C = \begin{bmatrix} 1 & -1 & 0 & 0 & 0 & 0 \\ 0 & 0 & 0 & 1 & -1 & 0 \end{bmatrix}$$

so that the unreachable but observable part is precisely the one described by the two state variables ε_1 and ε_2. The eigenvalues of the relevant dynamic matrix are both zero so that this part is not asymptotically stable. Consistent with this result, Problem 3.3 was found not to admit a solution for each initial state.

Remark 3.12 *(Decomposition of the ARE)* If the triple (A, B, C) is not *minimal*, the ARE to be taken into account simplifies a lot. In fact, the canonical decomposition of the triple induces a decomposition of the equation as well, thus enabling us to set some parts of its solution to zero. More precisely, assume that A, B and C are already in canonical form, namely

$$A = \begin{bmatrix} A_1 & A_2 & A_3 & A_4 \\ 0 & A_5 & 0 & A_6 \\ 0 & 0 & A_7 & A_8 \\ 0 & 0 & 0 & A_9 \end{bmatrix}, \quad B = \begin{bmatrix} B_1 \\ B_2 \\ 0 \\ 0 \end{bmatrix}$$

$$C = \begin{bmatrix} 0 & C_1 & 0 & C_2 \end{bmatrix}$$

and partition matrix P accordingly by letting

$$P := \begin{bmatrix} P_1 & P_2 & P_3 & P_4 \\ P_2' & P_5 & P_6 & P_7 \\ P_3' & P_6' & P_8 & P_9 \\ P_4' & P_7' & P_9' & P_{10} \end{bmatrix}$$

From the differential equations for the P_i's it follows that $P_i(\cdot, t_f) = 0$, $i = 1, 2, 3, 4, 6, 8, 9$ while the remaining blocks solve the three equations

$$\begin{aligned}
\dot{P}_5 &= -P_5A_5 - A_5'P_5 + P_5B_2R^{-1}B_2'P_5 - C_1'C_1, \\
\dot{P}_7 &= -P_7A_9 - (A_5' - P_5B_2R^{-1}B_2')P_7 - P_5A_6 - C_1'C_2, \\
\dot{P}_{10} &= -P_{10}A_9 - A_9'P_{10} - P_7'A_6 - A_6'P_7 + P_7'B_2R^{-1}B_2'P_7 - C_2'C_2,
\end{aligned}$$

which sequentially can be managed. The only nonlinear equation is the first one which actually is the Riccati equation for Problem 3.3 relative to the reachable and observable part of the triple (A, B, C). The two remaining equations are linear. Thus

$$\bar{P} = \begin{bmatrix} 0 & 0 & 0 & 0 \\ 0 & \bar{P}_5 & 0 & \bar{P}_7 \\ 0 & 0 & 0 & 0 \\ 0 & \bar{P}_7' & 0 & \bar{P}_{10} \end{bmatrix}$$

where $\bar{P}_i$, $i = 5, 7, 10$, are the limiting values (as $t_f \to \infty$) of the solutions of the above equations with boundary conditions $P_i(t_f, t_f) = 0$. These matrices are solutions of the algebraic equations which are obtained from the differential ones by setting the derivatives to zero and substituting for P_5 and P_7 their limiting values. The next section will show that $\bar{P}_5$ is such that $A_5 - B_2R^{-1}B_2'\bar{P}_5$ is *stable* (all its eigenvalues have negative real part): this fact implies that the two linear algebraic equations which determine $\bar{P}_7$ and $\bar{P}_{10}$ admit a unique solution. Indeed both of them are of the form $XF + GX + H = 0$ with F and G stable. Thus the solution of Problem 3.3 (when it exists relative to any initial state) can be found by first computing $\bar{P}_5$, solution of the ARE (in principle, by exploiting Theorem 3.4, actually by making reference to the results in Chapter 5) and subsequently determining the (unique) solutions $\bar{P}_7$ and $\bar{P}_{10}$ of the remaining two linear equations.

Finally, if the given triple (A, B, C) is not in canonical form (resulting from a change of variables defined by a nonsingular matrix T) the solution of the problem relies on $P_{or} := T'\bar{P}T$. The check of this claim is straightforward.

Remark 3.13 *(Tracking problem over an infinite horizon)* The optimal tracking problem presented in Remark 3.6 with reference to a finite control interval, can be stated also for an infinite time horizon. This extension is particularly easy if the problem at hand is time-invariant (the matrices which define both the system and the performance index are constant) and the signal to be tracked is the output of a linear time-invariant system. Under these circumstances the optimal control problem is specified by

$$\begin{aligned}\dot{x}(t) &= Ax(t) + Bu(t),\\ y(t) &= Cx(t),\\ x(0) &= x_0\end{aligned}$$

and

$$J = \int_0^\infty \{[y'(t) - \mu'(t)]\hat{Q}[y(t) - \mu(t)] + u'(t)Ru(t)\}dt$$

where μ is the output of the dynamic system

$$\begin{aligned}\dot{\vartheta}(t) &= F\vartheta(t),\\ \mu(t) &= H\vartheta(t),\\ \vartheta(0) &= \vartheta_0.\end{aligned}$$

As in Remark 3.6, $\hat{Q} = \hat{Q}' > 0$ and $R = R' > 0$. Further, due to self-explanatory motivations, the pair (F, H) is assumed to be observable so that if x_0 and ϑ_0 are generic though given, asymptotic stability of F must be required. Under these circumstances it is not difficult to verify that the solution of the problem exists for each x_0 and ϑ_0 if and only if the observable but unreachable part of the triple (A, B, C) is asymptotically stable. The solution can be deduced by noticing that the problem at hand can be given the form of Problem 3.3 provided that the new system

$$\dot{\xi} = W\xi + Vu$$

and the performance index

$$J = \int_0^\infty [\xi'\Theta\xi + u'Ru]dt$$

are considered, where

$$\xi := \begin{bmatrix} x \\ \vartheta \end{bmatrix},\ W := \begin{bmatrix} A & 0 \\ 0 & F \end{bmatrix},\ V := \begin{bmatrix} B \\ 0 \end{bmatrix},\ \Theta := \begin{bmatrix} C'\hat{Q}C & -C'\hat{Q}H \\ -H'\hat{Q}C & H'\hat{Q}H \end{bmatrix}$$

Thus the optimal control law is

$$u_c^o(x,t) = -R^{-1}B'(\bar{P}_1 x + \bar{P}_2\vartheta(t))$$

where $\bar{P}_1$ solves the ARE $0 = PA + A'P - PBR^{-1}B'P + C'\hat{Q}C$ and is such that $\bar{P}_1 = \lim_{t_f\to\infty} P(t,t_f)$, $P(t,t_f)$ being the solution of the DRE $\dot{P} = -PA - A'P + PBR^{-1}B'P - C'\hat{Q}C$ satisfying the boundary condition $P(t_f,t_f) = 0$, while $\bar{P}_2$ solves the linear equation $0 = PF + (A - BR^{-1}B'\bar{P}_1)'P - C'\hat{Q}H$. Finally, the optimal value of the performance index is $J^o(x_0,\vartheta_0) = x_0'\bar{P}_1 x_0 + 2\vartheta_0'\bar{P}_2' x_0 + \vartheta_0'\bar{P}_3\vartheta_0$, where $\bar{P}_3$ is the solution of the Lyapunov equation $0 = PF + F'P - \bar{P}_2'BR^{-1}B'\bar{P}_2 + H'\hat{Q}H$.

Example 3.17 Consider an object with unitary mass which moves without friction along a straight line under the action of an external force u. If x_1 and x_2 denote the position and velocity, respectively, then $\dot{x}_1 = x_2$, $\dot{x}_2 = u$. The goal is to make the position behave as the damped sinusoidal signal μ which is the output of the system $\dot{\vartheta}_1 = -0.1\vartheta_1 + \vartheta_2$, $\dot{\vartheta}_2 = -\vartheta_1 - 0.1\vartheta_2$, $\mu = \vartheta_1$ when $\vartheta(0) = \begin{bmatrix} 1 & 1 \end{bmatrix}'$. Thus a meaningful performance index is

$$J = \int_0^\infty [q(x_1 - \mu)^2 + u^2]dt$$

where $q > 0$. For $q = 1$ we find

$$\bar{P}_1 = \begin{bmatrix} 1.41 & 1 \\ 1 & 1.41 \end{bmatrix},\ \bar{P}_2 = -\begin{bmatrix} 0.70 & 0.87 \\ 0.06 & 0.61 \end{bmatrix},\ \bar{P}_3 = \begin{bmatrix} 1.60 & 0.34 \\ 0.34 & 1.50 \end{bmatrix}$$

while for $q = 10$ we get

$$\bar{P}_1 = \begin{bmatrix} 7.95 & 3.16 \\ 3.16 & 2.51 \end{bmatrix},\ \bar{P}_2 = -\begin{bmatrix} 6.83 & 3.53 \\ 1.83 & 2.05 \end{bmatrix},\ \bar{P}_3 = \begin{bmatrix} 8.25 & 2.50 \\ 2.50 & 4.00 \end{bmatrix}$$

In Fig. 3.18 (a) the time-plots of μ and x_1^o are shown corresponding to these values of q and $x^o(0) = 0$. Note that x_1^o more closely tracks μ as q increases. In Fig. 3.18 (b) the time-plots of μ and x_1^o are reported when $q = 1$ and $x(0) = 0$ or $x(0) = x_0^o = \begin{bmatrix} 1.55 & -0.62 \end{bmatrix}'$, x_0^o being the optimal initial state (recall the discussion in Remark 3.6 concerning the selection of the initial state and observe that it applies also in the case of an infinite time horizon). The apparent improvement resulting from the second choice is also witnessed by the values of the performance index, $J^o(0) = 3.77$ and $J^o(x_0^o) = 1.75$, respectively.

Suppose that the velocity x_2 is now to mimic the signal μ, so that the term $q(x_2 - \mu)^2$ takes the place of the term $q(x_1 - \mu)^2$ in the performance index. We obtain

$$\bar{P}_1 = \begin{bmatrix} 0 & 0 \\ 0 & \sqrt{q} \end{bmatrix},\ \bar{P}_2 = -q\begin{bmatrix} 0 & 0 \\ \frac{\sqrt{q}+0.1}{(\sqrt{q}+0.1)^2+1} & \frac{1}{(\sqrt{q}+0.1)^2+1} \end{bmatrix}$$

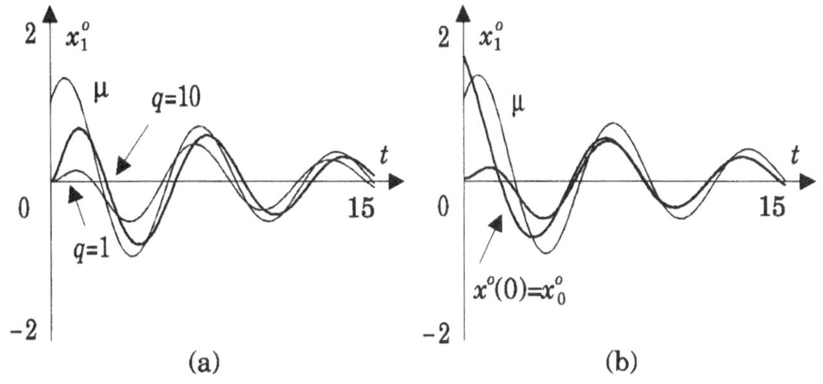

Figure 3.18: Example 3.17: responses of x_1^o for some values of q and $x^o(0) = 0$ (a) or $q = 1$, $x^o(0) = x_0^o$ and $x^o(0) = 0$ (b).

and the optimal initial state (no longer unique since $\bar{P}_1$ is singular) is any vector of the form $x_0^o = \left[\begin{array}{cc} \alpha & \sqrt{q}(\sqrt{q}+1.1)/[(\sqrt{q}+0.1)^2+1] \end{array} \right]'$ where α is an arbitrary real number (notice that $\bar{P}_1^\dagger = \text{diag}[0, 1/\sqrt{q}]$).

Remark 3.14 *(Penalties on the control derivative)* The discussion in Remark 3.8 is still valid in the case $t_f = \infty$ even if some care must be paid to existence of the solution. With the notation adopted there, let Λ_{nr} and $\hat{\Lambda}_{nr}$ be the spectra of the unreachable parts of the pairs (A, B) and $(\hat{A}, \hat{B})$, respectively. Then, $\Lambda_{nr} = \hat{\Lambda}_{nr}$. In fact, if T_r is a nonsingular matrix which performs the canonical decomposition of the pair (A, B) into the reachable and unreachable parts (see Section A.1 of Appendix A), namely a matrix such that

$$T_r A T_r^{-1} = \begin{bmatrix} A_{1r} & A_{2r} \\ 0 & A_{3r} \end{bmatrix}, \; T_r B = \begin{bmatrix} B_{1r} \\ 0 \end{bmatrix}, \; (A_{1r}, B_{1r}) = \text{reachable},$$

then Λ_{nr} is the spectrum of A_{3r}. By letting

$$\hat{T}_r := \begin{bmatrix} T_r & 0 \\ 0 & I \end{bmatrix}$$

we obtain

$$\hat{T}_r \hat{A} \hat{T}_r^{-1} = \begin{bmatrix} A_{1r} & A_{2r} & B_{1r} \\ 0 & A_{3r} & 0 \\ 0 & 0 & 0 \end{bmatrix}, \; \hat{T}_r \hat{B} = \begin{bmatrix} 0 \\ 0 \\ I \end{bmatrix}$$

so that $\Lambda_{nr} \subseteq \hat{\Lambda}_{nr}$, since the spectrum of A_{3r} is a subset of $\hat{\Lambda}_{nr}$. It is not difficult to verify that

$$\left(\begin{bmatrix} A_{1r} & B_{1r} \\ 0 & 0 \end{bmatrix}, \begin{bmatrix} 0 \\ I \end{bmatrix} \right) = \text{reachable}$$

from which $\hat{\Lambda}_{nr} = \Lambda_{nr}$. Indeed, if such a pair is not reachable, then, in view of the *PBH* test (see Theorem A.1 of Section A.1 of Appendix A) it follows that,

$$\begin{bmatrix} A'_{1r} & 0 \\ B'_{1r} & 0 \end{bmatrix} \begin{bmatrix} x \\ u \end{bmatrix} = \lambda \begin{bmatrix} x \\ u \end{bmatrix}, \begin{bmatrix} x \\ u \end{bmatrix} \neq 0,$$

$$\begin{bmatrix} 0 & I \end{bmatrix} \begin{bmatrix} x \\ u \end{bmatrix} = 0.$$

These equations imply that $u = 0$ and, again in view of the *PBH* test, the pair (A_{1r}, B_{1r}) should not be reachable.

Let now Λ_{no} and $\hat{\Lambda}_{no}$ be the spectra of the unobservable parts of the pairs (A, C) and $(\hat{A}, \hat{C})$, respectively, where $\hat{C}'\hat{C} = \hat{Q}$:=diag$[C'C, D'D]$, C and D being factorizations of Q and R, respectively. Then $\Lambda_{no} \subseteq \hat{\Lambda}_{no}$. In fact, let T_o be a nonsingular matrix which performs the canonical decomposition of the pair (A, C) into the observable and unobservable parts, namely a matrix such that

$$T_o A T_o^{-1} = \begin{bmatrix} A_{1o} & 0 \\ A_{2o} & A_{3o} \end{bmatrix}, \; C T_o^{-1} = \begin{bmatrix} C_{1o} & 0 \end{bmatrix}, \; (A_{1o}, C_{1o}) = \text{observable}.$$

Then, Λ_{no} is the spectrum of A_{3o}. If

$$\hat{T}_o := \begin{bmatrix} T_o & 0 \\ 0 & I \end{bmatrix}$$

we obtain

$$\hat{T}_o \hat{A} \hat{T}_o^{-1} = \begin{bmatrix} A_{1o} & 0 & B_{1o} \\ A_{2o} & A_{3o} & B_{2o} \\ 0 & 0 & 0 \end{bmatrix}, \; \hat{C}\hat{T}_o^{-1} = \begin{bmatrix} C_{1o} & 0 & 0 \\ 0 & 0 & D \end{bmatrix}$$

so that $\Lambda_{no} \subseteq \hat{\Lambda}_{no}$ since the spectrum of A_{3o} is a subset of $\hat{\Lambda}_{no}$.

Finally, denote with Λ_{nro} and $\hat{\Lambda}_{nro}$ the spectra of the unreachable but observable parts of the triples (A, B, C) and $(\hat{A}, \hat{B}, \hat{C})$. From the preceding discussion it can be concluded that $\hat{\Lambda}_{nro} \subseteq \Lambda_{nro}$.

It is now possible to state that if a solution of Problem 3.3, defined by the quadruple (A, B, Q, R) exists for each initial state $x(0)$, that is if all elements of Λ_{nro} lie in the open left half-plane (see Theorem 3.5), then a solution of Problem 3.3, defined by the quadruple $(\hat{A}, \hat{B}, \hat{Q}, \hat{R})$ (recall that $\hat{R}$ is the weighting matrix for $\dot{u}$ in the performance index), exists for each initial state $\begin{bmatrix} x'(0) & u'(0) \end{bmatrix}'$, since, necessarily, all elements of $\hat{\Lambda}_{nro}$ lie in the open left half-plane.

In the special case where rank(B) is maximum and equal to the number of columns, the optimal regulator can be given a form different from the one shown in Fig. 3.11 which, referring to a finite control interval can anyhow be adopted also in the present context, the only significant difference being the time-invariance of the system. Since $B'B$ is nonsingular, from the system equation $\dot{x} = Ax + Bu$ it follows that

$$u = (B'B)^{-1} B'(\dot{x} - Ax).$$

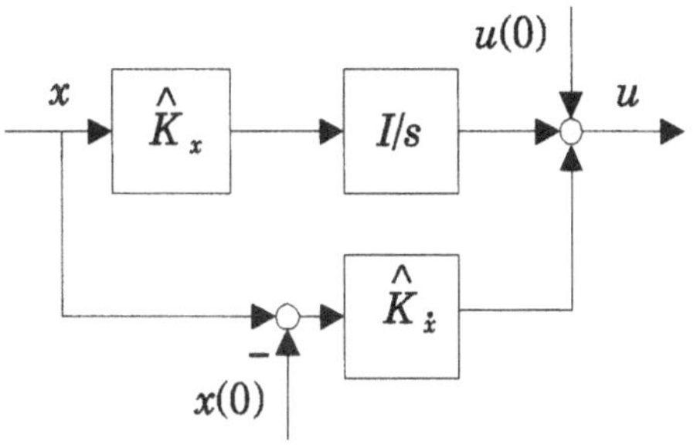

Figure 3.19: The optimal regulator (PI) when the performance index penalizes $\dot{u}$.

On the other hand, the solution of Problem 3.3 implies that $\dot{u} = K_x x + K_u u$ so that

$$\dot{u} = \hat{K}_{\dot{x}}\dot{x} + \hat{K}_x x$$

where $\hat{K}_{\dot{x}} := K_u(B'B)^{-1}B'$ and $\hat{K}_x := K_x - K_u(B'B)^{-1}B'A$. By performing the integration of the equation for $\dot{u}$ between the initial time 0 and a generic instant t we obtain

$$u(t) = \hat{K}_{\dot{x}}(x(t) - x(0)) + \int_0^t \hat{K}_x x(\tau)d\tau + u(0).$$

This is the control law enforced by the system in Fig. 3.19 which can be interpreted as a generalization of the PI controller to the multivariable case.

Remark 3.15 *(Performance evaluation in the frequency-domain)* The synthesis procedure based on the solution of Problem 3.3 can easily be exploited to account for requirements (more naturally) expressed in the frequency domain, as, for instance, those calling for a weak dependence of some variables of interest on others in a specified frequency range. In other words, the presence of harmonic components of some given frequencies in some state and/or control variables, must be avoided, or, equivalently, suitable penalties on them must be set. This can be done in a fairly easy way. Indeed, recall that thanks to Parceval's theorem

$$\int_0^\infty z'(t)z(t)dt = \frac{1}{2\pi}\int_{-\infty}^{\infty} Z^\sim(j\omega)Z(j\omega)d\omega$$

where z is a time function, Z is its Fourier transform and it has obviously been assumed that the written expression makes sense. Therefore, a penalty on some harmonic components in the signal $x(t)$ can be set by looking for the minimization of a performance index of the form

$$\frac{1}{2\pi}\int_{-\infty}^{\infty} X^\sim(j\omega)F_x^\sim(j\omega)F_x(j\omega)X(j\omega)d\omega$$

where F_x is a suitable matrix of shaping functions. If F_x is rational and proper (not necessarily strictly proper) it can be interpreted as the transfer function of a system with input x, $Z(j\omega) = F_x(j\omega)X(j\omega)$ is the Fourier transform of the output $z(t)$ and the integral of $z'z$ is the quantity to be evaluated. The usual performance index takes on the following (more general) form

$$J_f = \frac{1}{2\pi}\int_{-\infty}^{\infty}[X^{\sim}(j\omega)F_x^{\sim}(j\omega)F_x(j\omega)X(j\omega) + U^{\sim}(j\omega)F_u^{\sim}(j\omega)F_u(j\omega)U(j\omega)]d\omega$$

where F_x and F_u are proper rational matrices. This index has to be minimized subject to eq. (3.17). The resulting optimal control problem can be tackled by first introducing two (minimal) realizations of F_x and F_u. Let the quadruples (A_x, B_x, C_x, D_x) and (A_u, B_u, C_u, D_u) define such realizations, respectively, and note that

$$J_f = \int_0^{\infty}[x_A'(t)Q_Ax_A(t) + 2x_A'(t)Z_Au(t) + u'(t)R_Au(t)]dt$$

if $x_A := \begin{bmatrix} x' & z_x' & z_u' \end{bmatrix}'$ with $\dot{x}_A = A_Ax_A + B_Au$ where

$$A_A := \begin{bmatrix} A & 0 & 0 \\ B_x & A_x & 0 \\ 0 & 0 & A_u \end{bmatrix}, \quad B_A := \begin{bmatrix} B \\ 0 \\ B_u \end{bmatrix}$$

and

$$Q_A := \begin{bmatrix} D_x'D_x & D_x'C_x & 0 \\ C_x'D_x & C_x'C_x & 0 \\ 0 & 0 & C_u'C_u \end{bmatrix}, \quad Z_A := \begin{bmatrix} 0 \\ 0 \\ C_u'D_u \end{bmatrix}, \quad R_A := D_u'D_u.$$

Thus the problem with frequency-domain requirements has been restated as an LQ problem over an infinite horizon where a rectangular term is present in the performance index. The results for the optimal regulator problem can be exploited provided only that the quadratic form in u is positive definite, namely if $\text{rank}(D_u)$ equals the number of its columns. Indeed, the state weighting matrix is positive semidefinite since the form of J_f implies that $Q_{Ac} := Q_A - Z_AR_A^{-1}Z_A' \geq 0$ (see Remark 3.4). Assuming that $R_A > 0$, the solution of the problem exists for each initial state $x_A(0)$ if and only if (see Theorem 3.5) the observable but unreachable part of the triple (A_A, B_A, C_A) is asymptotically stable, where

$$C_A := \begin{bmatrix} D_x & C_x & 0 \\ 0 & 0 & C_u \end{bmatrix}$$

is a factorization of Q_A. In conclusion, the problem with frequency requirements can be viewed as a customary optimal regulator problem if F_x and F_u are such as to verify the above assumptions on sign and stability. Note that the resulting regulator is no longer purely algebraic. Indeed, the control variable u depends, through a constant matrix, upon the whole enlarged state vector x_A, so that $u = K_xx + K_{zx}z_x + K_{zu}z_u$. Thus the regulator is a dynamic system with state $\begin{bmatrix} z_x' & z_u' \end{bmatrix}'$, input x (the state of the controlled system) and output u (see also Fig. 3.20).

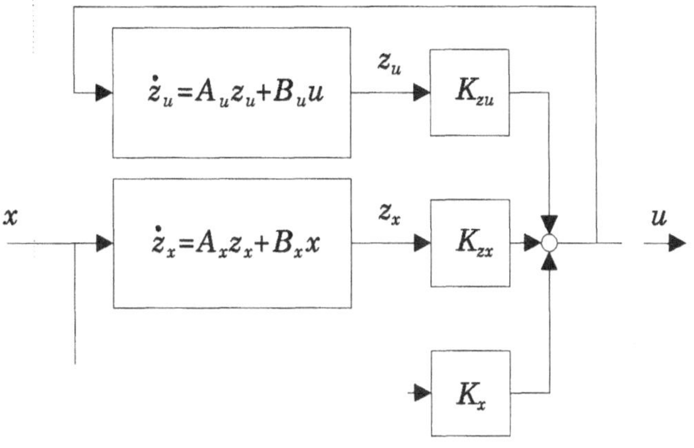

Figure 3.20: The optimal regulator when the performances are evaluated in the frequency-domain.

Example 3.18 Consider the mixer shown in Fig. 3.21 and assume that the concentration inside the tank, of constant section S, is uniform. Further let the outgoing flow q_u be proportional to the square root of the liquid level h. Then the system can be described by the equations

$$\dot{h} = \frac{1}{S}(q_1 + q_2 - \alpha\sqrt{h}),$$
$$\dot{c} = \frac{1}{Sh}[(c_1 - c)q_1 + (c_2 - c)q_2],$$

c_1 and c_2 being the concentrations of the two incoming flows q_1 and q_2, respectively. If the deviations of h, c, q_1, q_2, c_1 and c_2 about the equilibrium, characterized by $\bar{h}$, $\bar{c}$, $\bar{q}_1$, $\bar{q}_2$, $\bar{c}_1$ and $\bar{c}_2$, are denoted by x_1, x_2, u_1, u_2, d_1 and d_2, respectively, we obtain, by linearizing the above equations, $\dot{x} = Ax + Bu + Md$, where

$$A = \begin{bmatrix} -0.1 & 0 \\ 0 & -0.2 \end{bmatrix}, \quad B = \begin{bmatrix} 1 & 1 \\ 0.1 & -0.1 \end{bmatrix}, \quad M = \begin{bmatrix} 0 & 0 \\ 0.1 & 0.1 \end{bmatrix}$$

corresponding to a suitable choice of the physical parameters and the values of the variables at the equilibrium.

The first design is carried on in the usual way by selecting $Q = I$ and $R = 0.1I$, yielding the optimal control law $u = Kx$ with

$$K = -\begin{bmatrix} 2.19 & 1.45 \\ 2.19 & -1.45 \end{bmatrix}.$$

The second design is carried out in complying with the requirement of counteracting the effects on x_2 of a disturbance $d_1 = \sin(0.5t)$, accounting for fluctuations of the

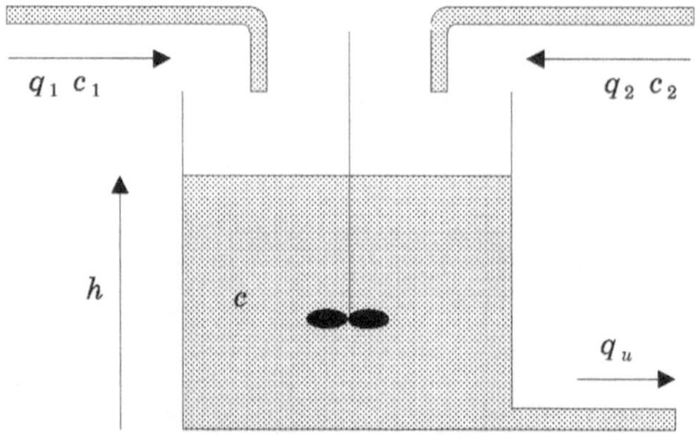

Figure 3.21: The mixer considered in Example 3.18.

concentration c_1. Choosing $F_u(s) = 0.1I$ and

$$F_x(s) = \begin{bmatrix} 1 & 0 \\ 0 & \dfrac{0.25(1+0.3s)^2}{s^2+0.01s+0.25} \end{bmatrix},$$

a realization of the (2,2)-element $\Phi(s)$ of $F_x(s)$ (the amplitude of its frequency-response is shown in Fig. 3.22) is

$$A_x = \begin{bmatrix} -0.01 & -0.25 \\ 1 & 0 \end{bmatrix}, \; B_x = \begin{bmatrix} 1 \\ 0 \end{bmatrix}, \; C_x = \begin{bmatrix} 0.15 & 0.24 \end{bmatrix}, \; D_x = 0.02.$$

Within this new framework the optimal regulator is the second order dynamic system described by $u = K_x x + K_{zx} z_x$ with $\dot{z}_x = A_x z_x + B_x x$ and

$$K_x = -\begin{bmatrix} 2.19 & 2.33 \\ 2.19 & -2.33 \end{bmatrix}, \; K_{zx} = -\begin{bmatrix} 0.99 & -0.22 \\ -0.99 & 0.22 \end{bmatrix}.$$

Figure 3.23 shows the responses of x_2^o and u_1^o when $d_1(t) = \sin(0.5t)$, corresponding to the implementation of the above control laws. Note the better performance of the second regulator as far as the second state variable is concerned and the consequent greater involvement of the control variable.

Remark 3.16 *(Stochastic control problem over an infinite horizon)* The discussion in Remark 3.9 can be suitably modified to cover the case of an unbounded control interval, provided that the material in Appendix A, Section A.5 is taken into consideration. Corresponding to the (time-invariant) system

$$\dot{x}(t) = Ax(t) + Bu(t) + v(t),$$
$$x(0) = x_0$$

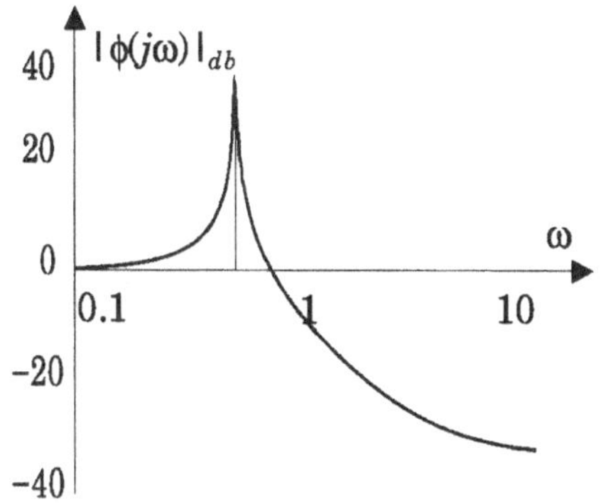

Figure 3.22: Example 3.18: frequency-response of $|\Phi(j\omega)|$.

where v and x_0 are as in Remark 3.9, reference can be made to either the performance index

$$J_{s1} = E\left[\int_0^{\infty} [x'(t)Qx(t) + u'(t)Ru(t)]dt\right]$$

when $v(\cdot) = 0$, or the performance index

$$J_{s2} = E\left[\lim_{T\to\infty}\frac{1}{T}\int_0^{T} [x'(t)Qx(t) + u'(t)Ru(t)]dt\right]$$

when $x_0 = 0$. In both cases the solution, if it exists, is constituted by the control law (3.19) defined in Theorem 3.3. By restraining our attention to the simple cases where $\Pi_0 + \bar{x}_0\bar{x}_0' > 0$ (performance index J_{s1}) and $V > 0$ (performance index J_{s2}), the solution exists if and only if the unreachable but observable part of the triple (A, B, C) is asymptotically stable, C being, as customary, such that $C'C = Q$.

The remaining part of this section is devoted to presenting some particular but important properties of the optimal feedback system and discussing the potentialities of design methods based on the minimization of quadratic indices.

3.4.1 Stability properties

The stability properties of system (3.17), (3.19), that is of system

$$\dot{x}(t) = (A - BR^{-1}B'\bar{P})x(t), \tag{3.22}$$

are now analyzed in detail. The fact that the control law guarantees a finite value of the performance index corresponding to any initial state suggests that system (3.22) should be asymptotically stable if every nonzero motion of the

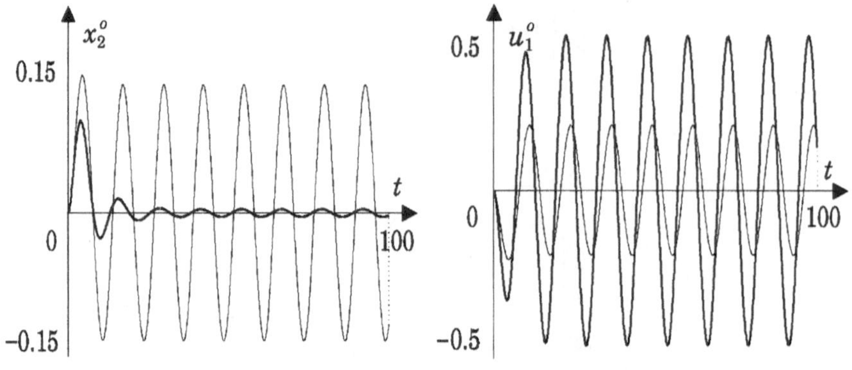

Figure 3.23: Example 3.18: response of x_2^o and u_1^o corresponding to a sinusoidal fluctuation of c_1 when frequency-domain requirements are taken (heavy line) or not taken (light line) into account.

state is *detected* by the performance index. The soundness of this statement is supported by the following simple example.

Example 3.19 Consider Problem 3.3 in the three cases specified by the triple

$$A = \begin{bmatrix} 0 & 1 & 0 \\ 0 & 0 & 1 \\ 0 & 0 & 0 \end{bmatrix}, \quad B = \begin{bmatrix} 0 \\ 0 \\ 1 \end{bmatrix}, \quad R = 1$$

and the matrices Q_i, $i = 1, 2, 3$ which have the element (i, i) equal to 1 and all others equal to 0. The solution of the problem exists in every case, since the pair (A, B) is reachable and is constituted by the control laws

$$\begin{aligned} u_{cs1}^o(x) &= K_1 x := -x_1 - 2x_2 - 2x_3, \ Q = Q_1, \\ u_{cs2}^o(x) &= K_2 x := -x_2 - \sqrt{2}x_3, \ Q = Q_2, \\ u_{cs3}^o(x) &= K_3 x := -x_3, \ Q = Q_3. \end{aligned}$$

The resulting closed loop system is asymptotically stable, simply stable and unstable, respectively, as could easily have been forecast. Indeed, the three matrices Q_i can be factorized as $Q_i = C_i' C_i$, where C_i is a 1×3 matrix having the $(1, i)$ element equal to 1, while the remaining elements are 0. Letting $y_i := C_i x$, the three performance indices are nothing but the integral of $y_i^2 + u^2$ and their boundedness requires the asymptotic zeroing of y_i which, in turn, implies the asymptotic zeroing of x (and hence asymptotic stability of the system since the initial state is arbitrary) only in the first case which is the one where the pair (A, C_i) is observable. In the remaining cases the unobservable part of the pair (A, C_i) has zero eigenvalues and is not stabilized by the control law which, nevertheless, is optimal.

The forthcoming Theorems 3.6 and 3.7 establish precise connections between the characteristics of the performance index and the stability properties of the resulting closed loop system. They rely on the following lemma.

Lemma 3.2 *Let the pair (A, B) be reachable and $Q = C'C$. Then the matrix $\bar{P}$ which specifies the optimal control law for Problem 3.3 is positive definite if and only if the pair (A, C) is observable.*

Proof. Assume that the pair (A, C) is observable and the matrix $\bar{P}$ is not positive definite so that $x_0'\bar{P}x_0 = 0$ for some suitable $x_0 \neq 0$. This implies that the performance index (3.13) is zero when $x(0) = x_0$ and the control law (3.19) is implemented. Since R is positive definite it follows that $u^o(\cdot) = 0$ and, if x_l is the state free motion originated in x_0, also $x_l'(\cdot)C'Cx_l(\cdot) = 0$, i.e., $Cx_l(\cdot) = 0$, contradicting the observability of (A, C). Vice versa, if the pair (A, C) is not observable, then, corresponding to $u(\cdot) = 0$, $Cx_l(\cdot) = 0$ for some suitable $x_0 \neq 0$ and the performance index (3.18) is zero. Because of optimality also $J^o(x_0) = 0$ and from eq. (3.21) it follows that $\bar{P}$ is not positive definite.

Theorem 3.6 *Let $Q = C'C$ and the triple (A, B, C) be minimal. Then the closed loop system resulting from the solution of Problem 3.3 is asymptotically stable.*

Proof. The proof is based on Krasowskii's criterion after $V(x) := \frac{1}{2}x'\bar{P}x$ has been chosen as a Lyapunov function. This function is positive definite thanks to Lemma 3.2. Then, recalling that $\bar{P}$ is symmetric and solves the ARE,

$$\begin{aligned}
\frac{dV(x^o(t))}{dt} &= x^{o\prime}(t)\bar{P}\frac{x^o(t)}{dt} = x^{o\prime}(t)\bar{P}(A - BR^{-1}B'\bar{P})x^o(t) \\
&= \frac{1}{2}x^{o\prime}(t)(\bar{P}A + A'\bar{P} - 2\bar{P}BR^{-1}B'\bar{P})x^o(t) \\
&= -\frac{1}{2}x^{o\prime}(t)(Q + \bar{P}BR^{-1}B'\bar{P})x^o(t).
\end{aligned}$$

Thus the time derivative of V evaluated along the system motion is negative semidefinite. Asymptotic stability follows if such a derivative is nonzero for each nonzero motion. In contradiction, assume that this happens when $x(0) = x_0 \neq 0$, so that

$$x^{o\prime}(t)(Q + \bar{P}BR^{-1}B'\bar{P})x^o(t) = 0, \ \forall t.$$

Then $B'\bar{P}x^o(\cdot) = 0$ and $Cx^o(\cdot) = 0$ which implies that x^o solves the equation $\dot{x} = Ax$ as well and the pair (A, C) is not observable.

Example 3.20 Consider the electric circuit described in Example 2.2 and assume that all electric parameters are equal to 1, for simplicity in notation. The system equations become $\dot{x}_1 = x_2$, $\dot{x}_2 = u - x_1$. As a performance index choose

$$J = \int_0^\infty (x_1^2 + x_2^2 + u^2)dt.$$

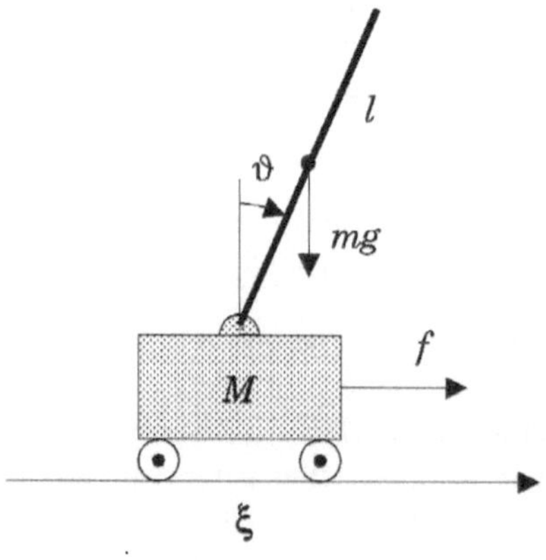

Figure 3.24: The inverted pendulum considered in Example 3.21.

The system is reachable and $Q > 0$, so that observability is ensured. The only symmetric and positive definite solution of the ARE is

$$\bar{P} = \begin{bmatrix} 1.91 & 0.41 \\ 0.41 & 1.35 \end{bmatrix}$$

and the eigenvalues of the closed loop system are $\lambda_{1,2} = -0.68 \pm j0.98$.

The result in Theorem 3.6 is of primary importance as it constitutes an efficient tool for the synthesis of a stabilizing controller. Indeed if Problem 3.3 is stated for the given process and a performance index with $Q > 0$, then the resulting optimal control law is stabilizing. Further, by suitably tuning the elements of Q and R it is also possible to achieve satisfactory transient responses.

Example 3.21 Consider the inverted pendulum shown in Fig. 3.24. The mass of the cart which can move along a straight trajectory under an external force f is M, while the homogeneous rod hinged on the cart is of length $2l$ and mass m. Neglecting all frictions and denoting with ϑ the angle between the rod and the vertical axis and with ξ the cart position, the system equations are

$$\begin{aligned}(M+m)\ddot{\xi} &= ml(\dot{\vartheta}^2 \sin(\vartheta) - \ddot{\vartheta}\cos(\vartheta)) + f,\\ (J+ml^2)\ddot{\vartheta} &= mgl\sin(\vartheta) - ml\ddot{\xi}\cos(\vartheta),\end{aligned}$$

where J is the rod inertia referred to its center of mass and g is the gravity constant. The system linearized equations corresponding to the equilibrium $\xi = 0$, $\dot{\xi} = 0$, $\vartheta = 0$,

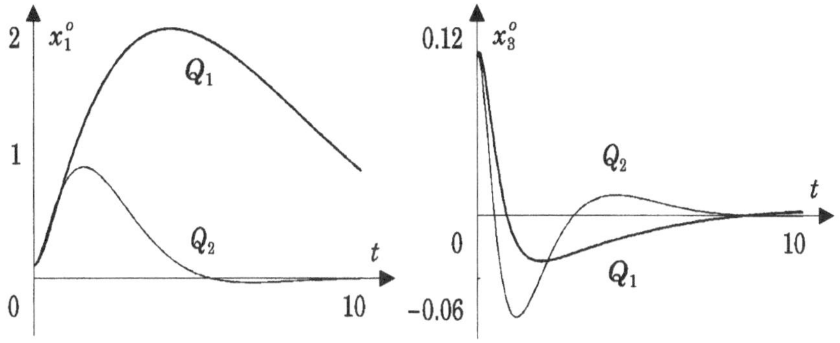

Figure 3.25: Example 3.21: responses of x_1^o and x_3^o.

$\dot{\vartheta} = 0$, $f = 0$ are characterized by the matrices

$$A = \begin{bmatrix} 0 & 1 & 0 & 0 \\ 0 & 0 & -\frac{9}{43}g & 0 \\ 0 & 0 & 0 & 1 \\ 0 & 0 & \frac{39}{43}g & 0 \end{bmatrix}, \quad B = \begin{bmatrix} 0 \\ \frac{52}{559} \\ 0 \\ -\frac{3}{43} \end{bmatrix}$$

with chosen values $m = 3$, $l = 1$, $M = 10$. The linearized system is reachable and unstable. First a performance index with $R = 1$ and $Q = Q_1 = I$ is selected. Then the (1,1) element of the state weighting matrix is increased to 100 in order to improve the transient response of the first state variable x_1 (the deviation of the cart position ξ), yielding the matrix Q_2. The responses of x_1^o and x_3^o (the deviation of the angle ϑ) are shown in Fig. 3.25 corresponding to the above choices for Q and $x(0) = \begin{bmatrix} 0.1 & 0.1 & 0.1 & 0.1 \end{bmatrix}'$.

In view of Theorems 3.5, 3.6 and Remark 3.12 it is straightforward to derive the following result which is therefore given without proof.

Theorem 3.7 *Assume that a solution of Problem 3.3 exists for each initial state. Then the optimal closed loop system is asymptotically stable if and only if the pair (A, Q) is detectable.*

Remark 3.17 *(Existence and stabilizing properties of the optimal regulator)* A summary of the discussion above concerning the existence and the stabilizing properties of the solution of Problem 3.3 is presented in Fig. 3.26 where reference is made to a canonical decomposition of the triple (A, B, C) and the notation of Remark 3.12 is adopted. Further, the term "stab". denotes asymptotic stability and the existence or inexistence of the solution has to be meant for an arbitrary initial state.

Remark 3.18 *(Optimal regulation with constant exogenous inputs)* The results concerning the optimal regulator can be exploited when the system has to be controlled

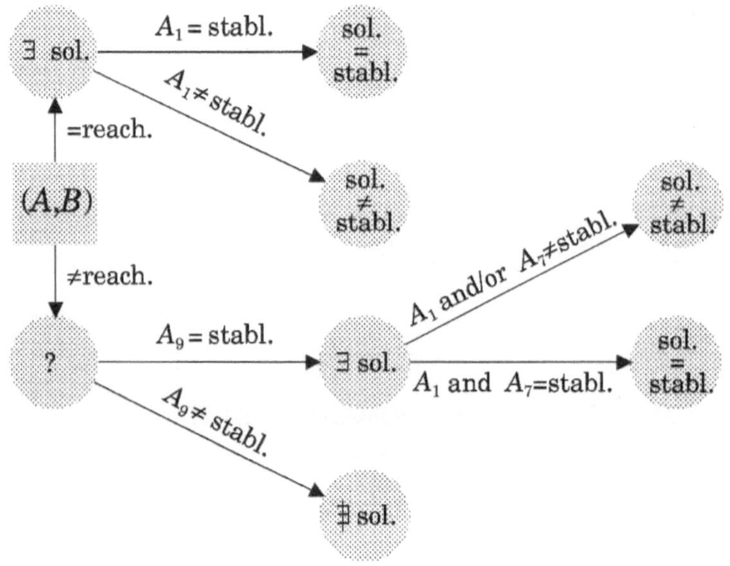

Figure 3.26: Optimal regulator problem: existence and stabilizing properties of the solution.

so as to achieve asymptotic zero error regulation in the presence of unknown inputs of polynomial type. Constraining the discussion (for the sake of conciseness) to the case of constant signals, the controlled system is described by

$$\dot{x}(t) = Ax(t) + Bu(t) + Md, \tag{3.23a}$$
$$y(t) = Cx(t) + Nd, \tag{3.23b}$$
$$x(0) = x_0,$$

and a regulator has to be designed so as to guarantee, for each constant signal y_s and d and each initial state x_0,

$$\lim_{t\to\infty} y(t) = y_s.$$

Within this framework y_s is the set point for y, while d accounts for the disturbances acting on the system input and output. In the present setting the triple (A, B, C) is minimal, the number of control variables equals the number of output variables and the state of the system is available to the controller. In view of Appendix B.4 the controller can be thought of as constituted by two subsystems: the first one is described by the equation

$$\dot{\xi}(t) = y_s - y(t) \tag{3.24}$$

while the second one has to generate the control variable u on the basis of x and ξ in such a way as to asymptotically stabilize the whole system. In designing this second subsystem it is no doubt meaningful to ask for small deviations of the state and control variables from their steady state values together with a fast zeroing of the error. Since the first variations of the involved variables for constant inputs are described by system Σ_v obtained from eqs. (3.23), (3.24) by setting $d = 0$ and $y_s = 0$, namely

$$\begin{aligned}\dot{\delta x}(t) &= A\delta x(t) + B\delta u(t),\\ \delta y(t) &= C\delta x(t),\\ \dot{\delta \xi}(t) &= -\delta y(t),\end{aligned}$$

a satisfactory answer is to let the second subsystem be constituted by the solution of Problem 3.3 for Σ_v and a suitable performance index. Thus, chosen

$$J = \int_0^\infty [\delta y'(t)Q_y\delta y(t) + \delta\xi'(t)Q_\xi\delta\xi(t) + \delta u'(t)R\delta u(t)]dt$$

with Q_ξ and R positive definite and Q_y positive semidefinite, the optimal control law, if it exists, will surely be stabilizing, since Σ_v is observable from ξ (the check can easily be performed via the *PBH* test) and given by

$$\delta u_{cs}^o(\delta x, \delta\xi) = K_1\delta x + K_2\delta\xi.$$

The existence of the solution is guaranteed by the reachability of Σ_v, namely by the fulfillment of the condition

$$\begin{aligned} n+m &= \mathrm{rank}(\begin{bmatrix} B & AB & A^2B & \cdots \\ 0 & -CB & -CAB & \cdots \end{bmatrix}) \\ &= \mathrm{rank}(\begin{bmatrix} A & B \\ -C & 0 \end{bmatrix}\begin{bmatrix} 0 & B & AB & \cdots \\ I & 0 & 0 & \cdots \end{bmatrix}) \end{aligned}$$

which in turn is equivalent to saying that system $\Sigma(A, B, C, 0)$ does not possess transmission zeros at the origin (actually invariant zeros, because of the minimality of Σ). In fact, reachability of the pair (A, B) implies that in the above equation the rank of the second matrix on the right-hand side be equal to $m+n$ and, in view of the already mentioned minimality of $\Sigma(A, B, C, 0)$, that there are transmission zeros at the origin if and only if $Ax+Bu = 0$ and $Cx = 0$ with x and/or u different from 0 (see Section A.3 of Appendix A). On the other hand, if $\Sigma(A, B, C, 0)$, which possesses as many inputs as outputs, has a transmission zero located at the origin, then it would follow that also $A'x + C'y = 0$ and $B'x = 0$ with x and/or y different from 0, which would in turn entail the existence of a zero eigenvalue in the unreachable part of Σ_v, thanks to the *PBH* test. Since this system is observable, we should conclude that no solution exists for Problem 3.3 when stated on such a system. Thus zero error regulation can be achieved in the presence of constant inputs only if none of the transmission zeros of $\Sigma(A, B, C, 0)$ is located at the origin.

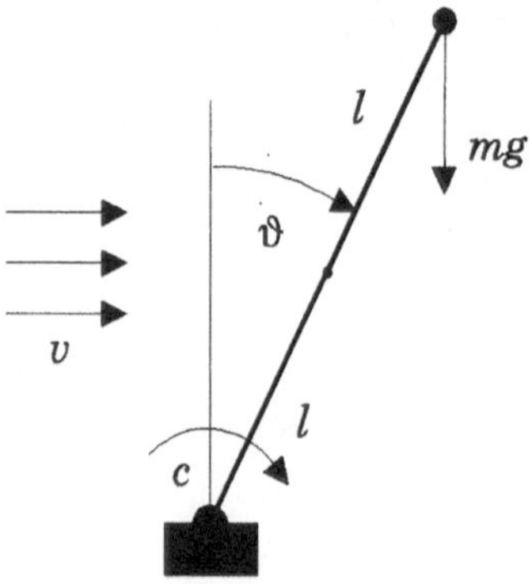

Figure 3.27: The inverted pendulum of Example 3.22.

Example 3.22 Consider the inverted pendulum shown in Fig. 3.27 and assume that the mass of the rod is negligable if compared to the mass m of the sphere located at its end. A torque c (control variable), a wind force and gravity act on the rod of length $2l$. If ϑ is the angle between the rod and the vertical axis and J is the inertia with respect to the hinge, the system equations are

$$J\ddot{\vartheta} = 2lmg\sin(\vartheta) + c + 2l^2(\cos(\vartheta))^2 v$$

where the last term accounts for the wind of intensity v. Letting the two components of x denote the deviations from 0 of the angle ϑ and its derivative, u the deviation of the torque c and d the deviation of the wind intensity from their nominal values, the linearized system is characterized, consistent with the notation in eqs. (3.23), (3.24), by

$$A = \begin{bmatrix} 0 & 1 \\ \frac{g}{2l} & 0 \end{bmatrix}, \quad B = \begin{bmatrix} 0 \\ \frac{1}{4ml} \end{bmatrix}, \quad M = \begin{bmatrix} 0 \\ \frac{1}{2m} \end{bmatrix}.$$

For $m = 1$, $l = 1$, $Q_v = 0$, $R = 1$, the responses of $e := x_{1s} - x_1$ and u when $Q_\xi = 10$ and $Q_\xi = 100$ are shown in Fig. 3.28 corresponding to a unitary set point variation at time 4 and a negative step of amplitude 10 in the wind intensity, occurring at time 0. Note the opposite effects of the changes in Q_ξ on the error and control variables.

Remark 3.19 *(Penalties on the control derivative)* The problem considered in Remarks 3.8 and 3.14 can be discussed further with reference to the stability properties of its solution.

With the same notation adopted in Remark 3.14, first recall that in view of Theorems 3.5 and 3.7, if a solution of Problem 3.3 stated for the quadruple (A, B, Q, R) exists for each (initial state) $x(0)$ and the resulting closed loop system is asymptotically stable, then the set $\Lambda_{no} \cup \Lambda_{nro}$ must be stable, i.e. all its elements must have negative real parts. Assuming that this set is such, it is possible to claim that Problem 3.3 stated for the quadruple $(\hat{A}, \hat{B}, \hat{Q}, \hat{R})$ admits a solution for each initial state

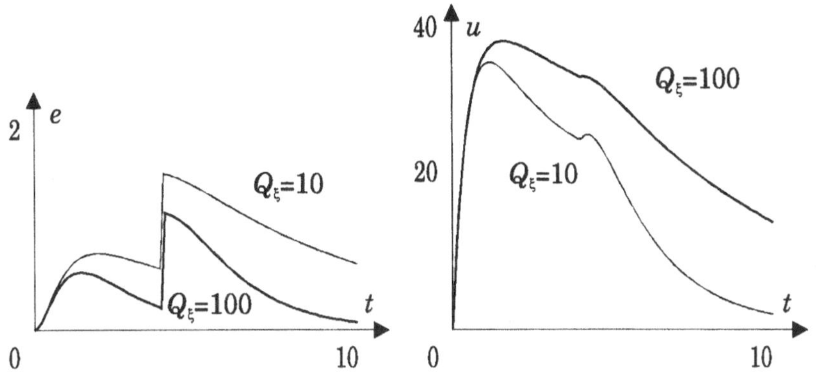

Figure 3.28: Example 3.22: responses of the error and control variables.

$\left[\begin{array}{cc} x'(0) & u'(0) \end{array}\right]'$ if the set $\hat{\Lambda}_{nro}$ is stable. Since $\hat{\Lambda}_{nro} \subseteq \Lambda_{nro}$ (see Remark 3.14), the set $\hat{\Lambda}_{nro}$ is stable if the set Λ_{nro} is such.

If stability of the resulting closed loop system has to be guaranteed as well, stability of the set $\hat{\Lambda}_{no}$ must be checked. To this end, observe that if $\lambda \in \hat{\Lambda}_{no}$, then, thanks to the *PBH* test, it must be

$$\hat{A}\left[\begin{array}{c} x \\ u \end{array}\right] = \lambda \left[\begin{array}{c} x \\ u \end{array}\right], \ \hat{C}\left[\begin{array}{c} x \\ u \end{array}\right] = 0, \ \left[\begin{array}{c} x \\ u \end{array}\right] \neq 0.$$

Since matrix R is positive definite, D (which is a factorization of it) is square and nonsingular without lack of generality, so that these equations imply $u = 0$ since $\hat{C}$ =diag$[C, D]$ and, again from the *PBH* test, $\lambda \in \Lambda_{no}$. Therefore, the set $\hat{\Lambda}_{no}$ is stable if the set Λ_{no} is stable.

Notice that the presence of a penalty term on the control derivative in the performance index allows one to relax the requirement on the sign of R: indeed it appears in matrix $\hat{Q}$ only and can be positive semidefinite. In this case, matrix D, if chosen of maximal rank, is no longer square and the above discussion about stability of the closed loop system has to be modified. In fact, assume as before that the set $\hat{\Lambda}_{nro}$ is stable in order to guarantee the existence of the solution for each initial state and, as for $\hat{\Lambda}_{no}$, note that the above equations (which amount to say $\lambda \in \hat{\Lambda}_{no}$) still imply $u = 0$ if $\lambda \neq 0$: thus such an eigenvalue is certainly stable if Λ_{no} is stable. If, on the other hand, $\lambda = 0$, those equations become $Ax + Bu = 0$, $Cx = 0$: hence a transmission zero of system $\Sigma(A, B, C, 0)$ is located at the origin since x and u cannot simultaneously be zero. Therefore, if R is not positive definite, the conclusion can be drawn that Problem 3.3, stated for the quadruple $(\hat{A}, \hat{B}, \hat{Q}, \hat{R})$, can not admit a stabilizing solution whenever system $\Sigma(A, B, C, 0)$ possesses transmission zeros located at the origin.

Example 3.23 Consider the system defined by the matrices

$$A = \begin{bmatrix} 0 & 1 \\ 1 & 1 \end{bmatrix}, \quad B = \begin{bmatrix} 0 \\ 1 \end{bmatrix}$$

and the performance index characterized by $Q = C'C$, $C = \begin{bmatrix} \alpha & 1 \end{bmatrix}$ and $R = \beta$, where α and β are either 0 or 1. The triple (A, B, C) is minimal, corresponding to both values of α. System $\Sigma(A, B, C, 0)$ has a transmission zero at $-\alpha$. Consider the LQ problem with the quadratic term $(\dot{u})^2$ in the performance index. The solution of this problem exists for all initial states and both values of the parameters α and β since the pair $(\hat{A}, \hat{B})$ is reachable, but, consistent with the discussion in Remark 3.19, the closed loop system is not always stable. Indeed, letting F be the dynamic matrix of the optimal feedback system, we obtain

$$\alpha = 0,\ \beta = 0: \quad \lambda_i(F) = \begin{cases} 0 \\ -1 \\ -\sqrt{2}, \end{cases}$$

$$\alpha = 0,\ \beta = 1: \quad \lambda_i(F) = \begin{cases} -0.50 \\ -1.40 + 0.27j \\ -1.40 - 0.27j, \end{cases}$$

$$\alpha = 1,\ \beta = 0: \quad \lambda_i(F) = \begin{cases} -1.52 \\ -0.70 + 0.40j \\ -0.70 - 0.40j, \end{cases}$$

$$\alpha = 1,\ \beta = 1: \quad \lambda_i(F) = \begin{cases} -1 \\ -1 \\ -\sqrt{2}. \end{cases}$$

The result presented in Theorem 3.6 makes precise the intuition that stability is achieved by the closed loop system if the value of the performance index, which is finite for all initial states, is affected by each nonzero motion of the state. A similar thought suggests that the solution of a suitable LQ problem over an infinite horizon might achieve any desired degree of exponential stability, that is make the system transients extinguish more rapidly than an arbitrarily given exponential function. In fact, if the standard quadratic function in the performance index is multiplied by a factor asymptotically unbounded, the existence of the solution of the optimal control problem (assured, for instance, by reachability of the pair (A, B)) requires that the rate the state approaches zero is faster than the rate such a term goes to infinity. However, this approach might be unsatisfactory since, at least *a priori*, a time varying control law should be expected, the performance index being no longer stationary. This unpleasant outcome can be avoided by a careful choice of the multiplicative factor, as shown in the forthcoming theorem which refers to a generalization of the optimal regulator problem.

Problem 3.4 (Optimal regulator problem with exponential stability) *For the time-invariant system*

$$\dot{x}(t) = Ax(t) + Bu(t), \qquad (3.25)$$
$$x(0) = x_0$$

where x_0 *is given, find a control which minimizes the performance index*

$$J = \int_0^\infty e^{2\alpha t}[x'(t)Qx(t) + u'(t)Ru(t)]dt. \qquad (3.26)$$

No constraints are imposed on the final state and further $Q = Q' \geq 0$, $R = R' > 0$, *while* α *is a given nonnegative real number.*

For this problem the following result holds.

Theorem 3.8 *Let the triple* (A, B, Q) *be minimal. Then the solution of Problem 3.4 exists for each initial state* x_0 *and each* $\alpha \geq 0$. *The solution is characterized by the control law*

$$u^o_{cs\alpha}(x) = -R^{-1}B'\bar{P}_\alpha x \qquad (3.27)$$

where $\bar{P}_\alpha$ *is the symmetric and positive definite solution of the algebraic Riccati equation*

$$0 = P(A + \alpha I) + (A + \alpha I)'P - PBR^{-1}B'P + Q \qquad (3.28)$$

such that $\bar{P}_\alpha = \lim_{t_f \to \infty} P_\alpha(t, t_f)$, $P_\alpha(t, t_f)$ *being the solution of the differential Riccati equation* $\dot{P} = -P(A+\alpha I) - (A+\alpha I)'P + PBR^{-1}B'P - Q$ *with boundary condition* $P_\alpha(t_f, t_f) = 0$.

Further, all eigenvalues of the closed loop system (3.25), (3.28) have real parts smaller than $-\alpha$.

Proof. For the control law (3.27) the proof is simple. In fact, by letting $\hat{x}(t) := e^{\alpha t}x(t)$ and $\hat{u}(t) := e^{\alpha t}u(t)$, Problem 3.4 becomes Problem 3.3 stated on the system

$$\hat{\Sigma}: \ \dot{\hat{x}} = (A + \alpha I)\hat{x} + B\hat{u}$$

and the performance index

$$J = \int_0^\infty [\hat{x}'(t)Q\hat{x}(t) + \hat{u}'(t)R\hat{u}(t)]dt.$$

Note that reachability of the pair (A, B) implies reachability of the pair $((A+\alpha I), B)$. Indeed, if this second pair is not reachable, i.e., (see the *PBH* test) if $(A+\alpha I)'x = \lambda x$ and $B'x = 0$ with $x \neq 0$, then $A'x = (\lambda - \alpha)x$ and also the first pair is not reachable.

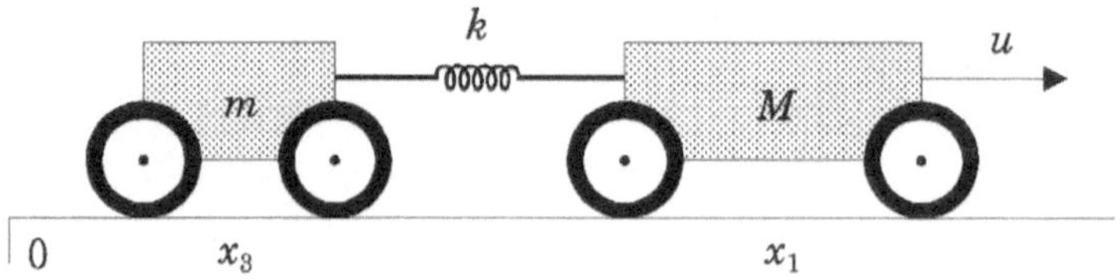

Figure 3.29: The system considered in Example 3.24.

Therefore, the solution of the optimal regulator problem stated on system $\hat{\Sigma}$ exists and is given by the control law

$$\hat{u}_{cs}^{o}(\hat{x}) = -R^{-1}B'\bar{P}_\alpha\hat{x} \tag{3.29}$$

where $\bar{P}_\alpha$ satisfies eq. (3.28) and possesses the relevant properties. By recalling the performed change of variables, eq. (3.27) is found.

For the eigenvalues location, observe that the *PBH* test implies that the pair $((A + \alpha I), Q)$ is observable and the control law (3.29) is stabilizing. Therefore all eigenvalues of $A + \alpha I - BR^{-1}B'\bar{P}_\alpha$ have negative real parts so that the eigenvalues of $A - BR^{-1}B'\bar{P}_\alpha$ have real parts less than $-\alpha$ (the eigenvalues of any matrix $T + aI$ are the eigenvalues of T increased by a).

Example 3.24 Consider the system shown in Fig. 3.29. Assume that both objects, the first one with mass M, the second one with mass m, can move without friction along a straight trajectory. The spring between them has stiffness k, while the external force u is the control variable. A regulator has to be designed along the lines of Theorem 3.8 to make the transients last, from a practical point of view, no more than 1 unit of time. If x is the vector of the positions and velocities of the two objects, the design can be carried on by stating Problem 3.4 on a system defined by

$$A = \begin{bmatrix} 0 & 1 & 0 & 0 \\ -k/M & 0 & k/M & 0 \\ 0 & 0 & 0 & 1 \\ k/m & 0 & -k/m & 0 \end{bmatrix}, \quad B = \begin{bmatrix} 0 \\ 1/M \\ 0 \\ 0 \end{bmatrix}$$

and a performance index characterized by $R = 1$, $Q = I$ and, as a first attempt, by $\alpha = 5$. The choice for α originates from the convention that the transients last 5 times the greatest time constant. Corresponding to $M = 10$, $m = 1$ and $k = 100$, the eigenvalues of the closed loop system are $\lambda_{1,2} = -10.00 \pm 10.49j$, $\lambda_{3,4} = -10.00 \pm 0.01j$ and the state response for $x(0) = \begin{bmatrix} 1 & 1 & 1 & 1 \end{bmatrix}'$ is the one depicted in Fig. 3.30. As a second attempt the choice $\alpha = 3$ is made, since the eigenvalues have unnecessarily been pushed far into the left half-plane. The new eigenvalues are $\lambda_{1,2} = -6.00 \pm 10.49j$ and $\lambda_{3,4} = -6.00 \pm 0.02j$ while the resulting transients are again shown in Fig. 3.30.

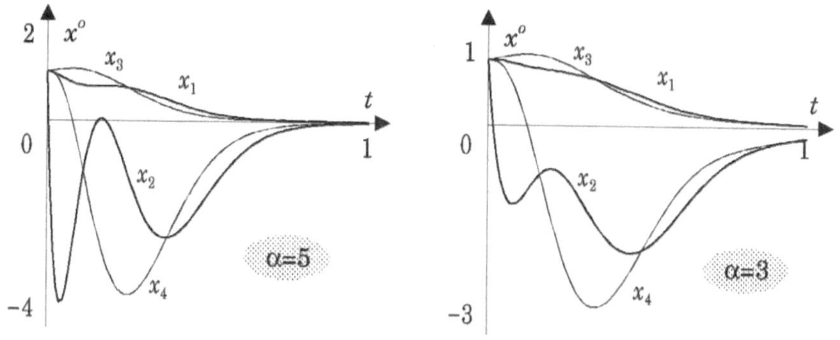

Figure 3.30: Example 3.24: responses of x^o for $\alpha = 5$ and $\alpha = 3$.

3.4.2 Robustness properties

The control law which is a solution of the optimal regulator problem has been shown to be stabilizing under suitable mild assumptions: however, this is not the only nice property that it possesses. First the following lemma (its proof is omitted as not particularly interesting) is needed: reference is made to the system

$$\dot{x}(t) = Ax(t) + Bu(t), \tag{3.30}$$
$$y(t) = Cx(t), \tag{3.31}$$
$$u(t) = -R^{-1}B'Px(t) := Kx(t) \tag{3.32}$$

where P is *any* symmetric solution of the ARE

$$0 = PA + A'P - PBR^{-1}B'P + Q \tag{3.33}$$

with $Q = C'C$ and $R = R' > 0$.

Lemma 3.3 *Let K be given by eqs. (3.32), (3.33). Then*

$$G^{\sim}(s)G(s) = I + H^{\sim}(s)QH(s) \tag{3.34}$$

where

$$G(s) = I - R^{\frac{1}{2}}KH(s)$$
$$H(s) = (sI - A)^{-1}BR^{-\frac{1}{2}}.$$

From eq. (3.34), letting $s = j\omega$, $\omega \in R$, it follows that $G'(-j\omega)G(j\omega) \geq I$ since its left-hand side is an *hermitian* matrix (actually it is the product of a complex

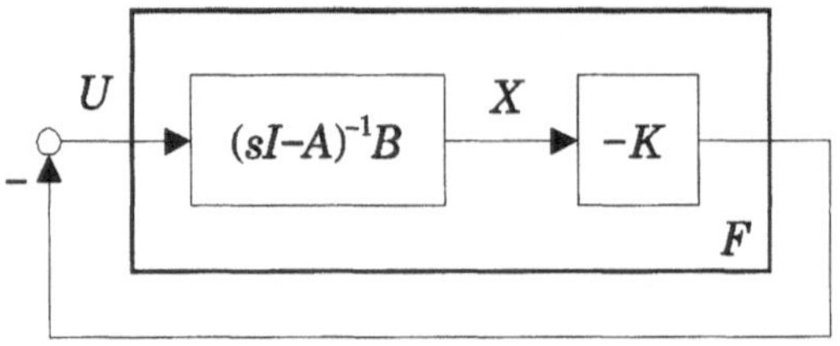

Figure 3.31: The closed loop system resulting from the solution of Problem 3.3.

matrix by its conjugate transpose), while its right-hand side is the sum of the identity matrix with an hermitian positive semidefinite matrix (recall that $Q \geq 0$). In the particular case of a scalar control variable, eq. (3.34) implies that

$$\left|1 - K(j\omega I - A)^{-1}B\right| \geq 1, \ \forall \omega \in R. \tag{3.35}$$

Based on this relation the following theorem states that the optimal closed loop system is *robust* in terms of both *phase and gain margin.*

Theorem 3.9 *Consider the system (3.30), (3.31) and assume that:*

i) *The input u is scalar;*
ii) *The triple (A, B, C) is minimal;*
iii) *In eq. (3.32) $P = \bar{P}$, the solution of eq. (3.33) relevant to Problem 3.3 defined by the quadruple $(A, B, C'C, R)$.*

Then the phase margin of the closed loop system is not less than $\pi/3$ while the gain margin is infinite.

Proof. Letting $F(s) := -\bar{K}(sI - A)^{-1}B$, where $\bar{K} := -R^{-1}B'\bar{P}$, the closed loop system, which is asymptotically stable in view of the above assumptions, can be given the structure shown in Fig. 3.31. The polar locus of the frequency response of F lies, in the complex plane, outside the unit circle centered at $(-1, j0)$ since, thanks to eq. (3.35), the polar locus of the frequency-response of $1 + F$ lies outside the unit circle centered at the origin. By the Nyquist stability criterion it is easy to conclude that the only possible shapes for the polar loci of the frequency-response of F are of the kind of those shown in Fig. 3.32 (a), where the diagrams labeled with a and b refer to the case in which at least one pole of F is in the open right half-plane, while those labeled with c and d refer to the case where no pole of F is located there, if the poles of F lying on the imaginary axis have been surrounded on the right in the Nyquist path. It is now obvious that the intersection of the polar locus of the frequency-response of F with the unit circle centered at the origin must occur within the sector delimited, on the left, by the straight lines r and s shown in Fig. 3.32 (b), which form an angle of $\pm 2\pi/3$ with the real axis. Therefore, the phase margin can

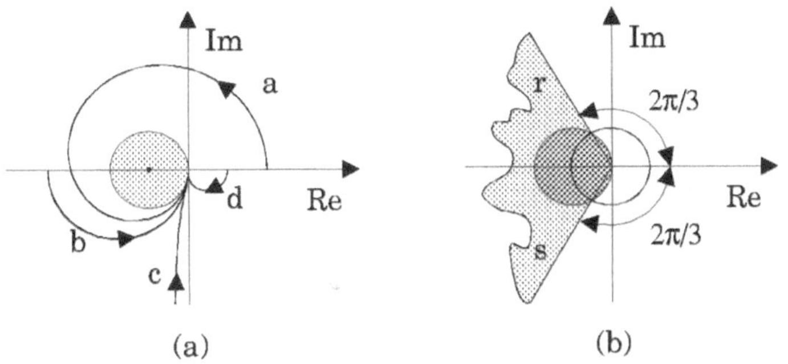

Figure 3.32: (a) Allowed shapes for the polar plot of the frequency-response of F; (b) Constraints on the phase margin.

not be less than $\pi/3$. Finally, the number of encirclements of the Nyquist locus of F around any point belonging to the interval $(-2, 0)$ of the real axis is always the same, and the number of encirclements of the Nyquist loci of μF and F around the point $(-1, j0)$ are equal, provided that $\mu \in (1/2, \infty)$. The gain margin is therefore unbounded since system stability is preserved even if the loop gain is multiplied by an arbitrarily large factor.

Example 3.25 Consider the unstable linear system defined by

$$A = \begin{bmatrix} 0 & 1 \\ 0.75 & -2 \end{bmatrix}, \; B = \begin{bmatrix} 0 \\ 1 \end{bmatrix}.$$

The state variables can be measured and a stabilizing regulator has to be designed to make all real parts of the eigenvalues not greater than -0.5. By exploiting pole-placement techniques (see Appendix B.1, Section B.2) we find that the control law $u = K_a x$ with $K_a = \begin{bmatrix} -1 & 1 \end{bmatrix}$ is such that the two eigenvalues are $\lambda_{1,2} = -0.5$. On the contrary, by resorting to the optimal regulator theory and choosing, after a few trials, $Q = \text{diag}(1, 0)$ and $R = 1$, we get a control law $u = K^o x$, $K^o = -\begin{bmatrix} 2.00 & 0.83 \end{bmatrix}$, and the two eigenvalues $\lambda_1 = -0.55$ and $\lambda_2 = -2.28$. The gain margin resulting from the first control law is quite small: indeed it suffices to multiply by 3 the loop gain in order to lose stability (the eigenvalues of $A + 3BK_a$ are $\lambda_{1,2} = 0.50 \pm j1.41$). On the contrary, the eigenvalues of $A + 3BK^o$ are $\lambda_{1,2} = -2.24 \pm j0.47$. As for the phase margin, the second design appears to be more satisfactory (see Fig. 3.33 (a) where the Nyquist loci of $F_a(s) := -K_a(sI - A)^{-1}B$ and $F^o(s) := -K^o(sI - A)^{-1}B$ are shown). Finally, the responses of the first state variable are plotted in Fig. 3.33 (b) when $u = Kx + v$, $v(t) = 1$, $t > 0$, $K = K_a$ or $K = K^o$. Since the two steady state conditions are different, in the second case the value of the state variable has been multiplied by the ratio β between the two steady state values. This allows a better comparison of the transient responses: specifically, the worst performance of the first design is enhanced.

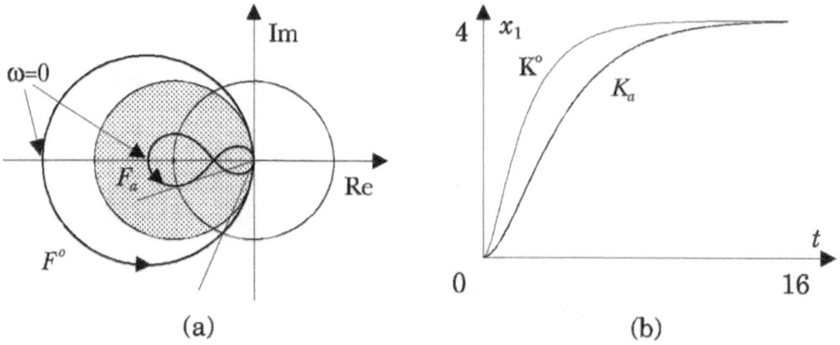

Figure 3.33: Example 3.25: (a) Nyquist loci of F_a and F^o; (b) step responses of x_1 (when $K = K^o$ the values are multiplied by β).

When the control variable is not scalar a similar result holds: it is stated in the following theorem, the proof of which can be found in the literature [12, 33, 35].

Theorem 3.10 *Consider Problem 3.3 and assume that the pair (A, B) is reachable, $Q > 0$ and $R > 0$ diagonal. Then each one of the m loops of the system resulting from the implementation of the optimal control law possesses a phase margin not smaller than $\pi/3$ and an infinite gain margin.*

The significance of this result can be better understood if reference is made to the block-scheme in Fig. 3.34. When the transfer functions $g_i(s)$ are equal to 1 it represents the system resulting from the implementation of the optimal control law $u = \bar{K}x = -R^{-1}B'\bar{P}x$. Let $F_i(s)$ be the transfer function obtained by opening the i-th loop (that is the one corresponding to the i-th control variable). We get

$$F_i(s) = [\bar{K}]_i(sI - (A + \sum_{\substack{j=1 \\ j\neq i}}^{m} [B]^j[\bar{K}]_j))^{-1}[B]^i.$$

Theorem 3.10 guarantees that the system remains stable if the functions $g_i(s)$ introduce a phase lag less than $\pi/3$ at the critical frequency of $F_i(s)$, or are purely nondynamic systems with an arbitrarily high gain.

Example 3.26 Consider the system shown in Fig. 3.35 which is constituted by a tank of constant section. The temperature ϑ_i of the incoming flow q_i is constant while the temperature ϑ of the outgoing flow q_u equals the temperature existing in each point of the tank. The external temperature is ϑ_e and Q_r is the heat supplied to the

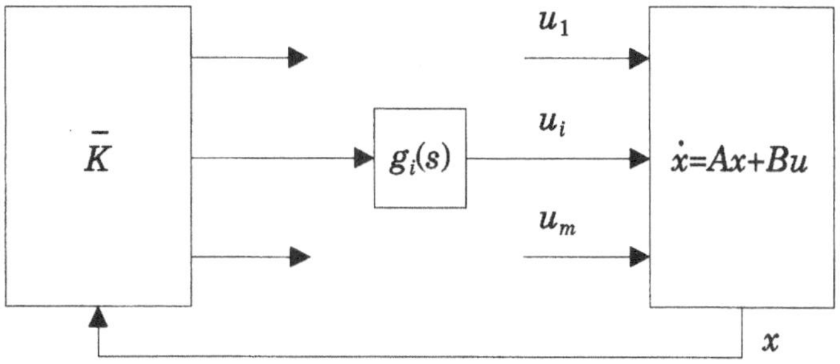

Figure 3.34: Block-scheme of the optimal control system.

liquid. If the thermal capacity of the tank is neglected, the system can be described by the equations

$$\dot{h} = \alpha_1(q_i - \beta\sqrt{h}),$$
$$h\dot{\vartheta} = \alpha_2 q_i(\vartheta_i - \vartheta) + \alpha_3 Q_r + \alpha_4 h(\vartheta_e - \vartheta)$$

where the value of the outgoing flow is assumed to be β times the square root of the liquid level while α_j, $j = 1, \ldots, 4$ are constants depending on the physical parameters of the system. By linearizing these equations and denoting with x_1, x_2, u_1, u_2, d_1, d_2 the deviations of h, ϑ, q_i, Q_r, β, ϑ_e, respectively, the system $\dot{x} = Ax + Bu + Md$ is obtained, where

$$A = \begin{bmatrix} -1/8 & 0 \\ -10/16 & -1/2 \end{bmatrix}, \quad B = \begin{bmatrix} 1 & 0 \\ -17/4 & 1/4000 \end{bmatrix}, \quad M = \begin{bmatrix} -2 & 0 \\ 0 & 1/4 \end{bmatrix},$$

corresponding to a particular equilibrium condition and suitable values of the parameters. The controller must attain asymptotic zero-error regulation of the liquid level and temperature in the face of step variations of the flow q_u and temperature ϑ_e (that is for arbitrary d_1 and d_2), while ensuring good transients. According to Remark 3.18 two integrators (with state ξ) are introduced together with a quadratic performance index characterized by $R = \text{diag}[1, 10^{-5}]$, $Q_x = I$ and $Q_\xi = 1000I$ (note that in this specific case the state of the controlled system coincides with the output to be regulated). The responses of x_1^o and x_2^o are reported in Fig. 3.36 corresponding to a step variation of d_1 (at $t = 0$) with amplitude 0.5 and d_2 (at $t = 5$) with amplitude 10. In the same figure the responses corresponding to the same inputs are reported when also the actuators dynamics are taken into account. If the transfer functions of these devices (controlling the incoming flow q_i and the heat Q_r) are $g_1(s) = 1/(1+0.05s)$ and $g_2(s) = 1/(1+s)$, respectively, one can check that they add a phase lag of approximately 45° at the critical frequencies of the two feedback loops. The figure proves that the control system designed according to the LQ approach behaves satisfactorily, even in the presence of the neglected dynamics.

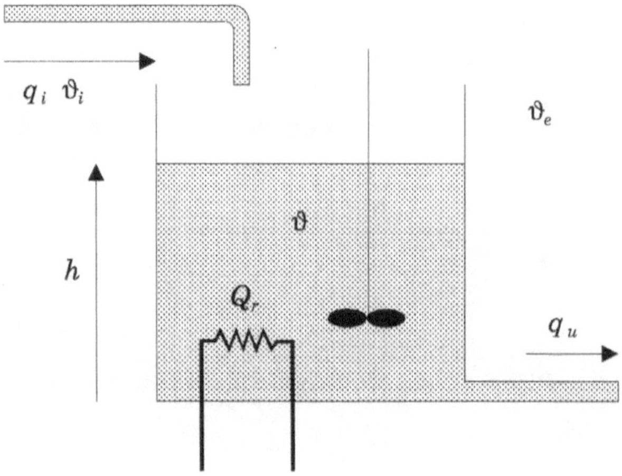

Figure 3.35: The system considered in Example 3.26.

3.4.3 The cheap control

The behaviour of the solution of the optimal regulator problem is now analyzed when the penalty set on the control variable becomes less and less important, that is when it is less and less mandatory to keep the input values at low levels. Obviously, when the limit situation is reached where no cost is associated to the use of the control variable (the control has become a *cheap* item) the input can take on arbitrarily high values and the ultimate capability of the system to follow the desired behaviour (as expressed by the performance index) is put into evidence.

With reference to the system

$$\dot{x}(t) = Ax(t) + Bu(t), \tag{3.36a}$$

$$y(t) = Cx(t), \tag{3.36b}$$

$$x(0) = x_0, \tag{3.36c}$$

which is assumed to be both reachable and observable, consider the performance index

$$J(\rho) = \frac{1}{2}\int_0^{\infty} [y'(t)y(t) + \rho u'(t)Ru(t)]dt \tag{3.37}$$

where $R = R' > 0$ is given and $\rho > 0$ is a scalar. Obviously, the desired behaviour for the system is $y(\cdot) = 0$.

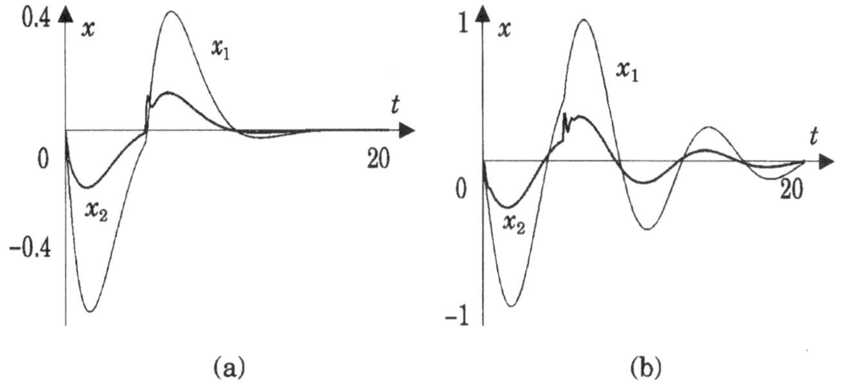

Figure 3.36: Example 3.26: state responses in design conditions (a) and when the actuators dynamics are inserted (b).

Under the assumptions above, the solution of Problem 3.3 defined by eqs. (3.36)–(3.37) exists for each initial state and $\rho > 0$ and is specified by the control law

$$u^o_{cs}(x,\rho) = -\frac{1}{\rho}R^{-1}B'P(\rho)x \tag{3.38}$$

where $P(\rho)$ is the (unique) positive definite solution of the ARE (see Theorems 5.8 and 5.10 in Section 5.3 of Chapter 5)

$$0 = PA + A'P - \frac{1}{\rho}PBR^{-1}B'P + C'C.$$

A preliminary result concerning the asymptotic properties of $P(\rho)$, is given in the following lemma where $Q := C'C$, $x^o(t,\rho)$ is the solution of eqs. (3.36a), (3.36c), (3.38) and $u^o(t,\rho)$ is the relevant control.

Lemma 3.4 *The limit of $P(\rho)$ as $\rho \to 0^+$ exists and is denoted by P_0.*

Proof. For $\rho_1 > \rho_2 > 0$ the following relations hold:

$$\begin{aligned}
x_0'P(\rho_1)x_0 &= \int_0^\infty [x^{o\prime}(t,\rho_1)Qx^o(t,\rho_1) + \rho_1 u^{o\prime}(t,\rho_1)Ru^o(t,\rho_1)]dt \\
&\geq \int_0^\infty [x^{o\prime}(t,\rho_1)Qx^o(t,\rho_1) + \rho_2 u^{o\prime}(t,\rho_1)Ru^o(t,\rho_1)]dt \\
&\geq \int_0^\infty [x^{o\prime}(t,\rho_2)Qx^o(t,\rho_2) + \rho_2 u^{o\prime}(t,\rho_2)Ru^o(t,\rho_2)]dt \\
&= x_0'P(\rho_2)x_0
\end{aligned}$$

since the pair $(x^o(t,\rho_1), u^o(t,\rho_1))$, which is optimal when $\rho = \rho_1$, is no longer such, in general, when $\rho = \rho_2 \neq \rho_1$. Therefore, $x_0'P(\rho)x_0$ is a monotonic nondecreasing function, bounded from below (because it is nonnegative) and admits a limit when $\rho \to 0^+$. Since x_0 is arbitrary, also the limit of $P(\rho)$ exists.

A meaningful measure of how similar the system response is to the desired one (that is how $y(\cdot)$ is close to zero) is supplied by the quantity

$$J_x(x_0,\rho) := \frac{1}{2}\int_0^\infty x^{o\prime}(t,\rho)Qx^o(t,\rho)dt,$$

the limiting value of which is given in the following theorem.

Theorem 3.11 *Let P_0 be the limit value of $P(\rho)$. Then*

$$\lim_{\rho\to 0+} J_x(x_0,\rho) = \frac{1}{2}x_0'P_0x_0.$$

Proof. The form of the performance index (3.37) makes obvious that the value of J_x is a monotonic nondecreasing function of ρ, bounded from below since it is nonnegative. Further, $J_x(x_0,\rho) \leq J^o(x_0,\rho)$, $\forall\rho$, so that

$$\lim_{\rho\to 0+} J_x(x_0,\rho) = \frac{1}{2}x_0'P_0x_0 - \varepsilon,\ \varepsilon \geq 0.$$

Let $\varepsilon > 0$ and assume that there exist $\bar{\varepsilon} > 0$ and $\bar{\rho} > 0$ such that $2J_x(x_0,\bar{\rho}) = x_0'P_0x_0 - 2\bar{\varepsilon}$. By choosing

$$\hat{\rho} = \frac{\bar{\varepsilon}}{2}J_u^{-1}(x_0,\bar{\rho}) \quad \text{where} \quad J_u(x_0,\rho) := \frac{1}{2}\int_0^\infty u^{o\prime}(t,\rho)Ru^o(t,\rho)dt,$$

it follows that

$$\begin{aligned} x_0'P_0x_0 - \bar{\varepsilon} &= 2(J_x(x_0,\bar{\rho}) + \hat{\rho}J_u(x_0,\bar{\rho})) \\ &= 2\int_0^\infty [x^{o\prime}(t,\bar{\rho})Qx^o(t,\bar{\rho}) + \hat{\rho}u^{o\prime}(t,\bar{\rho})Ru^o(t,\bar{\rho})]dt \\ &\geq 2\int_0^\infty [x^{o\prime}(t,\hat{\rho})Qx^o(t,\hat{\rho}) + \hat{\rho}u^{o\prime}(t,\hat{\rho})Ru^o(t,\hat{\rho})]dt = x_0'P(\hat{\rho})x_0 \geq x_0'P_0x_0 \end{aligned}$$

because *i)* the pair $(x^o(t,\bar{\rho}), u^o(t,\bar{\rho}))$ is optimal when $\rho = \bar{\rho}$ but it is not so when $\rho = \hat{\rho}$; *ii)* it has been proved (Lemma 3.4) that $x_0'P(\rho)x_0$ is a monotonic nondecreasing function the limit of which, as ρ goes to 0, is $x_0'P_0x_0$. The last equation implies $\bar{\varepsilon} = 0$.

Thanks to this theorem matrix P_0 supplies the required information about the maximum achievable accuracy in attaining the desired behaviour $(y(\cdot) = 0)$. Apparently, the best result is $P_0 = 0$: the circumstances under which this happens are specified in the forthcoming theorem, the proof of which can be found in the literature. For the sake of simplicity, matrices B and C (which appear in eqs. (3.36a), (3.36b) and have dimensions $n \times m$ and $p \times n$) are assumed to have rank equal to m and p, respectively.

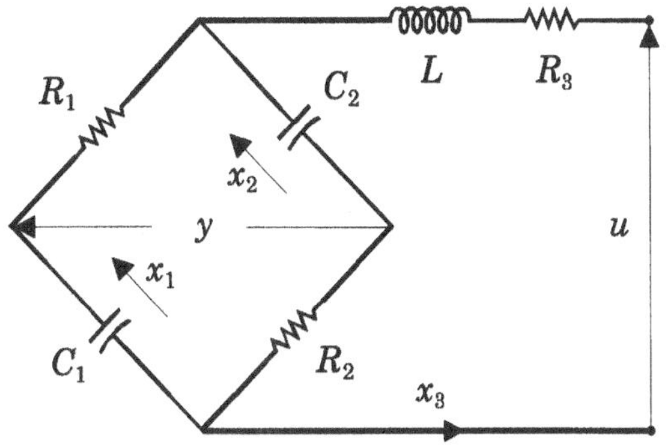

Figure 3.37: The electric circuit considered in Example 3.28.

Theorem 3.12 *Let system (3.36a), (3.36b) be both reachable and observable. The following conclusions hold:*

i) *If $m < p$, then $P_0 \neq 0$;*

ii) *If $m = p$, then $P_0 = 0$ if and only if the transmission zeros of the system have nonpositive real parts;*

iii) *If $m > p$ and there exists a full rank $m \times p$ matrix M such that the transmission zeros of system $\Sigma(A, BM, C, 0)$ have nonpositive real parts, then $P_0 = 0$.*

Example 3.27 Consider a system $\Sigma(A, B, C, 0)$ where the state, input and output are scalars and $B \neq 0$, $C \neq 0$. This system does not possess finite transmission zeros, since its transfer function is $F(s) = BC/(s - A)$: hence $P_0 = 0$. Indeed, the relevant ARE is $0 = 2AP - P^2B^2/(\rho R) + C^2$, the only positive solution of which is

$$P(\rho) = \frac{1}{B^2}\left(\rho AR + \sqrt{(\rho AR)^2 + \rho B^2 QR}\right)$$

and $\lim_{\rho \to 0^+} P(\rho) = 0$.

Example 3.28 Consider the electric circuit shown in Fig. 3.37 which, under the assumption of ideal components, can be described by the equations

$$\begin{aligned}
C_1\dot{x}_1 &= -\frac{1}{R_1 + R_2}(x_1 - x_2 - R_2x_3),\\
C_2\dot{x}_2 &= \frac{1}{R_1 + R_2}(x_1 - x_2 + R_1x_3),\\
L\dot{x}_3 &= u - \frac{1}{R_1 + R_2}(R_2x_1 + R_1x_2 + (R_1R_2 + R_1R_3 + R_2R_3)x_3),\\
y &= \frac{1}{R_1 + R_2}(R_1x_1 + R_2x_2 - R_1R_2x_3).
\end{aligned}$$

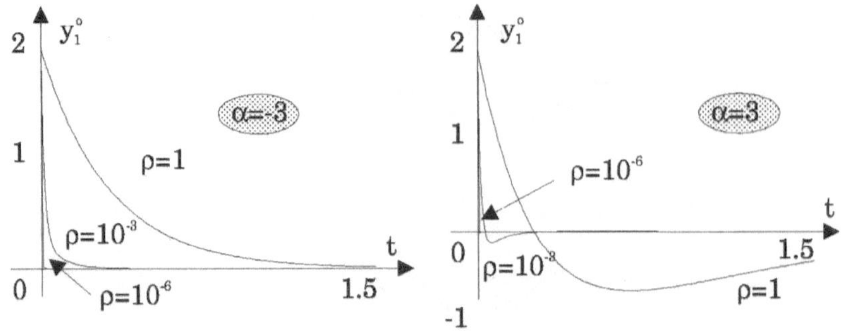

Figure 3.38: Example 3.29: responses of y_1^o for some values of ρ and α.

Letting $R_1 = R_2 = 1$, $R_3 = 0.5$, $C_1 = 1$, $C_2 = 0.5$, $L = 1$, it is easy to verify that the transfer function from u to y is

$$G(s) = \frac{2 - s^2}{2s^3 + 5s^2 + 4.5s + 2}$$

which exhibits a zero in the right half-plane. Thus, if the performance index is the integral of $y^2 + \rho u^2$, it follows that $P_0 \neq 0$. In fact, for $\rho = 10^{-10}$ we get

$$P(\rho) = \begin{bmatrix} 0.49 & 0.34 & 0.00 \\ 0.34 & 0.24 & 0.00 \\ 0.00 & 0.00 & 0.00 \end{bmatrix}.$$

Example 3.29 Consider the system $\Sigma(A, B, C, 0)$ where

$$A = \begin{bmatrix} -2 & 0 & 0 \\ 0 & -1 & 0 \\ 0 & 0 & \alpha \end{bmatrix}, \ B = \begin{bmatrix} 1 & 0 \\ 0 & 1 \\ 0 & 1 \end{bmatrix}, \ C = \begin{bmatrix} 1 & 1 & 1 \\ 0 & 1 & 0 \end{bmatrix},$$

α being a parameter the value of which is either -3 or 0 or 3. For these values of α, Σ is observable and reachable, thus the transmission zeros coincide with the zeros of $\det[G(s)] := \det[C(sI - A)^{-1}B]$ (see Section A.3 of Appendix A) provided that it is expressed as

$$\det[G(s)] = \frac{\varphi(s)}{\det[sI - A]}.$$

In the case at hand, $\det[sI - A] := \psi(s) = (s + 2)(s + 1)(s - \alpha)$ and $\det[G(s)] = (s - \alpha)/\psi(s)$, so that there exists a transmission zero with positive real part if $\alpha = 3$. Consistent with this, for $R = I$ in the performance index, we get, when $\rho = 10^{-10}$

$$P(\rho) = 10^{-1} \begin{bmatrix} 0.00 & 0.00 & 0.00 \\ 0.00 & 3.75 & -3.75 \\ 0.00 & -3.75 & 3.75 \end{bmatrix} \quad \text{if } \alpha = 3$$

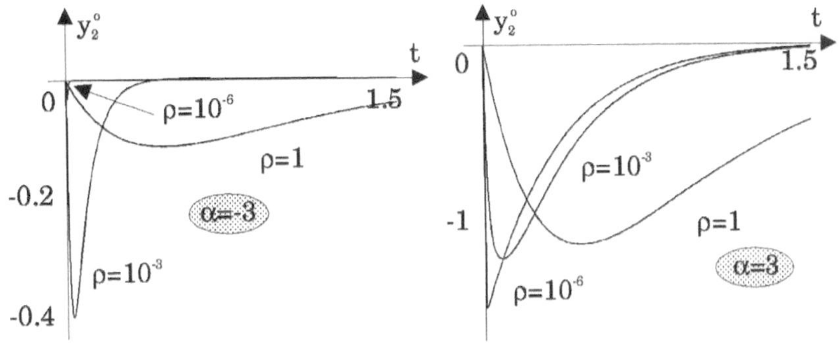

Figure 3.39: Example 3.29: responses of y_2^o for some values of ρ and α.

and

$$P(\rho) = 10^{-4}\begin{bmatrix} 0.07 & 0.00 & 0.07 \\ 0.00 & 0.34 & -0.20 \\ 0.07 & -0.20 & 0.27 \end{bmatrix}, \; P(\rho) = 10^{-4}\begin{bmatrix} 0.07 & 0.00 & 0.07 \\ 0.00 & 0.14 & 0.00 \\ 0.07 & 0.00 & 0.07 \end{bmatrix}$$

if $\alpha = 0$ and $\alpha = -3$, respectively. The responses of y_1^o and y_2^o are shown in Figs. 3.38 and 3.39 for various values of ρ and α, corresponding to $x(0) = \begin{bmatrix} 1 & 0 & 1 \end{bmatrix}'$. Note that when the transmission zero is in the right half-plane, $y_2(\cdot)$ does not tend to 0 as ρ goes to 0. On the contrary, this happens when no transmission zeros are located there.

Example 3.30 Consider the system $\Sigma(A, B, C, 0)$ where

$$A = \begin{bmatrix} 0 & 0 & 0 \\ 0 & 0 & 1 \\ 1 & 0 & 0 \end{bmatrix}, \; B = \begin{bmatrix} 1 & 0 \\ 0 & 0 \\ 0 & 1 \end{bmatrix}, \; C = \begin{bmatrix} 0 & \alpha & 1 \end{bmatrix},$$

α being a real parameter. Letting $M = \begin{bmatrix} a & b \end{bmatrix}'$, we find

$$C(sI - A)^{-1}BM = \frac{(s+\alpha)(bs+a)}{s^3}$$

so that the zeros of system $\Sigma(A, BM, C, 0)$ do not lie in the open right half-plane if $\alpha \geq 0$ and a and b have been suitably selected. Consistent with this, we get, for $R = 1$, $\alpha = 1$ and $\rho = 10^{-10}$

$$P(\rho) = 10^{-5}\begin{bmatrix} 0.00 & 0.00 & 0.00 \\ 0.00 & 1.00 & 1.00 \\ 0.00 & 1.00 & 1.00 \end{bmatrix}.$$

3.4.4 The inverse problem

The *inverse* optimal control problem consists in finding, for a given system and *control law*, a performance index with respect to which such a control law is optimal. The discussion as well as the solution of this seemingly useless problem allows us to clarify the ultimate properties of a control law in order that it can be considered optimal and to precisely evaluate the number of degrees of freedom which are actually available when designing a control law via an LQ approach.

We will deal only with the case of scalar control variable and time invariant system and control law: thus the inverse problem is stated on the linear dynamical system

$$\dot{x}(t) = Ax(t) + Bu(t) \tag{3.39}$$

and the control law

$$u(x) = Kx \tag{3.40}$$

where the control variable u is scalar, while the state vector x has n components. The problem consists in finding a matrix $Q = Q' \geq 0$ such that the control law (3.40) is optimal relative to the system (3.39) and the performance index

$$J = \int_0^\infty [x'(t)Qx(t) + u^2(t)]dt. \tag{3.41}$$

Since u is a scalar, there is no lack of generality in taking $R = 1$.

For the above problem a few results are available, among which the most significant one is stated in the following theorem where K is assumed to be nonzero. Should this not be the case, the solution of the problem would be trivial, namely $Q = 0$.

Theorem 3.13 *With reference to the system (3.39) and the control law (3.40) where $K \neq 0$, let the following assumptions be assumed:*

(a1) *The pair (A, B) is reachable;*

(a2) *The system (3.39), (3.40) is asymptotically stable;*

(a3) $\mu(j\omega) := |1 - K(j\omega I - A)^{-1}B| \geq 1$, $\forall \omega$ *real*, $\mu(\cdot) \neq 1$;

(a4) *The pair (A, K) is observable.*

Then there exists $Q = Q' \geq 0$ such that the control law (3.40) is optimal for the LQ problem defined on the system (3.39) and the performance index (3.41).

Proof. The assumptions are independent of the particular choice of the state variables, therefore the pair (A, B) can be written in the canonical form

$$A = \begin{bmatrix} 0 & 1 & 0 & \cdots & 0 \\ 0 & 0 & 1 & \cdots & 0 \\ \vdots & \vdots & \vdots & \cdots & \vdots \\ 0 & 0 & 0 & \cdots & 1 \\ -a_1 & -a_2 & -a_3 & \cdots & -a_n \end{bmatrix}, \quad B = \begin{bmatrix} 0 \\ 0 \\ \vdots \\ 0 \\ 1 \end{bmatrix}.$$

The characteristic polynomial of A is $\psi(s) := s^n + \sum_{i=1}^n a_i s^{i-1}$. If k_i, $i = 1, 2, \ldots, n$, denotes the i-th component of K, it follows that

$$K(sI - A)^{-1}B = \frac{\sum_{i=1}^n k_i s^{i-1}}{\psi(s)}$$

while the characteristic polynomial of $A + BK$ is

$$\varphi(s) := s^n + \sum_{i=1}^n (a_i - k_i)s^{i-1}.$$

From assumption *(a3)* it follows that, $\forall \omega$ *real,*

$$[1 - K(-j\omega I - A)^{-1}B][1 - K(j\omega I - A)^{-1}B] = \frac{\varphi^\sim(j\omega)\varphi(j\omega)}{\psi^\sim(j\omega)\psi(j\omega)} \geq 1 \tag{3.42}$$

and, since $\psi^\sim(j\omega)\psi(j\omega) \geq 0$, also

$$\pi(j\omega) := \varphi^\sim(j\omega)\varphi(j\omega) - \psi^\sim(j\omega)\psi(j\omega) \geq 0.$$

By recalling that $\mu(\cdot) \neq 1$, the last inequality ensures that there exists a nonzero polynomial $\sigma(s) := \sum_{i=1}^n \vartheta_i s^{i-1}$ of degree not greater than $n - 1$ having all roots in the closed left half-plane and such that

$$\varphi^\sim(j\omega)\varphi(j\omega) - \psi^\sim(j\omega)\psi(j\omega) = \sigma^\sim(j\omega)\sigma(j\omega). \tag{3.43}$$

In fact if z is a root of π also $-z$ is a root of it and both polynomials φ and ψ are monic and of degree n. By collecting into a vector ϑ the coefficients of the polynomial σ we obtain $\vartheta'(sI - A)^{-1}B = \sigma(s)/\psi(s)$ and, in view of eqs. (3.42), (3.43), it follows that

$$\begin{aligned} [1 - K(-j\omega I - A)^{-1}B][1 - K(j\omega I - A)^{-1}B] &= 1 + \frac{\sigma^\sim(j\omega)\sigma(j\omega)}{\psi^\sim(j\omega)\psi(j\omega)} \\ &= 1 + B'(-j\omega I - A')^{-1}\vartheta\vartheta'(j\omega I - A)^{-1}B. \end{aligned} \tag{3.44}$$

From Lemma 3.3, if the control law (3.40) is optimal with respect to the performance index (3.41), then it follows that

$$\begin{aligned} &[1 - K(-j\omega I - A)^{-1}B][1 - K(j\omega I - A)^{-1}B] \\ &\quad = 1 + B'(-j\omega I - A')^{-1}Q(j\omega I - A)^{-1}B. \end{aligned} \tag{3.45}$$

This lemma justifies the requirement $\mu(\cdot) \neq 1$ (see assumption *(a3)*). In fact, if this is not the case, then $Q := C'C = 0$, since

$$B'(-j\omega I - A')^{-1}C'C(j\omega I - A)^{-1}B = 0, \ \forall\omega$$

would imply $C(j\omega I - A)^{-1}B = 0$, $\forall\omega$ and, in view of assumption (a1), $C = 0$. But, with $Q = 0$, the optimal control law is $u = 0$, against the assumption $K \neq 0$.

The comparison of eq. (3.44) with eq. (3.45) suggests that the control law (3.40) could be optimal with respect to the performance index (3.41) with $Q = \vartheta'\vartheta$. Let $\hat{K} := \begin{bmatrix} \hat{k}_1 & \hat{k}_2 & \cdots \hat{k}_n \end{bmatrix}$ be the matrix which defines the optimal control law for such a Q, so that, in view of Lemma 3.3,

$$\begin{aligned}&[1 - \hat{K}(-j\omega I - A)^{-1}B][1 - \hat{K}(j\omega I - A)^{-1}B] \\ &= 1 + B'(-j\omega I - A')^{-1}Q(j\omega I - A)^{-1}B\end{aligned}$$

and (recall eq. (3.45))

$$\begin{aligned}&[1 - K(-j\omega I - A)^{-1}B][1 - K(j\omega I - A)^{-1}B] \\ &= [1 - \hat{K}(-j\omega I - A)^{-1}B][1 - \hat{K}(j\omega I - A)^{-1}B].\end{aligned}$$

The characteristic polynomial of $A + B\hat{K}$ is $\hat{\varphi}(s) := s^n + \sum_{i=1}^{n}(a_i - \hat{k}_i)s^{i-1}$, hence the last equation, together with what has been shown above, imply

$$\hat{\varphi}^{\sim}(j\omega)\hat{\varphi}(j\omega) = \varphi^{\sim}(j\omega)\varphi(j\omega) = \psi^{\sim}(j\omega)\psi(j\omega) + \sigma^{\sim}(j\omega)\sigma(j\omega). \tag{3.46}$$

If it were possible to conclude from this relation that $\hat{\varphi} = \varphi$, then it would also be possible to conclude that $\hat{K} = K$ since the two polynomials are identical functions of the elements of these matrices. Thanks to the assumption *(a2)*, φ is uniquely determined from the polynomial on the right-hand side of eq. (3.46), since all its roots must have negative real parts.

It is now shown that all the roots of $\hat{\varphi}$ have negative real parts, thus entailing that $\hat{\varphi} = \varphi$. If the pair (A, ϑ') is observable, such a property is guaranteed by Theorem 3.6. By contradiction assume that this is not true: the transfer function $F_\vartheta(s) := \vartheta'(sI - A)^{-1}B$ would have the degree of the denominator *less* than n if expressed as the ratio of two coprime polynomials. Since $F_\vartheta(s) = \sigma(s)/\psi(s)$, the polynomials σ and ψ would possess at least one common root, say z, with nonpositive real part because of the definition of σ. If eq. (3.43) is taken into account, then z must also be a root of φ, since all the roots of $\varphi^{\sim}$ have positive real parts. As a consequence, the degree of the denominator of the transfer function $F_K(s) := K(sI - A)^{-1}B$, which is the ratio of the polynomials $\psi - \varphi$ and ψ, would be *strictly less* than n after the suitable cancellations have been performed. In view of assumption *(a1)* the observability of the pair (A, K) (assumption *(a4)*) is then violated. In conclusion, the pair (A, ϑ') is observable, all roots of the polynomial $\hat{\varphi}$ have negative real parts, $\varphi = \hat{\varphi}$, $K = \hat{K}$ and the control law (3.40) is optimal with respect to the performance index (3.41) with $Q = \vartheta\vartheta'$.

Example 3.31 Consider the inverse problem defined by

$$A = \begin{bmatrix} 0 & 1 & 0 \\ 0 & 0 & 1 \\ 0 & 0 & 0 \end{bmatrix}, \; B = \begin{bmatrix} 0 \\ 0 \\ 1 \end{bmatrix}, \; K = \begin{bmatrix} -1 & -2 & -2 \end{bmatrix}.$$

It is straightforward to verify that the assumptions in Theorem 3.13 hold and $\pi(s) = 1$, so that $\sigma(s) = 1$, $\vartheta = \begin{bmatrix} 1 & 0 & 0 \end{bmatrix}'$ and Q =diag[1, 0, 0]. This result can easily be checked by computing the solution of the optimal regulator problem stated on the quadruple $(A, B, Q, 1)$.

Remark 3.20 *(Unnecessity of assumption (a4))* The fourth assumption in Theorem 3.13 can be removed. Indeed, let the pair (A, K) be not observable and the unobservable part of it be already put into evidence, so that

$$A = \begin{bmatrix} A_1 & 0 \\ A_2 & A_3 \end{bmatrix}, \; B = \begin{bmatrix} B_1 \\ B_2 \end{bmatrix}, \; K = \begin{bmatrix} K_1 & 0 \end{bmatrix}$$

with the pair (A_1, K_1) observable and the pair (A_1, B_1) reachable (this last claim follows from the reachability assumption of the pair (A, B)). Theorem 3.13 can be applied to the subsystem Σ_1 described by $\dot{x}_o = A_1 x_o + B_1 u$ and the control law $u = K_1 x_o$. In fact, assumption *(a2)* is verified for the triple (A_1, B_1, K_1) if it holds for the triple (A, B, K), since

$$A + BK = \begin{bmatrix} A_1 + B_1 K_1 & 0 \\ A_2 + B_2 K_1 & A_3 \end{bmatrix}.$$

Furthermore, since $K(sI - A)^{-1}B = K_1(sI - A_1)^{-1}B_1$, if condition *(a3)* holds for the triple (A, B, K) it holds also for the triple (A_1, B_1, K_1). Thus a matrix Q_1 can be found which defines a performance index relative to which $K_1 x$ is an optimal control law for the subsystem Σ_1. It is then obvious that matrix Q =diag$[Q_1, 0]$ specifies a performance index corresponding to which the given control law is optimal (the state of the subsystem $\dot{x}_{no} = A_2 x_o + A_3 x_{no} + B_2 u$ affects neither in a direct nor in an indirect way the performance index: thus it should not contribute to the current value of the control variable).

Example 3.32 Consider the inverse problem defined by the triple

$$A = \begin{bmatrix} 0 & 1 & 0 \\ 0 & 0 & 0 \\ 1 & 1 & -1 \end{bmatrix}, \; B = \begin{bmatrix} 0 \\ 1 \\ 1 \end{bmatrix}, \; K = \begin{bmatrix} -1 & -\sqrt{2} & 0 \end{bmatrix}.$$

By exploiting Remark 3.20 we find $Q_1 = \text{diag}[1, 0]$ and, consequently,

$$Q = \text{diag}[1, 0, 0].$$

Remark 3.21 *(Degrees of freedom in the choice of the performance index)* The proof of Theorem 3.13 points out a fact which is of some interest in exploiting the LQ theory in the synthesis of control systems. Indeed, within this framework only the structure of the performance index should be seen as given, the relevant free parameters being selected (through a sequence of rationally performed trials) so as to eventually specify a satisfactory control law. If the control is scalar, the number of these design parameters is substantially less than $1 + n(n+1)/2$, the number of elements in R and Q. In fact, on the one hand, it has already been shown that R can be set to 1 without loss of generality, while, on the other hand, it is apparent that, under the (mild) assumptions of reachability of (A, B) and stability of the feedback system, conditions *(a1)–(a3)* of Theorem 3.13 are satisfied whenever the control law results from the solution of the LQ problem corresponding to an arbitrary $Q \geq 0$. These conditions are sufficient (thanks to Remark 3.20) to ensure the existence of the solution of the inverse problem. Thus the same control law must also result from a Q expressed as the product of an n-vector by its transpose and the *truly free* parameters are n.

Example 3.33 Consider the optimal regulator problem defined by $Q = I$, $R = 1$ and

$$A = \begin{bmatrix} 0 & 1 \\ 0 & 0 \end{bmatrix}, \quad B = \begin{bmatrix} 0 \\ 1 \end{bmatrix}.$$

The optimal control law is specified by $K = \begin{bmatrix} -1 & -\sqrt{3} \end{bmatrix}$. The inverse problem for the triple (A, B, K) gives $\vartheta = \begin{bmatrix} 1 & 1 \end{bmatrix}'$ and a different state weighting matrix, namely

$$\bar{Q} = \vartheta\vartheta' = \begin{bmatrix} 1 & 1 \\ 1 & 1 \end{bmatrix} \neq Q.$$

Conditions *(a1)–(a3)* in Theorem 3.13 can further be weakened, up to becoming necessary and sufficient. These new conditions are presented in the following theorem where the triple (A, B, K) has already undergone the canonical decomposition, thus exhibiting the form

$$A = \begin{bmatrix} A_1 & A_2 & A_3 & A_4 \\ 0 & A_5 & 0 & A_6 \\ 0 & 0 & A_7 & A_8 \\ 0 & 0 & 0 & A_9 \end{bmatrix}, \quad B = \begin{bmatrix} B_1 \\ B_2 \\ 0 \\ 0 \end{bmatrix}, \quad K = \begin{bmatrix} 0 & K_1 & 0 & K_2 \end{bmatrix} \tag{3.47}$$

with the pair (A_r, B_r) reachable, the pair (A_o, K_o) observable and

$$A_r := \begin{bmatrix} A_1 & A_2 \\ 0 & A_5 \end{bmatrix}, \quad B_r := \begin{bmatrix} B_1 \\ B_2 \end{bmatrix}$$

$$A_o := \begin{bmatrix} A_5 & A_6 \\ 0 & A_9 \end{bmatrix}, \quad K_o := \begin{bmatrix} K_1 & K_2 \end{bmatrix}.$$

Theorem 3.14 *With reference to the system (3.39) and the control law (3.40) there exists a matrix $Q = Q' \geq 0$ such that the optimal regulator problem* *defined on that system and the performance index (3.41) admits a solution for each initial state, the solution being specified by the given control law, if and only if*

(a1) *All the eigenvalues of A_9 have negative real part,*

(a2) *All eigenvalues of $A_5 + B_2 K_1$ have negative real part,*

(a3) *One of the two following conditions holds:*

  (a31) $K = 0$

  (a32) $\mu(j\omega) := |1 - K_1(j\omega I - A_5)^{-1} B_2| \geq 1$, $\forall \omega$ *real*, $\mu(\cdot) \neq 1$.

Proof. Necessity. Let K specify the optimal control law corresponding to a given $Q := C'C$. Then Lemma 3.3 implies that condition *(a32)* holds or, if $\mu(\cdot) = 1$, $C(sI - A)^{-1} B = 0$, which in turn causes the solution of the optimal control problem to be $u(\cdot) = 0$, that is $K = 0$. Note that the unobservable part of the triple (A, B, C) is included in the unobservable part of the triple (A, B, K): in fact, if the first triple has already undergone the canonical decomposition (in general, the relevant canonical form differs from the one given in eq. (3.47)), it is easy to verify that the columns of K corresponding to the zero columns of C are also zero (recall Remark 3.12). Since the control law ensures that all eigenvalues not belonging to the unobservable part have negative real parts, conditions *(a1)* and *(a2)* are satisfied.

Sufficiency. If condition *(a31)* holds, then $Q = 0$ is the (trivial) solution of the inverse problem. Thus the case $K \neq 0$ is left and, in view of condition *(a32)*, $K_1 \neq 0$, necessarily. By following the procedure outlined in the proof of Theorem 3.13, a polynomial σ_1 with roots in the closed left half-plane can be determined from the triple (A_5, B_2, K_1). The vector ϑ_1 which collects the coefficients of σ_1 is such that $Q_1 := \vartheta_1 \vartheta_1'$ is a solution of the inverse problem relevant to the triple (A_5, B_2, K_1) and the pair (A_5, ϑ_1) is observable (see the proof of Theorem 3.13). Now it will be shown that a vector ϑ_2 with the same number of rows as matrix A_9 can be determined so that the n-dimensional vector $\vartheta := \begin{bmatrix} 0 & \vartheta_1' & 0 & \vartheta_2' \end{bmatrix}'$ defines a solution of the inverse problem relevant to the triple (A, B, K). This solution is constituted by the matrix $Q := \vartheta\vartheta'$. Note that a solution of the optimal regulator problem defined by the quadruple $(A, B, Q, 1)$ exists whatever the vector ϑ_2 might be, thanks to assumption *(a1)*. In view of Remark 3.12 in Section 3.4 it can be stated that the nonzero blocks in the solution of the relevant ARE are only four, precisely $\bar{P}_5$, $\bar{P}_7$, $\bar{P}_{10}$, $\bar{P}_7'$, and satisfy the equations

$$\begin{aligned} 0 &= P_5 A_5 + A_5' P_5 - P_5 B_2 B_2' P_5 + Q_1 \\ 0 &= P_7 A_9 + (A_5 - B_2 B_2' P_5)' P_7 + P_5 A_6 + Q_{12} \\ 0 &= P_{10} A_9 + A_9' P_{10} + P_7' A_6 + A_6' P_7 - P_7' B_2 B_2' P_7 + Q_2 \end{aligned} \tag{3.48}$$

where $Q_{12} := \vartheta_1 \vartheta_2'$, $Q_2 := \vartheta_2 \vartheta_2'$. Corresponding to this, the optimal control law is specified by a matrix possessing the same structure as K (recall eq. (3.47)).

The remaining part of the proof aims at showing that a suitable choice of ϑ_2 implies that the solution $\bar{P}_7$ of eq. (3.48) with $P_5 = \bar{P}_5$ is such that $K_2 = -B_2' \bar{P}_7$.

The solution of eq. (3.48) is

$$\bar{P}_7 = \int_0^\infty e^{A'_{5c}t}(\bar{P}_5 A_6 + Q_{12})e^{A_9 t}dt$$

where $A_{5c} := A_5 - B_2 B_2' \bar{P}_5$ (the check can easily be done by resorting to integration by parts and exploiting the fact that all eigenvalues of matrices A_9 and A_{5c} have negative real parts). Since Q_{12} is a function of ϑ_2, we obtain

$$\hat{K}_2'(\vartheta_2) := -\bar{P}_7'(\vartheta_2)B_2 = -\xi - \Lambda\vartheta_2$$

where

$$\xi := \int_0^\infty e^{A'_9 t} A'_6 \bar{P}_5 e^{A_{5c}t} B_2 dt, \ \Lambda := \int_0^\infty (\vartheta_1' e^{A_{5c}t} B_2) e^{A'_9 t} dt$$

are a vector and a matrix, respectively, which are known because $\bar{P}_5$ (and thus also A_{5c}) are uniquely determined once Q_1 has been selected as specified above. It should be clear that if Λ is nonsingular, then ϑ_2 can be selected so that $\hat{K}_2 = K_2$. Therefore the proof ends by checking that Λ is nonsingular. Without loss of generality matrix A_9 can be assumed to be in Jordan canonical form, hence $e^{A_9 t}$ and Λ are triangular. In order that this last matrix be nonsingular it is sufficient to ascertain that all its diagonal elements are different from 0. If α_i is the i-th eigenvalue of A_9, the i-th diagonal element of Λ is

$$\beta_i := \int_0^\infty (\vartheta_1' e^{A_{5c}t} B_2) e^{\alpha_i t} dt = \frac{1}{2\pi}\int_{-\infty}^\infty \frac{\sigma_1(j\omega)}{\varphi_1(j\omega)} \frac{1}{-j\omega - \alpha_i^\sim} d\omega,$$

having exploited Parceval's theorem and denoted by φ_1 the characteristic polynomial of A_{5c}. The evaluation of β_i can be performed by resorting to well-known properties of analytic functions, yielding

$$\beta_i = \frac{\sigma_1(-\alpha_i^\sim)}{\varphi_1(-\alpha_i^\sim)}$$

which is not 0 because $\mathrm{Re}[-\alpha_i^\sim] > 0$ (assumption *(a1)*) and all roots of σ_1 have nonpositive real parts.

Example 3.34 Consider the inverse problem defined by the matrices

$$A = \begin{bmatrix} 0 & 1 & 0 \\ 0 & 0 & 2 \\ 0 & 0 & -1 \end{bmatrix}, \ B = \begin{bmatrix} 0 \\ 1 \\ 0 \end{bmatrix}, \ K = \begin{bmatrix} -2 & -3 & -1 \end{bmatrix}$$

so that, if the notations in Theorem 3.14 and its proof are adopted, we have

$$A_5 = \begin{bmatrix} 0 & 1 \\ 0 & 0 \end{bmatrix}, \ A_6 = \begin{bmatrix} 0 \\ 2 \end{bmatrix}, \ B_2 = \begin{bmatrix} 0 \\ 1 \end{bmatrix}, \ K_1 = \begin{bmatrix} -2 & -3 \end{bmatrix}, \ A_9 = K_2 = -1$$

and all assumptions are verified (in *(a3)* $\mu_1(\cdot) \neq 1$). Then we find that $\vartheta_1 = \begin{bmatrix} 2 & \sqrt{5} \end{bmatrix}'$, $\xi = 5/3$, $\Lambda = (2+\sqrt{5})/6$, $\vartheta_2 = 4(2-\sqrt{5})$, since

$$\bar{P}_5 = \begin{bmatrix} 6-2\sqrt{5} & 2 \\ 2 & 3 \end{bmatrix}, \ e^{A_{5c}t} = \begin{bmatrix} 2e^{-t} - e^{-2t} & e^{-t} - e^{-2t} \\ 2e^{-2t} - 2e^{-t} & 2e^{-2t} - e^{-t} \end{bmatrix}.$$

3.5 Problems

Problem 3.5.1 Show that the solution of the DRE

$$\frac{dP}{dt} = -PA - A'P + PBR^{-1}B'P, \; P(t_f) = S$$

with $R(t) > 0$, $S > 0$ is positive definite over any finite interval $[t_0, t_f]$ and that $\Pi := P^{-1}$ solves the *linear* equation

$$\frac{d\Pi}{dt} = A\Pi + \Pi A' - BR^{-1}B', \; \Pi(t_f) = S^{-1}.$$

Problem 3.5.2 Find the solution of the tracking problem discussed in Remark 3.7 when the term $[y(t_f) - \mu(t_f)]'\hat{S}[y(t_f) - \mu(t_f)]/2$, $\hat{S} = \hat{S}' > 0$ is added to J.

Problem 3.5.3 Find an example of the problem discussed in Remark 3.3 where the optimization with respect to the initial state cannot be performed.

Problem 3.5.4 Find a solution of the LQ problem (with known exogenous input) defined by $\dot{x}(t) = Ax(t) + Bu(t) + Md(t)$, $y(t) = Cx(t) + Nd(t)$, $x(0) = x_0$,

$$J = \int_0^T \{y'(t)y(t) + u'(t)u(t)\} \, dt.$$

where x_0, T, $d(\cdot)$ are given and the state is available to the controller.

Problem 3.5.5 Consider the LQ problem over an infinite horizon where $\dot{x}(t) = tu(t)$, $x(t_0) = x_0$ and

$$J = \int_{t_0}^{\infty} [(tx(t))^2 + u^2(t)]dt$$

Does this problem admit a solution for each x_0 and t_0? If the answer is affirmative, find such a solution.

Problem 3.5.6 Find the solution of the (generalization of the) LQ problem defined by $\dot{x}(t) = A(t)x(t) + B(t)u(t)$, $x(t_0) = x_0$ and

$$J = \int_0^{T_f} [x'(t)Q(t)x(t) + u'(t)R(t)u(t)]dt + x(T_i)'Sx(T_i)$$

where $0 < T_i < T_f < \infty$, x_0, T_f and T_i are given and $Q(t) = Q'(t) \geq 0$, $S = S' \geq 0$, $R(t) = R'(t) > 0$.

Problem 3.5.7 Consider the regulator problem defined on the quadruple

$$(A, B, C'C, 1), \quad \text{where} \quad A = \begin{bmatrix} 0 & 1 \\ 0 & 1 \end{bmatrix}, \; B = \begin{bmatrix} \alpha \\ 1 \end{bmatrix}, \; C = \begin{bmatrix} \beta & 1 \end{bmatrix}$$

and $-\infty < \alpha < \infty$, $-\infty < \beta < \infty$. Discuss the existence of the solution for all initial states and the stability properties of the resulting closed loop system.

Problem 3.5.8 Assume that the solution of the optimal regulator problem defined by a quadruple of constant matrices $(A, B, Q = Q' \geq 0, R = R' > 0$, exists for each initial state. Can the relevant ARE admit only the positive semidefinite solutions

$$P_1 = \begin{bmatrix} 1 & 1 \\ 1 & 1 \end{bmatrix}, \quad P_2 = \begin{bmatrix} 1 & 0 \\ 0 & 1 \end{bmatrix}?$$

Problem 3.5.9 Consider the tracking problem defined on the system $\dot{x}(t) = Ax(t) + Bu(t)$, $y(t) = Cx(t)$, $x(0) = x_0$ and the performance index

$$J = \int_0^\infty \left[(y(t) - \mu(t))'(y(t) - \mu(t)) + \rho u'(t)u(t) \right] dt$$

where $\mu(t) = \vartheta(t)$, $\dot{\vartheta}(t) = F\vartheta(t)$, $\vartheta(0) = \vartheta_0$, $F < 0$, $\rho > 0$ and

$$A = \begin{bmatrix} 0 & 1 \\ 0 & 0 \end{bmatrix}, \quad B = \begin{bmatrix} 0 \\ 1 \end{bmatrix}, \quad C = \begin{bmatrix} \alpha & 1 \end{bmatrix}.$$

With reference to the notation in Remark 3.13, check that $\bar{P}_i \to 0$, $i = 1, 2, 3$ when $\rho \to 0^+$ if $\alpha \geq 0$, while this is not so when $\alpha < 0$. Could this outcome have been forecast?

Problem 3.5.10 Find a solution of the LQ problem defined on the system $\dot{x}(t) = Ax(t) + Bu(t) + E\eta(t)$, $y(t) = Cx(t) + D\eta(t)$, $x(0) = x_0$ and the performance index

$$J = \int_0^\infty \left[y'(t)y(t) + u'(t)u(t) \right] dt$$

where the pair (A, B) is reachable, $\eta(t) = H\xi(t)$, $\dot{\xi}(t) = F\xi(t)$, $\xi(0) = \xi_0$, F is an asymptotically stable matrix. The state x is available to the controller and ξ_0 is known.

Chapter 4

The LQG problem

4.1 Introduction

The discussion in this chapter will be embedded into a not purely deterministic framework and focused on two problems which, at first glance, might seem to be related to each other only because they are concerned with the same stochastic system. Also the connection between these problems and the previously presented material is not apparent from the very beginning while, on the contrary, it will be shown to be very tight. Reference is made to a stochastic system described by

$$\dot{x}(t) = A(t)x(t) + B(t)u(t) + v(t), \tag{4.1a}$$

$$y(t) = C(t)x(t) + w(t), \tag{4.1b}$$

$$x(t_0) = x_0, \tag{4.1c}$$

where, as customary, A, B, C are continuously differentiable functions. In eqs. (4.1a), (4.1b), $\begin{bmatrix} v' & w' \end{bmatrix}'$ is a *zero-mean, gaussian, stationary,* $(n+p)$-dimensional *stochastic process* (n and p are the dimensions of the state and output vectors, respectively) which is assumed to be a white noise. In eq. (4.1c) the initial state is an n-dimensional *gaussian* random variable independent from $\begin{bmatrix} v' & w' \end{bmatrix}'$. The uncertainty on system (4.1) is thus specified by

$$E[\begin{bmatrix} v(t) \\ w(t) \end{bmatrix}] = 0, \ \forall t, \tag{4.2}$$

$$E[x_0] = \bar{x}_0, \tag{4.3}$$

and

$$E[\begin{bmatrix} v(t) \\ w(t) \end{bmatrix}\begin{bmatrix} v'(\tau) & w'(\tau) \end{bmatrix}] = \begin{bmatrix} V & Z \\ Z' & W \end{bmatrix}\delta(t-\tau)$$

$$:= \Xi\delta(t-\tau),\ \forall t,\tau, \tag{4.4}$$

$$E[x_0\begin{bmatrix} v'(t) & w'(t) \end{bmatrix}] = 0,\ \forall t, \tag{4.5}$$

$$E[(x_0-\bar{x}_0)(x_0-\bar{x}_0)'] = \Pi_0, \tag{4.6}$$

where the quantities $\bar{x}_0$, V, Z, W, Π_0 are given and δ is the impulsive function. Moreover, matrices V, W, Ξ, Π_0 are symmetric and positive semidefinite.

The two problems under consideration are concerned with the optimal *estimate* of the state of system (4.1) and the optimal (*stochastic*) control of it. The first problem is to determine the optimal (in a sense to be specified) approximation $\hat{x}(t_f)$ of $x(t_f)$, relying on all available information, namely the time history of the control and output variables (u and y, respectively) on the interval $[t_0, t_f]$ and the uncertainty characterization provided by eqs. (4.2)–(4.6). The second problem is to design a regulator with input y which generates the control u so as to minimize a suitable performance criterion.

Remark 4.1 *(Different system models)* Sometimes the system under consideration is not described by eq. (4.1a) but rather by the equation

$$\dot{x}(t) = A(t)x(t) + B(t)u(t) + B^*v^*(t) \tag{4.7}$$

where v^* is a zero mean white noise independent from x_0 and characterized by

$$E[\begin{bmatrix} v^*(t) \\ w(t) \end{bmatrix}\begin{bmatrix} v^{*\prime}(\tau) & w'(\tau) \end{bmatrix}] = \Xi\delta(t-\tau)$$

with

$$\Xi' = \Xi = \begin{bmatrix} V^* & Z^* \\ Z^{*\prime} & W \end{bmatrix} \geq 0,\ V^* > 0.$$

In this case the previously presented formulation can still be adopted by defining the stochastic process $v := B^*v^*$, which apparently verifies eqs. (4.1a), (4.2), (4.4), (4.5), with $V := B^*V^*B^{*\prime}$, $Z := B^*Z^*$.

At other times, eqs. (4.1a), (4.1b) are replaced by eq. (4.7) and

$$y(t) = C(t)x(t) + C^*v^*(t), \tag{4.8}$$

v^* being a zero mean white noise independent of x_0 with intensity $V^* > 0$. Letting, as above, $v := B^*v^*$ and $w := C^*v^*$, it is straightforward to get back to eqs. (4.1)–(4.6) with $V := B^*V^*B^{*\prime}$, $Z := B^*Z^*$, $W := C^*V^*C^{*\prime}$.

Finally, note that in eqs. (4.7), (4.8) matrices B^* and C^* could be continuously differentiable functions of time. In such a case the discussion above is still valid but the process $\begin{bmatrix} v' & w' \end{bmatrix}'$ is no longer stationary.

4.2 The Kalman filter

The problem of the optimal estimate or filtering of the state of system (4.1)–(4.6) is considered in the present section. The adopted performance criterion for the performed estimate is the expected value of the square of the error undergone in evaluating an arbitrarily given linear combination of the state components. Thus the problem to be discussed can be formally described as follows.

Problem 4.1 (Optimal estimate of $b'x(t_f)$) *Given an arbitrary vector $b \in R^n$, determine, on the basis of the knowledge of $y(t)$ and $u(t)$, $t_0 \leq t \leq t_f$, a scalar β such that the quantity*

$$J_b := E[(b'x(t_f) - \beta)^2] \tag{4.9}$$

is minimized with reference to the system (4.1)–(4.6).

The first case to be examined is the so-called *normal* case where the matrix W (the intensity of the output noise w) is positive definite, while the so-called *singular* case in which W is positive semidefinite only, will subsequently be dealt with. These two situations differ significantly from each other not only from a technical point of view, but also because of the meaning of the underlying problem. This aspect can be put into evidence by recalling that the symmetry of W allows us setting $W := T'DT$ where T is an orthogonal matrix, $D := \text{diag}[d_1, d_2, \ldots, d_r, 0, \ldots, 0]$ and $\text{rank}(W) = r$. By letting $y^* := Ty$, it follows that $y^* = TCx + Tw := C^*x + w^*$ with $E[w^*(t)] = 0$ and $E[w^*(t)w^{*\prime}(\tau)] = TE[w(t)w'(\tau)]T' = D$, so that it is possible to conclude that the last $p - r$ components of y^* are not corrupted by noise. In other words, the assumption $W > 0$ amounts to assuming that no outputs or linear combinations of them are noise free.

4.2.1 The normal case

The optimal state estimation problem is now considered under the assumption that the matrix W is positive definite: first the observation interval is supposed to be finite, subsequently the case of an unbounded interval will also be tackled. Thus, letting $-\infty < t_0 \leq t_f < \infty$, Problem 4.1 is discussed by adding a seemingly unmotivated constraint to the scalar β which is asked to *linearly* depend on y, according to the equation

$$\beta = \int_{t_0}^{t_f} \vartheta'(t)y(t)dt \tag{4.10}$$

where the function ϑ must be selected so as to minimize the value of the criterion (4.9). However, it is possible to prove that the choice (4.10) for the form of the estimate of $b'x(t_f)$ does not actually cause any loss in optimality, since in the adopted stochastic framework the estimate which minimizes J_b is indeed of that form. With reference to the selection of ϑ the following result holds.

Theorem 4.1 *Consider eq. (4.10): the function ϑ^o which solves Problem 4.1 relative to system (4.1)-(4.6) when the observation interval is finite, $u(\cdot) = 0$, $\bar{x}_0 = 0$ and $Z = 0$, is given by*

$$\vartheta^o(t) = W^{-1}C(t)\Pi(t)\alpha^o(t) \tag{4.11}$$

where Π is the (unique, symmetric, positive semidefinite) solution of the differential Riccati equation

$$\dot{\Pi}(t) = \Pi(t)A'(t) + A(t)\Pi(t) - \Pi(t)C'(t)W^{-1}C(t)\Pi(t) + V(t) \tag{4.12}$$

satisfying the boundary condition

$$\Pi(t_0) = \Pi_0, \tag{4.13}$$

while α^o is the (unique) solution of the linear equation

$$\dot{\alpha}(t) = -[A(t) - \Pi(t)C'(t)W^{-1}C(t)]'\alpha(t) \tag{4.14}$$

satisfying the boundary condition

$$\alpha(t_f) = b. \tag{4.15}$$

Proof. Consider, over the interval $[t_0, t_f]$, the dynamical system

$$\dot{\alpha} = -A'\alpha + C'\vartheta, \tag{4.16a}$$

$$\alpha(t_f) = b. \tag{4.16b}$$

In view of eqs. (4.1a), (4.1b) it follows that

$$\frac{d(\alpha' x)}{dt} = \vartheta' y - \vartheta' w + \alpha' v.$$

By integrating both sides of this equation between t_0 and t_f we get, in view of eqs. (4.10), (4.15),

$$b'x(t_f) - \beta = \alpha'(t_0)x(t_0) - \int_{t_0}^{t_f} \vartheta'(t)w(t)dt + \int_{t_0}^{t_f} \alpha'(t)v(t)dt.$$

By squaring both sides of this equation, performing the expected value operation, exploiting the linearity of the operator E and the identity $(r's)^2 = r'ss'r$, which holds for arbitrary elements r, s of R^n, we get

$$\begin{aligned} E[(b'x(t_f)-\beta)^2] &= E[\alpha'(t_0)x(t_0)x'(t_0)\alpha(t_0)] \\ &+E\left[\int_{t_0}^{t_f}\vartheta'(t)w(t)dt\int_{t_0}^{t_f}w'(\tau)\vartheta(\tau)d\tau\right] \\ &+E\left[\int_{t_0}^{t_f}\alpha'(t)v(t)dt\int_{t_0}^{t_f}v'(\tau)\alpha(\tau)d\tau\right] \\ &-E\left[2\alpha'(t_0)x(t_0)\int_{t_0}^{t_f}w'(t)\vartheta(t)dt\right] \\ &+E\left[2\alpha'(t_0)x(t_0)\int_{t_0}^{t_f}v'(t)\alpha(t)dt\right] \\ &-E\left[2\int_{t_0}^{t_f}\vartheta'(t)w(t)dt\int_{t_0}^{t_f}v'(\tau)\alpha(\tau)d\tau\right]. \end{aligned}$$

Note that

$$\begin{aligned} E\left[\int_{t_0}^{t_f}\vartheta'(t)w(t)dt\int_{t_0}^{t_f}w'(\tau)\vartheta(\tau)d\tau\right] &= \int_{t_0}^{t_f}\vartheta'(t)\int_{t_0}^{t_f}E[w(t)w'(\tau)]\vartheta(\tau)d\tau dt \\ &= \int_{t_0}^{t_f}\vartheta'(t)\int_{t_0}^{t_f}W\delta(t-\tau)\vartheta(\tau)d\tau dt \\ &= \int_{t_0}^{t_f}\vartheta'(t)W\vartheta(t)dt. \end{aligned}$$

Thus, by exploiting similar arguments for the remaining terms and taking into account eqs. (4.2)–(4.6), it follows that

$$J_b = \alpha'(t_0)\Pi_0\alpha(t_0) + \int_{t_0}^{t_f}[\alpha'(t)V\alpha(t)+\vartheta'(t)W\vartheta(t)]dt, \tag{4.17}$$

and selecting ϑ^o so as to minimize J_b amounts to solving the optimal control problem defined by the linear system (4.16) and the quadratic performance index (4.17), i.e., an LQ problem (see Section 3.2 of Chapter 3.1) where, however, the roles of the final and initial times have been interchanged. This fact can be managed by letting, for any function f of t, $\hat{f}(\hat{t}) := f(t)$, with $\hat{t} := -t$: then the problem at hand is equivalent to the problem stated on the system

$$\begin{aligned} \frac{d\hat{\alpha}}{d\hat{t}} &= \hat{A}'\hat{\alpha} - \hat{C}'\hat{\vartheta}, \\ \hat{\alpha}(\hat{t}_f) &= b, \end{aligned}$$

and the performance index

$$J_b = \int_{\hat{t}_f}^{\hat{t}_0}[\hat{\alpha}'(\hat{t})V\hat{\alpha}(\hat{t})+\hat{\vartheta}'(\hat{t})W\hat{\vartheta}(\hat{t})]d\hat{t} + \hat{\alpha}'(\hat{t}_0)\Pi_0\hat{\alpha}(\hat{t}_0)$$

where $\hat{t}_f := -t_f$ and $\hat{t}_0 := -t_0$. This is a standard LQ problem because the inequality $t_f > t_0$ implies that $\hat{t}_f < \hat{t}_0$. Therefore, it follows that

$$\hat{\vartheta}^o(\hat{t}) = W^{-1}\hat{C}(\hat{t})\hat{\Pi}(\hat{t})\hat{\alpha}^o(\hat{t})$$

where $\hat{\Pi}$ is the solution of the DRE

$$\frac{d\hat{\Pi}}{d\hat{t}} = -\hat{\Pi}\hat{A}' - \hat{A}\hat{\Pi} + \hat{\Pi}\hat{C}'W^{-1}\hat{C}\hat{\Pi} - V$$

with $\hat{\Pi}(\hat{t}_0) = \Pi_0$ and $\hat{\alpha}^o$ is the solution of the equation

$$\frac{d\hat{\alpha}}{d\hat{t}} = (\hat{A} - \hat{\Pi}\hat{C}'W^{-1}\hat{C})'\hat{\alpha}$$

with $\alpha(\hat{t}_f) = b$. The theorem is proved once the inverse transformation from the time $\hat{t}$ to the time t has been carried out.

Remark 4.2 *(Meaning of β^o)* Within the particular framework into which Theorem 4.1 is embedded, both $x(t)$ and $y(t)$ are, for each t, zero-mean random variables because $\bar{x}_0 = 0$ and v and w are zero-mean white noises. Therefore β^o is a zero-mean random variable as well and its value, as given in Theorem 4.1, is the one which minimizes the variance of the estimation error of $b'x(t_f)$.

Remark 4.3 *(Variance of the estimation error)* The proof of Theorem 4.1 allows us to easily conclude that the optimal value of the performance criterion is

$$J_b^o = b'\Pi(t_f)b$$

which is the (minimal) variance of the estimation error at time t_f: thus the variance depends on the value of the matrix Π at that time. Note that in the proof of Theorem 4.1 both the final time t_f and the initial time t_0 are finite and given but *generic*: therefore $b'\Pi(t)b$ is the (minimal) variance of the estimation error of $b'x(t)$.

Example 4.1 A constant quantity has to be estimated on the basis of a series of measurements performed with continuity over a finite time interval $[0, t_f]$. Letting x be the deviation of the value of this quantity from its expected value, the problem at hand is nothing but Problem 4.1 relative to the (first order) system

$$\begin{aligned}\dot{x} &= 0,\\ y &= x + w,\\ x(0) &= x_0,\end{aligned}$$

where the dependence of the measurement (as supplied by the instrument) on the value of the quantity has been assumed to be purely algebraic (see also Fig. 4.1 (a)). Letting $\Pi_0 > 0$ and $W > 0$ denote the variance of $x(0)$ and the intensity of the white noise w, respectively, it is easy to check that the solution of the relevant DRE is

$$\Pi(t) = \frac{\Pi_0 W}{W + \Pi_0 t}$$

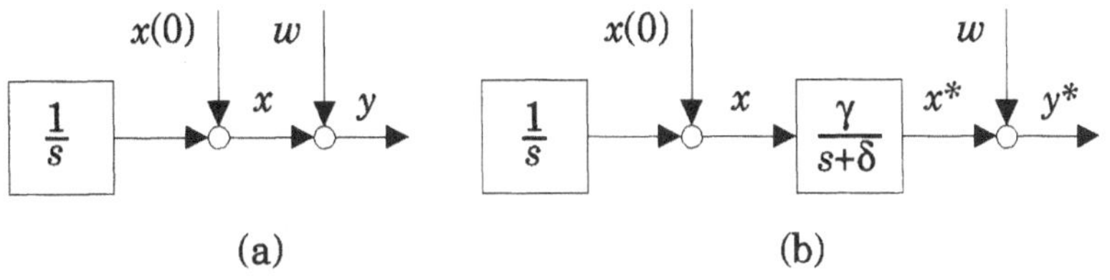

Figure 4.1: The problem discussed in Example 4.1 when the instrument behaves in a purely algebraic (a) or dynamic way (b).

so that, with the obvious choice $b = 1$,

$$\alpha^o(t) = \frac{W + \Pi_0 t}{W + \Pi_0 t_f}.$$

Therefore it follows that

$$\beta^o = \frac{\Pi_0}{W + \Pi_0 t_f} \int_0^{t_f} y(t)dt$$

since

$$\vartheta^o(t) = \frac{\Pi_0}{W + \Pi_0 t_f}.$$

Note that when $\Pi_0 t_f \gg W$, i.e., when the observation interval is sufficiently large or the confidence in the (*a priori* available) expected value of the quantity to be measured is sufficiently smaller than the confidence on the reliability of the instrument ($\Pi_0 \gg W$), the optimal estimate amounts, in essence, to averaging the collected measurements. Finally, observe that when the observation interval becomes unbounded ($t_f \to \infty$), the variance of the estimation error tends to 0 because $\lim_{t\to\infty} \Pi(t) = 0$.

Assume now that the data y^* supplied by the instrument results from a first order filtering process so that it is given by the equations

$$\begin{aligned} \dot{x}^* &= -\delta x^* + \gamma x, \\ y^* &= x^* + w \end{aligned}$$

with δ and γ both positive (see also Fig. 4.1 (b)). Suppose that $x^*(0) = 0$: to be consistent let $\Pi_0 = \mathrm{diag}[1, 0]$ and consider the combinations of values for the parameters δ, γ, W which are shown in Tab. 4.1 together with the corresponding values of the variance of the estimation error $x(1)$, i.e., the values of $\Pi_{11}(1)$ (the (1,1)-element of Π in the new framework). The comparison between the second and the fourth case points out the expected negative effect that an increase in the output noise causes in the performances, while the comparison between the first and second case emphasizes the benefits entailed in the adoption of an instrument with the same gain γ/δ but wider band δ, i.e., a faster instrument. The comparison between the second and third case confirms the prediction of the unfavourable effect of a gain reduction that occurs while keeping the band unaltered. Finally, the comparison

case	γ	δ	W	$\Pi_{11}(1)$
1	1	1	1	0.86
2	10	10	1	0.54
3	1	10	1	0.99
4	10	10	10	0.92

Table 4.1: Example 4.1: variance $\Pi_{11}(1)$ of the estimation error of $x(1)$.

between the first and third case shows that, while keeping the product gain×band constant, the negative effect entailed in a gain reduction is not compensated, at least for sufficiently short observation intervals, by increasing the band.

Remark 4.4 *(Correlated noises)* The proof of Theorem 4.1 suggests a way of dealing with the case where v and w are correlated noises, i.e., the case where $Z \neq 0$. In fact it is easy to check that eq. (4.17) becomes

$$J_b = \alpha'(t_0)\Pi_0\alpha(t_0) + \int_{t_0}^{t_f} [\alpha'(t)V\alpha(t) - 2\alpha'(t)Z\vartheta(t) + \vartheta'(t)W\vartheta(t)]dt$$

so that the estimation problem reduces to an LQ problem with a rectangular term which can be managed as shown in Remark 3.4. However, observe that matrix $V_c := V - ZW^{-1}Z'$ is positive semidefinite, since $V_c = T\Xi T'$ where $T := \begin{bmatrix} I & -ZW^{-1} \end{bmatrix}$ and, from eq. (4.4), $\Xi \geq 0$. Thus, Theorem 4.1 holds with V, $A(t)$ replaced by V_c, $A_c(t) := A(t) - ZW^{-1}C(t)$, respectively, and eq. (4.11) replaced by

$$\vartheta^o(t) = W^{-1}(C(t)\Pi(t) + Z')\alpha^o(t).$$

In view of this discussion there is no true loss of generality in considering only the case $Z = 0$.

Example 4.2 Again consider the problem presented in Example 4.1 and assume that the instrument is a first order dynamical system with unitary gain. A feedback action is performed on the instrument: a term proportional to the measure is added to the input of the instrument itself. The equations for the whole system are

$$\begin{aligned} \dot{x} &= 0, \\ \dot{x}^* &= \delta x + \delta(k-1)x^* + kw, \\ y^* &= x^* + w \end{aligned}$$

with $\delta > 0$ and k real. With the usual notations we get

$$A = \begin{bmatrix} 0 & 0 \\ \delta & \delta(k-1) \end{bmatrix}, \; \Xi = \begin{bmatrix} 0 & 0 & 0 \\ 0 & \delta^2k^2W & \delta kW \\ 0 & \delta kW & W \end{bmatrix}.$$

It is easy to check that A_c and V_c are independent of k. Thus also the solution of the relevant DRE is independent of k and the conclusion can be drawn that such a feedback action has no effect on the quality of the estimate within the framework at hand.

The importance of Theorem 4.1 is greater than might appear from its statement. Indeed, it allows us to devise the structure of a dynamical system the state of which, $\hat{x}(t)$, is, for each $t \in [t_0, t_f]$, the optimal estimate of the state of system (4.1)–(4.6). This fact is presented in the next theorem.

Theorem 4.2 *Consider the system (4.1)–(4.6) with $u(\cdot) = 0$, $\bar{x}_0 = 0$ and $Z = 0$. Then, for each $b \in R^n$ and for $-\infty < t_0 \le t \le t_f < \infty$ the optimal estimate of $b'x(t)$ is $b'\hat{x}(t)$, $\hat{x}(t)$ being the state, at time t, of the system*

$$\dot{\hat{x}}(t) = [A(t) + L(t)C(t)]\hat{x}(t) - L(t)y(t), \tag{4.18a}$$

$$\hat{x}(t_0) = 0 \tag{4.18b}$$

where $L(t) := -\Pi(t)C'(t)W^{-1}$ and Π is the solution (unique, symmetric and positive semidefinite) of the differential Riccati equation (4.12), satisfying the boundary condition (4.13).

Proof. For any given, generic, finite interval $[t_0, t_f]$ and an arbitrary $b \in R^n$ it follows, from Theorem 4.1, that

$$\begin{aligned}\beta^o &= \int_{t_0}^{t_f} \alpha^{o\prime}(t)\Pi(t)C'(t)W^{-1}y(t)dt \\ &= b' \int_{t_0}^{t_f} \Psi'(t, t_f)\Pi(t)C'(t)W^{-1}y(t)dt\end{aligned}$$

where $\Psi(t, \tau)$ is the transition matrix associated to $-(A + LC)'$. By exploiting well-known properties of the transition matrix (see Section A.2 of Appendix A) and denoting with $\Phi(t, \tau)$ the transition matrix associated to $(A + LC)$, we get

$$\beta^o = -b' \int_{t_0}^{t_f} \Phi(t_f, t)L(t)y(t)dt = b'\hat{x}(t_f)$$

where eqs. (4.18) have been taken into account. The proof is complete since the discussion is obviously independent of the actual value of t_0, t_f and b.

The above results can easily be generalized to cope with the case where $u(\cdot) \ne 0$ and $\bar{x}_0 \ne 0$ since the linearity of the system allows us to independently evaluate the effects on x of the deterministic input u and the time propagation of the expected value of the initial state. The presence of the deterministic input is taken into account by simply adding the term Bu to the equation for $\dot{\hat{x}}$, while the propagation of the state expected value is correctly performed by giving the value $\bar{x}_0$ to $\hat{x}(t_0)$. In so doing we get

$$[\dot{\hat{x}}(t) - \dot{x}(t)] = [A(t) + L(t)C(t)][\hat{x}(t) - x(t)] - v(t) - L(t)w(t),$$
$$\hat{x}(t_0) - x(t_0) = \bar{x}_0 - x_0.$$

Recalling that $E[\hat{x}(t_0) - x(t_0)] = 0$, we can conclude that, in view of the material in Section A.5 of Appendix A, $E[\hat{x}(t) - x(t)] = 0$, $\forall t$, and $b'\hat{x}(t)$ is still the estimate of $b'x(t)$ which entails an error with minimal variance. In short, $b'\hat{x}(t)$ is said to be the optimal or *minimal variance* estimate of $b'x(t)$, thus justifying the commonly adopted terminology according to which Problem 4.1 is the minimal variance estimation problem. These remarks are collected in the following theorem.

Theorem 4.3 *Consider the system (4.1)–(4.6) with $Z = 0$. Then, for each $b \in R^n$ and $-\infty < t_0 \leq t \leq t_f < \infty$ the minimal variance estimate of $b'x(t)$ is $b'\hat{x}(t)$, $\hat{x}(t)$ being the state, at time t, of the dynamical system*

$$\dot{\hat{x}}(t) = [A(t) + L(t)C(t)]\hat{x}(t) - L(t)y(t) + B(t)(u(t), \tag{4.19a}$$
$$\hat{x}(t_0) = \bar{x}_0 \tag{4.19b}$$

where $L(t) := -\Pi(t)C'(t)W^{-1}$ and Π is the solution (unique, symmetric and positive semidefinite) of the differential Riccati equation (4.12) satisfying the boundary condition (4.13).

It is interesting to observe that the system given by eqs. (4.19), usually referred to as optimal filter or *Kalman filter*, possesses the same structure as a state observer, briefly discussed in Sections B.2, B.3 of Appendix B.1, with reference to time invariant systems (see also Fig. 4.2, where a block scheme of the filter is presented). The *gain* L of the filter is determined on the basis of the adopted characterization of the system uncertainty.

Remark 4.5 *(Not strictly proper system)* In view of the discussion preceding Theorem 4.3, it is quite obvious how the case where a term $D(t)u(t)$ appears in eq. (4.1b) can be handled. Indeed it is sufficient to add to $\hat{y}$, which is the optimal estimate of y, the term Du, so that eq. (4.19a) becomes

$$\dot{\hat{x}} = (A + LC)\hat{x} - Ly + (B + LD)u.$$

Remark 4.6 *(Meaning of $\Pi(t)$)* By making reference to the proof of Theorem 4.1, Remark 4.3 and Theorem 4.3 it is easy to conclude that

$$b'\Pi(t)b = E[(b'x(t) - b'\hat{x}(t))^2] = b'E[(x(t) - \hat{x}(t))(x(t) - \hat{x}(t))']b.$$

Since b is arbitrary, the matrix $\Pi(t)$ is the variance of the optimal estimation error at time t: therefore, any norm of it, for instance its *trace,* constitutes, when evaluated at some time τ, a meaningful measure of how good is the estimate performed on the basis of the data available up to τ.

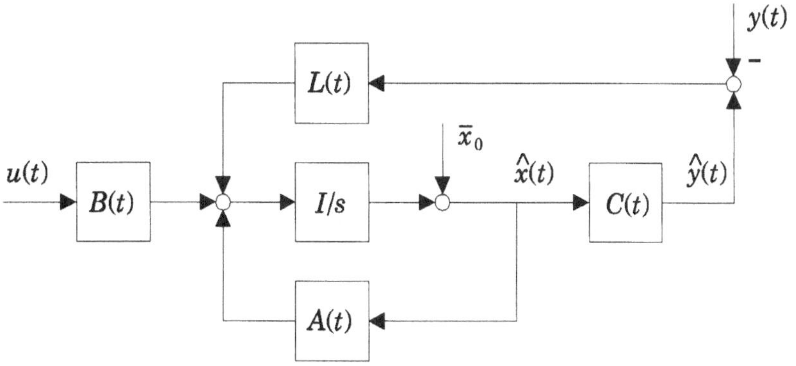

Figure 4.2: The Kalman filter.

Example 4.3 Consider the system (4.1)–(4.6) where

$$A = \begin{bmatrix} -1 & 1 \\ 0 & -1 \end{bmatrix}, \; \Pi_0 = I, \; C = \begin{bmatrix} 1 & 0 \end{bmatrix}, \; W = 1, \; V = 0.$$

The inverse of Π (recall the discussion in Example 3.2) is

$$M(t) = \frac{1}{4} \begin{bmatrix} 2 + 2e^{2t} & 1 - e^{2t} - 2te^{2t} \\ 1 - e^{2t} - 2te^{2t} & 1 + 3e^{2t} + 2te^{2t} + 2t^2e^{2t}. \end{bmatrix}$$

Therefore it follows that

$$\det[M(t)] = \frac{1}{16}(1 + 8e^{2t} + 6te^{2t} + 4t^2e^{2t} + 5e^{4t})$$

and

$$\lim_{t\to\infty} \Pi(t) = \lim_{t\to\infty} M^{-1}(t) = 0$$

from which one can conclude that the variance of the estimation error vanishes as the observation interval increases.

Remark 4.7 *(Incorrelation between the estimation error and the filter state)* An interesting property of the Kalman filter is put into evidence by the following discussion. Let $e := x - \hat{x}$ and consider the system with state $\begin{bmatrix} e' & \hat{x}' \end{bmatrix}'$ which is described by the equations

$$\begin{bmatrix} \dot{e} \\ \dot{\hat{x}} \end{bmatrix} = \begin{bmatrix} A + LC & 0 \\ -LC & A \end{bmatrix} \begin{bmatrix} e \\ \hat{x} \end{bmatrix} + \begin{bmatrix} I & L \\ 0 & -L \end{bmatrix} \begin{bmatrix} v \\ w \end{bmatrix} + \begin{bmatrix} 0 \\ B \end{bmatrix} u.$$

By denoting with $\bar{e}(t)$ and $\bar{\hat{x}}(t)$ the expected values of $e(t)$ and $\hat{x}(t)$, respectively, and letting

$$E[\begin{bmatrix} e(t) - \bar{e}(t) \\ \hat{x}(t) - \bar{\hat{x}}(t) \end{bmatrix} \begin{bmatrix} e'(t) - \bar{e}'(t) & \hat{x}'(t) - \bar{\hat{x}}'(t) \end{bmatrix}] := \begin{bmatrix} \Pi_{11}(t) & \Pi_{12}(t) \\ \Pi_{12}'(t) & \Pi_{22}(t) \end{bmatrix},$$

it follows that the matrices Π_{ij}, $i = 1, j$, $j = 1, 2$, satisfy the differential equations

$$\dot{\Pi}_{11} = \Pi_{11}(A + LC)' + (A + LC)\Pi_{11} + V + LWL', \tag{4.20}$$
$$\dot{\Pi}_{12} = (A + LC)\Pi_{12} + \Pi_{12}A' - \Pi_{11}C'L' - LWL', \tag{4.21}$$
$$\dot{\Pi}_{22} = -LC\Pi_{12} - \Pi_{12}'C'L' + A\Pi_{22} + \Pi_{22}A' + LWL'$$

with the boundary conditions

$$\Pi_{11}(t_0) = \Pi_0, \tag{4.22}$$
$$\Pi_{12}(t_0) = 0, \tag{4.23}$$
$$\Pi_{22}(t_0) = 0.$$

Here the material in Section A.5 of Appendix A has been exploited. By recalling that $L = -\Pi C'W^{-1}$, it is straightforward to check that eq. (4.20) coincides with eq. (4.12) and that eq. (4.22) is indeed eq. (4.13), so that $\Pi_{11}(\cdot) = \Pi(\cdot)$. From this identity it follows that in eq. (4.21) it is $-\Pi_{11}C'L' - LWL' = 0$: thus, $\Pi_{12}(\cdot) = 0$ solves such an equation with the relevant boundary condition (4.23). This fact proves that the stochastic processes e and $\hat{x}$ are uncorrelated. By exploiting Remark 4.4 the same arguments can easily be extended to the case where v and w are correlated ($Z \neq 0$).

The proof of Theorem 4.1 suggests which results pertaining to LQ problems are useful in the case of an unbounded observation interval, that is when $t_0 = -\infty$. By referring to Section 3.3 of Chapter 3.1, the initial state of the system is supposed to be known and equal to zero, so that $\bar{x}_0 = 0$, $\Pi_0 = 0$ and a suitable *reconstructability* assumption is introduced (recall that this property is dual to controllability). On this basis the forthcoming theorem can be stated without requiring any formal proof.

Theorem 4.4 *Let the pair $(A(t), C(t))$ be reconstructable for $t \leq t_f$. Then the problem of the optimal state estimation for the system (4.1)–(4.6) with $Z = 0$, $\bar{x}_0 = 0$, $\Pi_0 = 0$ admits a solution also when $t_0 \to -\infty$. For each $b \in R^n$ and $\tau \leq t_f$ the optimal estimate of $b'x(\tau)$ is given by $b'\hat{x}_\infty(\tau)$, where $\hat{x}_\infty(\tau)$ is the limit approached by the solution, evaluated at τ, of the equation*

$$\dot{\hat{x}}(t) = [A(t) + L(t)C(t)]\hat{x}(t) - L(t)y(t) + Bu(t), \tag{4.24}$$
$$\bar{x}(t_0) = 0$$

when $t_0 \to -\infty$. In eq. (4.24) $L(t) := -\bar{\Pi}(t)C'(t)W^{-1}$ and, for all t, $\bar{\Pi}$ is a symmetric and positive semidefinite matrix given by

$$\bar{\Pi}(t) = \lim_{t_0 \to -\infty} \Pi(t, t_0),$$

$\Pi(t, t_0)$ being the solution (unique, symmetric and positive semidefinite) of the differential Riccati equation (4.12) satisfying the boundary condition $\Pi(t_0, t_0) = 0$.

Thus, the apparatus which supplies the optimal estimate possesses the structure shown in Fig. 4.2 (with $\bar{x}_0 = 0$ and, if it is the case, the term Du added to $\hat{y}$) also when the observation interval is unbounded.

In a similar way it is straightforward to handle filtering problems over an unbounded observation interval when the system is time-invariant: indeed, it is sufficient to mimic the results relevant to the optimal regulator problem (see Section 3.4 of Chapter 3.1) in order to state the following theorem which refers to the time-invariant system

$$\dot{x}(t) = Ax(t) + Bu(t) + v(t), \tag{4.25a}$$

$$y(t) = Cx(t) + w(t). \tag{4.25b}$$

Theorem 4.5 *Consider the system (4.25), (4.1c)–(4.6) with $Z = 0$, $\bar{x}_0 = 0$, $\Pi_0 = 0$ and the pair (A, C) observable. Then the problem of the optimal state estimation admits a solution also when $t_0 \to -\infty$. For each $b \in R^n$ and $\tau \leq t_f$ the optimal estimate of $b'x(\tau)$ is given by $b'\hat{x}_\infty(\tau)$, where $\hat{x}_\infty(\tau)$ is the limit approached by the solution, evaluated at τ, of the equation*

$$\dot{\hat{x}}(t) = (A + \bar{L}C)\hat{x}(t) - \bar{L}y(t) + Bu(t), \tag{4.26}$$
$$\bar{x}(t_0) = 0$$

when $t_0 \to -\infty$. In eq. (4.26) $\bar{L} := -\bar{\Pi}C'W^{-1}$, $\bar{\Pi}$ being a constant matrix, symmetric and positive semidefinite, which solves the algebraic Riccati equation

$$0 = \Pi A' + A\Pi - \Pi C'W^{-1}C\Pi + V \tag{4.27}$$

and is such that

$$\bar{\Pi} = \lim_{t_0 \to -\infty} \Pi(t, t_0),$$

$\Pi(t, t_0)$ being the solution (unique, symmetric and positive semidefinite) of the differential Riccati equation (4.12) with the boundary condition $\Pi(t_0, t_0) = 0$.

Example 4.4 Suppose that an estimate is wanted of a quantity which is slowly varying in a totally unpredictable way. Measures of this quantity are available over an unbounded interval. The problem can be viewed as a filtering problem relative to the first order system

$$\dot{x} = v,$$

$$y = x + w$$

where $\begin{bmatrix} v & w \end{bmatrix}'$ is a white noise with intensity $\Xi = \text{diag}[V, W]$, $V \neq 0$, $W \neq 0$. The ARE relevant to this problem admits a unique nonnegative solution $\bar{\Pi} = \sqrt{VW}$: thus the filter gain is $\bar{L} = -\sqrt{V/W}$. Note that the transfer function of the filter from the input y to the state $\hat{x}$ is

$$G(s) = \frac{1}{1 + s\tau}$$

where $\tau = \sqrt{W/V}$. Therefore the Kalman filter is a low-pass filter the band of which depends in a significant way on the noise intensities: the band increases when the input noise is stronger than the output noise and, vice versa, it reduces in the opposite case.

Obviously, when Theorem 4.5 applies the Kalman filter is a time-invariant system, the stability properties of which can be analyzed as done within the framework of the optimal regulator problem (Section 3.4.1 of Chapter 3.1). All the results there are still valid, provided the necessary modifications have been brought. As an example, the particularly meaningful result concerning asymptotic stability can be stated as shown in the following theorem.

Theorem 4.6 *Consider the system (4.25) and let the triple (A, F, C) be minimal, F' being any factorization of V. Then the Kalman filter relevant to an unbounded observation interval is asymptotically stable, i.e., all the eigenvalues of the matrix $A + \bar{L}C$ have negative real parts.*

Example 4.5 The filter designed in Example 4.4 is a stable system since its unique eigenvalue is $\lambda = \bar{L}$. Indeed, the pair $(0, 1)$ is observable and the pair $(0, \sqrt{V})$ is reachable.

4.2.2 The singular case

A possible way of dealing with the filtering problem in the singular case is now presented with reference to the time-invariant system described by eqs. (4.25). Thus the intensity of the output noise is a matrix W which is not positive definite, i.e., $W \geq 0$, $\det[W] = 0$ and, for the sake of simplicity, the rank of matrix C (eq. (4.25b)) is assumed to be equal to the number p of its rows.

Denote with $T := \begin{bmatrix} T_1' & T_2' \end{bmatrix}'$ an orthogonal matrix such that

$$TWT' = \begin{bmatrix} \Omega & 0 \\ 0 & 0 \end{bmatrix}$$

where Ω is a nonsingular matrix of dimensions $p_1 < p$. Letting $y^*(t) := Ty(t)$, it follows that

$$y^*(t) := \begin{bmatrix} y_d(t) \\ y_c(t) \end{bmatrix} = \begin{bmatrix} T_1Cx(t) \\ T_2Cx(t) \end{bmatrix} + \begin{bmatrix} T_1w(t) \\ T_2w(t) \end{bmatrix}.$$

In view of the fact that the intensity of the white noise T_2w is zero, this relation can be rewritten as

$$y_d(t) = C_dx(t) + w^*(t), \tag{4.28a}$$
$$y_c(t) = C_cx(t) \tag{4.28b}$$

where $C_d := T_1 C$, $C_c := T_2 C$ and $w^*(t) := T_1 w(t)$. The vector y_d (with p_1 components) is thus constituted by those output variables which are actually affected by the noise w, while the vector y_c (with $p - p_1$ components) accounts for those output variables which are not affected by it. Therefore the noise-free information carried by y_c should be exploited in tackling the state estimation problem. In this regard, let C^* be an $n - (p - p_1) \times n$ matrix such that $\begin{bmatrix} C_c' & C^{*\prime} \end{bmatrix}'$ is nonsingular and denote with $\begin{bmatrix} \Gamma_c & \Gamma^* \end{bmatrix}$ the inverse of this last matrix. It follows that

$$x(t) = \Gamma_c y_c(t) + \Gamma^* x^{(1)}(t) \tag{4.29}$$

where

$$x^{(1)}(t) := C^* x(t). \tag{4.30}$$

In principle, the time derivative of the noise-free function y_c can be computed so that from eqs. (4.25a), (4.16a), (4.28)–(4.30) it follows that

$$\begin{aligned} \dot{y}_c(t) &= C_c[Ax(t) + Bu(t) + v(t)] \\ &= C_c A \Gamma_c y_c(t) + C_c A \Gamma^* x^{(1)}(t) + C_c B u(t) + C_c v(t), \qquad (4.31a) \\ \dot{x}^{(1)}(t) &= C^*[Ax(t) + Bu(t) + v(t)] \\ &= C^* A \Gamma_c y_c(t) + C^* A \Gamma^* x^{(1)}(t) + C^* B u(t) + C^* v(t), \qquad (4.31b) \\ y_d(t) &= C_d \Gamma_c y_c(t) + C_d \Gamma^* x^{(1)}(t) + w^*(t). \qquad (4.31c) \end{aligned}$$

Equations (4.31) define a dynamical system with state $x^{(1)}$, known inputs u and y_c, unknown inputs (noises) v and w^*, outputs y_d and $\dot{y}_c$. More concisely, eqs. (4.31) become

$$\begin{aligned} \dot{x}^{(1)}(t) &= A^{(1)} x^{(1)}(t) + B^{(1)} u^{(1)}(t) + v^{(1)}(t), \qquad (4.32a) \\ y^{(1)}(t) &= C^{(1)} x^{(1)}(t) + D^{(1)} u^{(1)}(t) + w^{(1)}(t) \qquad (4.32b) \end{aligned}$$

where $v^{(1)}(t) := C^* v(t)$,

$$u^{(1)}(t) := \begin{bmatrix} y_c(t) \\ u(t) \end{bmatrix}, \ y^{(1)}(t) := \begin{bmatrix} y_d(t) \\ \dot{y}_c(t) \end{bmatrix}, \ w^{(1)}(t) := \begin{bmatrix} w^*(t) \\ C_c v(t) \end{bmatrix}$$

and $A^{(1)} := C^* A \Gamma^*$, $B^{(1)} := \begin{bmatrix} C^* A \Gamma_c & C^* B \end{bmatrix}$,

$$C^{(1)} := \begin{bmatrix} C_d \Gamma^* \\ C_c A \Gamma^* \end{bmatrix}, \ D^{(1)} := \begin{bmatrix} C_d \Gamma_c & 0 \\ C_c A \Gamma_c & C_c B \end{bmatrix}.$$

The intensity of the white noise $\begin{bmatrix} v^{(1)\prime} & w^{(1)\prime} \end{bmatrix}'$ is

$$\Xi^{(1)} := \begin{bmatrix} C^* V C^{*\prime} & C^* Z T_1' & C^* V C_c' \\ T_1 Z' C^{*\prime} & \Omega & T_1 Z' C_c' \\ C_c V C^{*\prime} & C_c Z T_1' & C_c V C_c' \end{bmatrix}.$$

If the intensity of $w^{(1)}$, i.e., the matrix

$$W^{(1)} := \begin{bmatrix} \Omega & T_1 Z' C_c' \\ C_c Z T_1' & C_c V C_c' \end{bmatrix}$$

is positive definite, the filtering problem relative to the system (4.32) is normal and the results of the preceding section can be applied, provided that the probabilistic characterization of $x^{(1)}(t_0)$ could be performed on the basis of *all* available information. If, on the other hand, $W^{(1)}$ is not positive definite, the above outlined procedure can again be applied. Note that the dimension of the vector $x^{(1)}$ is strictly less than n: thus, the procedure can be iterated only a finite number of times, and either the situation is reached where $W^{(i)} > 0$ or n noise-free outputs are available and suited to exactly estimate x.

Assuming, for the sake of simplicity, that $W^{(1)}$ is positive definite, the expected value and variance of the initial state of system (4.32a) has to be computed, given $y_c(t_0) = C_c x(t_0)$. If $x(t_0)$ is not a *degenerate* gaussian random variable, i.e., if $\Pi_0 > 0$, it can be shown that the following relations hold

$$\begin{aligned} \bar{x}_0^{(1)} &:= E[x^{(1)}(t_0)|y_c(t_0)] \\ &= C^*[\bar{x}_0 + \Pi_0 C_c'(C_c \Pi_0 C_c')^{-1}[y_c(t_0) - C_c \bar{x}_0]], \\ \Pi_0^{(1)} &:= E[(x^{(1)}(t_0) - \bar{x}_0^{(1)})(x^{(1)}(t_0) - \bar{x}_0^{(1)})'|y_c(t_0)] \\ &= C^*[\Pi_0 - \Pi_0 C_c'(C_c \Pi_0 C_c')^{-1} C_c \Pi_0] C^{*\prime}. \end{aligned}$$

Let $\Pi^{(1)}$ be the solution (unique, symmetric and positive semidefinite) of the DRE

$$\begin{aligned} \dot{\Pi}^{(1)}(t) = {} & A_c^{(1)} \Pi^{(1)}(t) + \Pi^{(1)}(t) A_c^{(1)\prime} + V_c^{(1)} \\ & - \Pi^{(1)}(t) C^{(1)\prime} (W^{(1)})^{-1} C^{(1)} \Pi^{(1)}(t) \end{aligned}$$

satisfying the boundary condition $\Pi^{(1)}(t_0) = \Pi_0^{(1)}$. In this equation $A_c^{(1)} := A^{(1)} - Z^{(1)}(W^{(1)})^{-1} C^{(1)}$, $Z^{(1)} := \begin{bmatrix} C^* Z T_1' & C^* V C - C' \end{bmatrix}$, $V_c^{(1)} := V^{(1)} - Z^{(1)}(W^{(1)})^{-1} Z^{(1)\prime}$ and $V^{(1)} := C^* V C^{*\prime}$. Then the Kalman filter for the system (4.32) when the uncertainty is specified as above and the observation interval is finite, possesses the structure shown in Fig. 4.3 where the *gains* L_c and L_d are given by the equation

$$\begin{bmatrix} L_d(t) & L_c(t) \end{bmatrix} = -[\Pi^{(1)}(t) C^{(1)\prime} + Z^{(1)}](W^{(1)})^{-1}. \tag{4.33}$$

Notice that these gains are time-varying.

Unlike what is shown in Fig. 4.3, the actual implementation of the last result does not need differentiation. In fact, the signal y_c, after differentiation

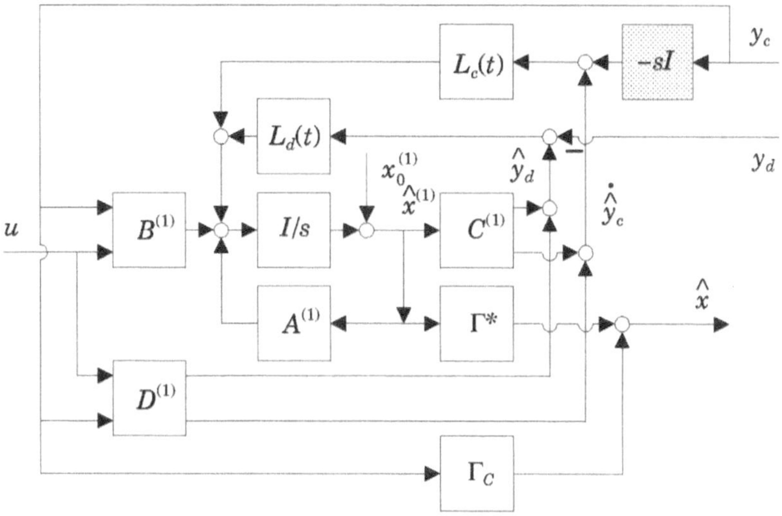

Figure 4.3: The singular Kalman filter.

and multiplication by $-L_c$, undergoes integration. Since $\Pi^{(1)}$ is a differentiable function, from eq. (4.33) it follows that also the function L_c is such, so that

$$\int L_c(t)\dot{y}_c(t)dt = L_c(t)y_c(t) - \int \dot{L}_c(t)y_c(t)dt$$

and the filter can be implemented according to the scheme in Fig. 4.4 where differentiation is no longer required since $\dot{L}_c$ can be evaluated in advance.

Example 4.6 Consider the system

$$\dot{x}_1 = x_2, \qquad y_1 = x_1 + w_1,$$
$$\dot{x}_2 = v_2, \qquad y_2 = x_2$$

where the zero-mean white noises v_2 and w_1 are uncorrelated and their intensities are $V_2 > 0$ and $\Omega > 0$, respectively. According to the previous notation, the system is affected by the noises v and w, their intensity Ξ being specified by the matrices $V = \text{diag}[0, V_2]$, $Z = 0$, $W = \text{diag}[\Omega, 0]$. The initial state is a zero-mean random variable with variance $\Pi_0 = I$. Consistent with the discussion above it follows that

$$C_c = \begin{bmatrix} 0 & 1 \end{bmatrix},\ C_d = C^* = \begin{bmatrix} 1 & 0 \end{bmatrix}$$

so that we obtain $v^{(1)} = 0$, $A^{(1)} = 0$, $B^{(1)} = 1$, $V^{(1)} = 0$, $\bar{x}_0^{(1)} = 0$, $\Pi_0^{(1)} = 1$, $Z^{(1)} = \begin{bmatrix} 0 & 0 \end{bmatrix}$, and

$$C^{(1)} = \begin{bmatrix} 1 \\ 0 \end{bmatrix},\ D^{(1)} = \begin{bmatrix} 0 \\ 0 \end{bmatrix},\ W^{(1)} = \begin{bmatrix} \Omega & 0 \\ 0 & V_2 \end{bmatrix}$$

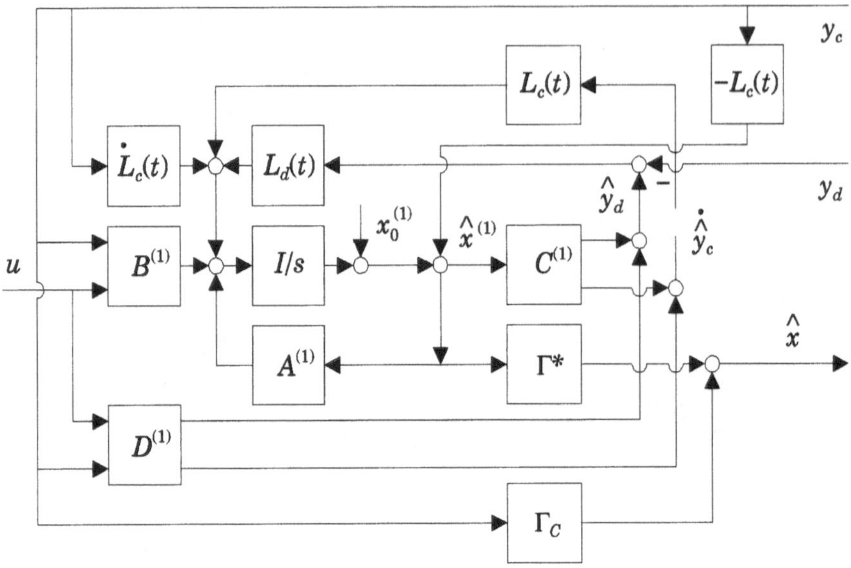

Figure 4.4: The singular Kalman filter without differentiator.

(recall that there is no control input u). Since $W^{(1)} > 0$ no further steps are required and it is easy to check that the solution of the relevant DRE is $\Pi^{(1)}(t) = \Omega/(t+\Omega)$. Therefore, $L_c(t) = \dot{L}_c(t) = 0$, $L_d(t) = -1/(t+\Omega)$ and the filter is defined by the equations

$$
\begin{aligned}
\dot{\xi} &= y_2 + \frac{1}{t+\Omega}(y_1 - \xi), \\
\hat{x} &= \begin{bmatrix} \xi \\ y_2 \end{bmatrix}.
\end{aligned}
$$

Note that the solution does not depend on the intensity of the input noise: this is a consequence of the fact that the state variable x_2 can be measured exactly.

Example 4.7 Consider the system

$$
\begin{aligned}
\dot{x}_1 &= x_2 + v_1, \\
\dot{x}_2 &= x_3 + v_2, \\
\dot{x}_3 &= v_3, \\
y_1 &= x_1 + w_1, \\
y_2 &= x_2 + w_2
\end{aligned}
$$

where the zero-mean white noises v and w have intensity Ξ specified by the matrices $V = \text{diag}[1, 0, 1]$, $Z = 0$, $W = \text{diag}[1, 0]$, so that both v_2 and w_2 can be considered zero. The initial state is a zero-mean random variable with variance $\Pi_0 = I$.

Consistent with the discussion above it follows that $y_c = y_2$, $y_d = y_1$ and hence we get

$$C_c = \begin{bmatrix} 0 & 1 & 0 \end{bmatrix},\ C_d = \begin{bmatrix} 1 & 0 & 0 \end{bmatrix},\ C^* = \begin{bmatrix} 1 & 0 & 0 \\ 0 & 0 & 1 \end{bmatrix}$$

and

$$\Gamma_c = \begin{bmatrix} 0 \\ 1 \\ 0 \end{bmatrix},\ \Gamma^* = \begin{bmatrix} 1 & 0 \\ 0 & 0 \\ 0 & 1 \end{bmatrix}.$$

Since there is no control input u, the following relations are obtained: $u^{(1)} = y_c$, $A^{(1)} = 0_{2\times 2}$, $C^{(1)} = I_2$, $D^{(1)} = 0_{2\times 1}$, $V^{(1)} = I_2$, $Z^{(1)} = 0_{2\times 2}$, $W^{(1)} = \text{diag}[1,0]$ and

$$x^{(1)} = \begin{bmatrix} x_1 \\ x_3 \end{bmatrix},\ v^{(1)} = \begin{bmatrix} v_1 \\ v_3 \end{bmatrix},\ w^{(1)} = \begin{bmatrix} w_1 \\ v_2 \end{bmatrix},\ y^{(1)} = \begin{bmatrix} y_d \\ \dot{y}_c \end{bmatrix}, B^{(1)} = \begin{bmatrix} 1 \\ 0 \end{bmatrix}.$$

The matrix $W^{(1)}$ is not positive definite and the procedure has to be further applied to the *second* order system

$$\begin{aligned} \dot{x}_1^{(1)} &= y_c + v_1^{(1)}, \\ \dot{x}_2^{(1)} &= v_2^{(1)}, \\ y_1^{(1)} &= x_1^{(1)} + w_1^{(1)} = y_d^{(1)}, \\ y_2^{(1)} &= x_2^{(1)} + w_2^{(1)} = y_c^{(1)}. \end{aligned}$$

Thus we obtain $C_c^{(1)} = \begin{bmatrix} 0 & 1 \end{bmatrix}$, $C_d^{(1)} = \begin{bmatrix} 1 & 0 \end{bmatrix} = C^{*(1)}$, and

$$\Gamma_c^{(1)} = \begin{bmatrix} 0 \\ 1 \end{bmatrix},\ \Gamma^{*(1)} = \begin{bmatrix} 1 \\ 0 \end{bmatrix}.$$

Consistent with this result, $x^{(2)} = x_1^{(1)}$, $v^{(2)} = v_1^{(1)}$, $V^{(2)} = 1$, $Z^{(2)} = 0_{1\times 2}$, $W^{(2)} = I_2$, $A^{(2)} = 0$, $D^{(2)} = 0_{2\times 2}$, $B^{(2)} = \begin{bmatrix} 0 & 1 \end{bmatrix}$,

$$w^{(2)} = \begin{bmatrix} w_1^{(1)} \\ v_2^{(1)} \end{bmatrix},\ C^{(2)} = \begin{bmatrix} 1 \\ 0 \end{bmatrix}.$$

Since the matrix $W^{(2)}$ is positive definite, the procedure ends at this stage and the state estimation problem concerns the *first* order system

$$\begin{aligned} \dot{x}^{(2)} &= u^{(1)} + v^{(2)} = y_c + v_1^{(1)}, \\ y_1^{(2)} &= x^{(2)} + w_1^{(1)}, \\ y_2^{(2)} &= v_2^{(1)}. \end{aligned}$$

By taking into account Fig. 4.3, the filter equation can easily be found, precisely

$$\dot{\hat{x}}^{(2)} = -L_d^{(1)}(y_1 - \hat{x}^{(2)}) + y_2 - L_c^{(1)} \frac{d^2 y_2}{dt^2}$$

with $\hat{x}^{(2)}(0) = \bar{x}_0^{(2)}$. By exploiting the discussion for the alternative scheme of Fig. 4.4 we get

$$\begin{aligned}\hat{x}^{(2)} &= -L_c^{(1)}\dot{y}_c^{(2)} + z,\\ \dot{z} &= -L_d^{(1)}(y_1 - \hat{x}^{(2)}) + y_2 + \dot{L}_c^{(1)}\dot{y}_2\end{aligned}$$

and a (first order) differentiation of the output y_2 is needed.

The quantities $L_c^{(1)}$ and $L_d^{(1)}$ are computed by resorting to the equation

$$\begin{bmatrix} L_d^{(1)} & L_c^{(1)} \end{bmatrix} = -\Pi^{(2)}C^{(2)\prime}(W^{(2)})^{-1}$$

where the matrix $\Pi^{(2)}$ is the solution of the DRE

$$\dot{\Pi}^{(2)} = A^{(2)}\Pi^{(2)} + \Pi^{(2)}A^{(2)\prime} - \Pi^{(2)}C^{(2)\prime}(W^{(2)})^{-1}C^{(2)}\Pi^{(2)} + V^{(2)}$$

satisfying the boundary condition $\Pi^{(2)}(0) = \Pi_0^{(2)}$. In the case at hand we find $\bar{x}_0^{(2)} = 0$ and $\Pi_0^{(2)} = 1$, so that $\Pi^{(2)}(t) = 1$, $L_d^{(1)}(t) = -1$, $L_c^{(1)}(t) = \dot{L}_c^{(1)}(t) = 0$. Finally, the estimate $\hat{x}$ of x is given by $\Gamma_c y_2 + \Gamma^*(\Gamma_c^{(1)}\dot{y}_2 + \Gamma^{*(1)}\hat{x}^{(2)})$.

The extension of the discussion above to the case of an unbounded observation interval can be performed in a straightforward way, and the conclusion is drawn that the solution of the DRE must be substituted by that solution of the ARE which is the limit of the solution of the DRE. Assuming that $W^{(i)}$ is positive definite, this limit is guaranteed to exist if the pair $(A^{(i)}, C^{(i)})$ is observable. In turn, this property holds whenever the pair (A, C) is observable. Indeed, let $W^{(1)} > 0$: if the pair $(A^{(1)}, C^{(1)})$ is not observable, there exists a vector $z \neq 0$ such that $C^*A\Gamma^*z = \lambda z$, $C_d\Gamma^*z = 0$, $C_cA\Gamma^*z = 0$ (see the *PBH* test in Section A.1 of Appendix A). Since the rank of Γ^* equals the number of its columns, it follows that $x := \Gamma^*z \neq 0$. Furthermore, it is also true that $C_c\Gamma^* = 0$ and $C^*\Gamma^* = I$. Therefore the three preceding equations are equivalent to $C^*Ax = \lambda C^*x$, $C_dx = 0$, $C_cAx = 0$, $C_cx = 0$, from which it follows that

$$\begin{bmatrix} C^* \\ C_c \end{bmatrix} Ax = \lambda \begin{bmatrix} C^* \\ C_c \end{bmatrix} x = 0.$$

These equations are, in turn, equivalent to $Ax = \lambda x$, $C_cx = 0$ which, together with the already found relation $C_dx = 0$, imply that the pair (A, C) is not observable. By iterating these arguments the truth of the claim above can be ascertained also in the case where $W^{(i)} > 0$, $i > 1$. Finally, notice that in the scheme of Fig. 4.4 the block $\dot{L}_c$ is no longer present since this function is actually constant.

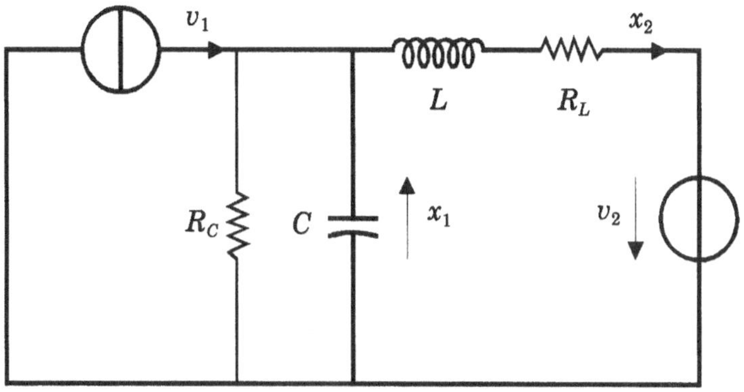

Figure 4.5: The electric circuit considered in Example 4.8.

Example 4.8 Consider the electric circuit in Fig. 4.5 where, for the sake of simplicity, all parameters are unitary. Therefore the system equations are

$$\begin{aligned}\dot{x}_1 &= -x_1 - x_2 + v_1,\\ \dot{x}_2 &= x_1 - x_2 + v_2,\\ y &= x_1\end{aligned}$$

under the assumption that the voltage across the condenser can exactly be measured. The signals v_1 and v_2 account for two sources of noise. For an unbounded observation interval the filter is sought corresponding to the intensity $V = \mathrm{diag}[\nu_1, \nu_2]$ of the stochastic process $\begin{bmatrix} v_1 & v_2 \end{bmatrix}$. By adopting the usual notation and letting $C^* = \begin{bmatrix} 0 & 1 \end{bmatrix}$, one finds $A^{(1)} = -1$, $B^{(1)} = 1$, $C^{(1)} = -1$, $D^{(1)} = -1$, $V^{(1)} = \nu_2$, $W^{(1)} = \nu_1$. The solution of the relevant ARE is

$$\bar{\Pi}^{(1)} = \sqrt{\nu_1^2 + \nu_1\nu_2} - \nu_1$$

so that the filter equations are (recall that $x^{(1)} = x_2$ and $L_c = \frac{-\bar{\Pi}^{(1)}}{\nu_1} = \sqrt{1+\nu_2/\nu_1} - 1$)

$$\begin{aligned}\dot{z} &= -(1+L_c)\hat{x}^{(1)} + (1-L_c)y,\\ \hat{x}^{(1)} &= z - L_c y,\\ \hat{x}_2 &= \hat{x}^{(1)},\end{aligned}$$

and the transfer function $G(s)$ from y to $\hat{x}_2$ is

$$G(s) = \frac{1 + L_c - L_c s}{1 + L_c + s}.$$

Note that the time constant of $G(s)$ approaches 1 when $\nu_2/\nu_1 \to 0$, while it approaches 0 when $\nu_2/\nu_1 \to \infty$.

4.3 The LQG control problem

The optimal control problem to be considered in this section refers to the system (4.1)–(4.6) and the performance index

$$J = E\left[\frac{1}{t_f - t_0}\int_{t_0}^{t_f}(x'(t)Q(t)x(t) + u'(t)R(t)u(t))dt\right] \tag{4.34}$$

where, as in the LQ context, $Q(t) = Q'(t) \geq 0$ and $R(t) = R'(t) > 0$, $\forall t$, are matrices of continuously differentiable functions and $Q(\cdot) \neq 0$ to avoid triviality. In the performance index (4.34) a term which is a nonnegative quadratic function of $x(t_f)$ could also be added: its presence, however, does not alter the essence of the problem but rather makes the discussion a little more involved. For the sake of simplicity it is not included here and likewise the intensity W of the noise w is assumed to be positive definite.

The Linear Quadratic Gaussian (LQG) optimal control problem under consideration is defined in the following way.

Problem 4.2 (LQG Problem) *Consider the system (4.1)–(4.6): Find the control which minimizes the performance index (4.34).*

In Problem 4.2 the control interval is given and may or may not be finite. In the first case it is obvious that the multiplicative factor in front of the integral is not important, while in the second case it is essential as far as the boundedness of the performance index is concerned.

4.3.1 Finite control horizon

The solution of Problem 4.2 is very simple and somehow obvious. In fact, according to it, the actual value of the control variable is made to depend on the optimal estimate of the state of the system (that is, the state of the Kalman filter) through a gain matrix which coincides with the matrix resulting from the minimization of the *deterministic* version of the performance index (4.34) in an LQ context, namely

$$J_d = \int_{t_0}^{t_f}[x'(t)Q(t)x(t) + u'(t)R(t)u(t)]dt.$$

The precise statement of the relevant result is given in the following theorem.

Theorem 4.7 *Let $Z = 0$, t_0 and t_f be given such that $-\infty < t_0 < t_f < \infty$. Then the solution of Problem 4.2 is*

$$u_c^o(\hat{x}, t) = K(t)\hat{x} \tag{4.35}$$

where $\hat{x}$ is the state of the Kalman filter

$$\dot{\hat{x}}(t) = [A(t) + L(t)C(t)]\hat{x}(t) + B(t)u_c^o(\hat{x}(t), t) - L(t)y(t) \tag{4.36}$$

with $\hat{x}(t_0) = \bar{x}_0$. In eqs. (4.35), (4.36), $K(t) = -R^{-1}(t)B'(t)P(t)$ and $L(t) = -\Pi(t)C'(t)W^{-1}$, where P and Π are the solutions (unique, symmetric and positive semidefinite) of the differential Riccati equations

$$\dot{P}(t) = -P(t)A(t) - A'(t)P(t) + P(t)B(t)R^{-1}(t)B'(t)P(t) - Q(t),$$
$$\dot{\Pi}(t) = \Pi(t)A'(t) + A(t)\Pi(t) - \Pi(t)C'(t)W^{-1}C(t)\Pi(t) + V$$

satisfying the boundary conditions $P(t_f) = 0$, $\Pi(t_0) = \Pi_0$, respectively.

Proof. Letting $e(t) := x(t) - \hat{x}(t)$, it follows that

$$E[x'(t)Q(t)x(t)] = E[e'(t)Q(t)e(t)] + 2E[e'(t)Q(t)\hat{x}(t)] + E[\hat{x}'(t)Q(t)\hat{x}(t)].$$

By exploiting known properties of the trace of a matrix and the material in Section 4.2.1 (in particular refer to Remarks 4.6 and 4.7) we get

$$E[e'(t)Q(t)e(t)] = \mathrm{tr}[Q(t)E[e(t)e'(t)]] = \mathrm{tr}[Q(t)\Pi(t)]$$

and

$$E[e'(t)Q(t)\hat{x}(t)] = \mathrm{tr}[Q(t)E[\hat{x}(t)e'(t)]] = 0.$$

Thus the problem of minimizing the performance index J is equivalent to the problem of minimizing the performance index

$$J_r := E\left[\int_{t_0}^{t_f} [\hat{x}(t)Q(t)\hat{x}(t) + u'(t)R(t)u(t)]dt\right] + \mathrm{tr}\left[\int_{t_0}^{t_f} Q(t)\Pi(t)dt\right].$$

The link between $\hat{x}$ and u can be established by means of eq. (4.36), yielding

$$\dot{\hat{x}}(t) = A(t)\hat{x}(t) + B(t)u(t) + L(t)(C(t)\hat{x}(t) - y(t)).$$

Hence the minimization of J_r implies solving a stochastic control problem (with accessible state) for this system where the random process $C(t)\hat{x}(t) - y(t)$ can be proved to be a white noise with intensity W. In view of Remark 3.9, the control law (4.35) solves the problem at hand.

Remark 4.8 *(Optimal value of the performance index)* The value of J resulting from the implementation of the control law given in Theorem 4.7 can easily be evaluated by computing the quantity J_r which is defined in the proof of the quoted theorem. By taking into account eq. (4.35) it follows that

$$J_r = \mathrm{tr}\left[\int_{t_0}^{t_f} Q(t)\Pi(t)dt\right] + E\left[\int_{t_0}^{t_f} \hat{x}'(t)[Q(t) + K'(t)R(t)K(t)]\hat{x}(t)dt\right].$$

In view of Remark 3.9 the second term is given by

$$\operatorname{tr}\left[P_K(t_0)\bar{x}_0\bar{x}_0' + \int_{t_0}^{t_f} L(t)WL'(t)P_K(t)dt\right]$$

where P_K is the solution of the equation

$$\dot{P}_K = -P_K(A+BK) - (A+BK)'P_K - (Q+K'RK)$$

with the boundary condition $P_K(t_f) = 0$. In writing down these relations the equation for $\hat{x}$, the relevant boundary condition and the circumstance that the noise intensity is W (see the proof of Theorem 4.7) have been taken into consideration. It is straightforward to check that the DRE for P (in the statement of the quoted theorem) reduces to the differential equation given above, provided that the terms $\pm PBK$, $\pm K'B'P$ are added to the DRE itself and the expression of K is taken into account. Therefore it follows that

$$J_r = \bar{x}_0'P(t_0)\bar{x}_0 + \operatorname{tr}\left[\int_{t_0}^{t_f} [Q(t)\Pi(t) + L(t)WL'(t)P(t)]dt\right].$$

Example 4.9 Consider the first order system

$$\dot{x} = u + v,$$
$$y = x + w$$

and the performance index

$$J = E\left[\int_0^{t_f} [Qx^2 + u^2]dt\right]$$

with $\bar{x}_0 = 0$ and $\Xi = \operatorname{diag}[V, W]$, $V > 0$, $W > 0$. Letting $\nu := \sqrt{VW}$, $\omega := \sqrt{V/W}$ and $\mu := \sqrt{Q}$, the solutions of the two DRE relevant to the problem are

$$P(t) = \mu\frac{1 - e^{2\mu(t-t_f)}}{1 + e^{2\mu(t-t_f)}},$$
$$\Pi(t) = \nu\frac{\Pi_0 + \nu + (\Pi_0 - \nu)e^{-2\omega t}}{\Pi_0 + \nu - (\Pi_0 - \nu)e^{-2\omega t}},$$

which imply $K = -P$ and $L = -\Pi/\sqrt{W}$. The responses of the system state and control input are shown in Fig. 4.6 when $Q = \Pi_0 = 1$, $V = 1$ and $W = 0.1$. Corresponding to the same sample of the noises, the responses of these variables are also presented in the quoted figure when the control law $u = Ky$ is implemented, that is when the Kalman filter is not utilized: the deterioration of the system behaviour is apparent. Similar conclusions can be drawn from the analysis of the same figure corresponding to different noises intensities, namely $V = 0.1$ and $W = 1$. The better performance of the control system which exploits a Kalman filter is put into particular evidence by the control transient.

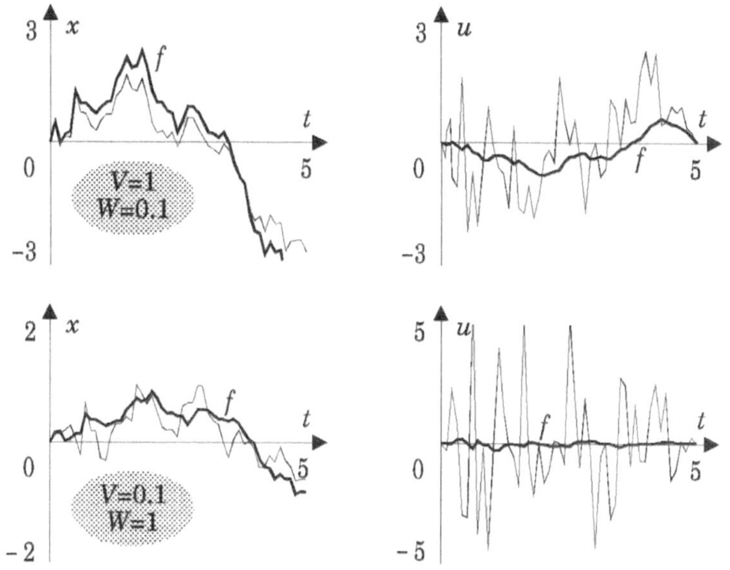

Figure 4.6: Example 4.9: responses of x and u when a Kalman filter is (f) or is not utilized.

4.3.2 Infinite control horizon

Unbounded control intervals can be dealt with also in a stochastic context and a particularly nice solution found when the problem at hand is stationary, thus paralleling the results of the LQ and filtering framework. Consistent with eq. (4.34), the performance index to be minimized is

$$J = E\left[\lim_{\substack{t_o \to -\infty \\ t_f \to \infty}} \frac{1}{t_f - t_0} \int_{t_0}^{t_f} [x'(t)Q(t)x(t) + u'(t)R(t)u(t)]dt \right] \tag{4.37}$$

where the matrices Q and R satisfy the usual continuity and sign definition assumptions. Moreover, the system (4.1)–(4.6) is controllable and reconstructable for all t, the initial state is zero, $Z = 0$ and $W > 0$. If these assumptions are satisfied the solutions of the two DRE relevant to Problem 4.2 can indefinitely be extended and the following theorem holds.

Theorem 4.8 *Assume that the above aasumptions hold. Then the solution of Problem 4.2 when the control interval is unbounded, that is when the perfor-*

mance index is given by eq. (4.37), is constituted by the control law

$$u_c^o(\hat{x},t) = \bar{K}(t)\hat{x}$$

where $\hat{x}$ *is the state of the Kalman filter*

$$\dot{\hat{x}}(t) = [A(t) + \bar{L}(t)C(t)]\hat{x}(t) + B(t)u_c^o(\hat{x}(t),t) - \bar{L}(t)y(t)$$

with $\bar{K}(t) = -R^{-1}(t)B'(t)\bar{P}(t)$ *and* $\bar{L}(t) = -\bar{\Pi}(t)C'(t)W^{-1}$*. The matrices* $\bar{P}$ *and* $\bar{\Pi}$ *are given by*

$$\bar{P}(t) = \lim_{t_f \to \infty} P(t,t_f),$$
$$\bar{\Pi}(t) = \lim_{t_0 \to -\infty} \Pi(t,t_0),$$

where $P(t,t_f)$ *and* $\Pi(t,t_0)$ *are the solutions of the differential Riccati equations specified in Theorem 4.7 with the boundary conditions* $P(t_f,t_f) = 0$ *and* $\Pi(t_0,t_0) = 0$.

Remark 4.9 *(Optimal value of the performance index)* If the LQG problem over an infinite interval admits a solution, the optimal value of the performance index can easily be evaluated by making reference to Remark 4.8. Thus

$$J^o = \lim_{\substack{t_o \to -\infty \\ t_f \to \infty}} \frac{1}{t_f - t_0} \mathrm{tr}\left[\int_{t_0}^{t_f} [Q(t)\bar{\Pi}(t) + \bar{L}(t)W\bar{L}'(t)\bar{P}(t)]dt\right].$$

The case when all the problem data are constant deserves particular attention: the most significant features are the constancy of matrices $\bar{P}$ and $\bar{\Pi}$ (and hence of matrices $\bar{K}$ and $\bar{L}$, too) and the fact that they satisfy the ARE resulting from setting to zero the derivatives in the DRE of the statement of Theorem 4.8. The importance of this particular framework justifies the formal presentation of the relevant result in the forthcoming theorem where (unnecessarily) restrictive assumptions are made in order to simplify its statement and some results concerning Riccati equations are exploited (see Chapter 5, Section 5.3).

Theorem 4.9 *Let the matrices* A, B, C, $Q := Q^{*\prime}Q^*$, R *be constant,* $Z = 0$, $V := V^*V^{*\prime}$, $W > 0$. *Moreover let the couples* (A,B) *and* (A,V^*) *be reachable and the couples* (A,C) *and* (A,Q^*) *observable. Then the solution of Problem 4.2 when the control interval is unbounded, i.e., when the performance index is given by eq. (4.37), is specified by the control law*

$$u_c^o(\hat{x}) = \bar{K}\hat{x}$$

where $\hat{x}$ *is the state of the Kalman filter*

$$\dot{\hat{x}}(t) = (A + \bar{L}C)\hat{x}(t) + Bu_c^o(\hat{x}(t)) - \bar{L}y(t)$$

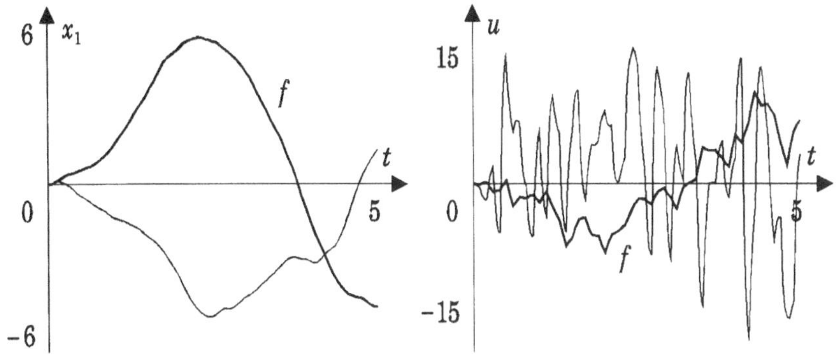

Figure 4.7: Example 4.10: responses of x_1 and u when the controller makes (f) or does not make use of a Kalman filter.

with $\bar{K} = -R^{-1}B'\bar{P}$ *and* $\bar{L} = -\bar{\Pi}C'W^{-1}$. *The matrices* $\bar{P}$ *and* $\bar{\Pi}$ *are the unique symmetric, positive definite solutions of the algebraic Riccati equations*

$$0 = PA + A'P - PBR^{-1}B'P + Q,$$
$$0 = \Pi A' + A\Pi - \Pi C'W^{-1}C\Pi + V.$$

Example 4.10 Consider the system

$$\dot{x}_1 = x_2,$$
$$\dot{x}_2 = u + v_2,$$
$$y = x_1 + w$$

with the noise intensities given by $V = \text{diag}[0, 10]$ and $W = 0.1$. Associated to this system is the performance index (4.37) where $Q = \text{diag}[1, 0]$ and $R = 1$. Each one of the two ARE relevant to the resulting LQG problem admits a unique positive semidefinite solution, namely

$$\bar{P} = \begin{bmatrix} \sqrt{2} & 1 \\ 1 & \sqrt{2} \end{bmatrix}, \quad \bar{\Pi} = \begin{bmatrix} \sqrt{0.2} & 1 \\ 1 & \sqrt{20} \end{bmatrix}$$

so that

$$\bar{K} = -\begin{bmatrix} 1 & \sqrt{2} \end{bmatrix}, \quad \bar{L} = -\begin{bmatrix} \sqrt{20} \\ 10 \end{bmatrix}.$$

Corresponding to the same noise samples, the responses of x_1 and u are shown in Fig. 4.7 when the above controller is used or a controller is adopted with the same structure and the matrix $\bar{L}$ replaced by $\hat{L} = -\begin{bmatrix} 20 & 100 \end{bmatrix}'$ (the eigenvalues of the error dynamic matrix are thus moved from $-2.24(1 \pm j)$ to -10). The control variable effort is apparently smaller when the result in Theorem 4.9 is exploited.

Remark 4.10 *(Optimal value of the performance index in the time-invariant case)* In view of Remark 4.9, the optimal value of the performance index when the LQG problem is time-invariant and the control interval is unbounded is given simply by

$$J^o = \mathrm{tr}[Q\bar{\Pi} + \bar{P}\bar{L}W\bar{L}'].$$

This expression implies that $J^o \geq \mathrm{tr}[Q\bar{\Pi}]$ since the second term is nonnegative (the eigenvalues of the product of two symmetric positive semidefinite matrices are nonnegative). This inequality holds independently of the actual matrix R: therefore, even if the control cost becomes negligible (i.e., when $R \to 0$), the value of the performance index cannot be less than $\mathrm{tr}[Q\bar{\Pi}]$ which, in a sense, might be seen as the *price* to be paid because of the *imprecise* knowledge of the system state. Since it can also be proved that

$$J^o = \mathrm{tr}[\bar{P}V + \bar{\Pi}\bar{K}'R\bar{K}],$$

the conclusion can be drawn that $J^o \geq \mathrm{tr}[\bar{P}V]$ and again, even when the output measurement becomes arbitrarily accurate ($W \to 0$), the optimal value of the performance index cannot be less than $\mathrm{tr}[\bar{P}V]$ which, in a sense, might be seen as the *price* to be paid because of the *presence* of the input noise.

Remark 4.11 *(Stabilizing properties of the LQG solution)* When the assumptions in Theorem 4.10 hold, the resulting control system is asymptotically stable. Indeed, since the Kalman filter which is the core of the controller has the structure of a state observer, it follows, from Theorem B.2 (see Section B.3 of Appendix B.1) that the eigenvalues of the control system are those of matrices $A+B\bar{K}$ and $A+\bar{L}C$. All these eigenvalues have negative real parts because the solutions of the ARE (from which the matrices $\bar{K}$ and $\bar{L}$ originate) are stabilizing (recall the assumptions in Theorem 4.10 and see Theorem 5.10 of Section 5.3 of Chapter 5).

The solution of the optimal regulator problem has been proved to be *robust* in terms of phase and gain margins (see Subsection 3.4.2 of Section 3.4 in Chapter 3.1). The same conclusions hold in the filtering context because of the duality between the two problems. Thus one might conclude that the controller defined in Theorem 4.9 implies that the resulting control system is endowed with analogous robustness properties with regard to the presence of phase and gain uncertainties on the *control side* (point P_u in the block scheme of Fig. 4.8) and/or on the *output side* (point P_y in the same block scheme). This actually fails to be true, as shown in the following well-known example.

Example 4.11 Consider the LQG problem defined on a second order system Σ characterized by the matrices

$$A = \begin{bmatrix} 1 & 1 \\ 0 & 1 \end{bmatrix}, \; B = \begin{bmatrix} 0 \\ 1 \end{bmatrix}, \; V = \beta_v \begin{bmatrix} 1 & 1 \\ 1 & 1 \end{bmatrix}, \; Q = \beta_q \begin{bmatrix} 1 & 1 \\ 1 & 1 \end{bmatrix},$$

$$C = \begin{bmatrix} 1 & 0 \end{bmatrix}, \; W = 1, \; R = 1$$

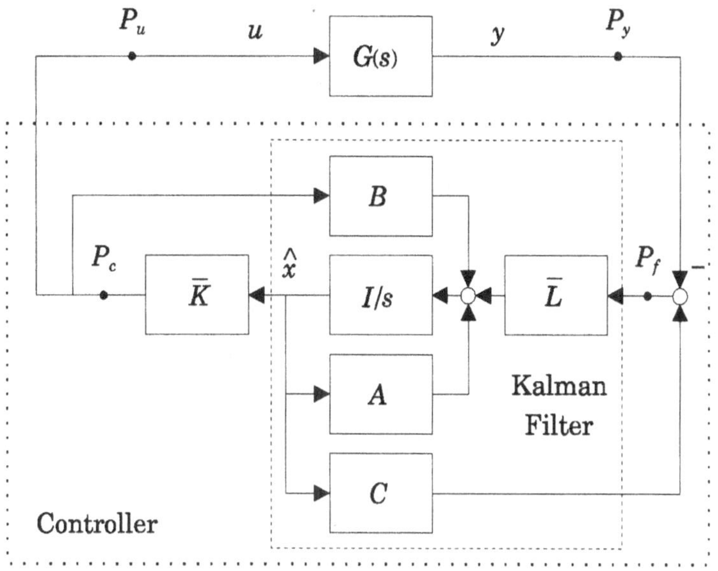

Figure 4.8: The LQG control system: $G(s) = C(sI - A)^{-1}B$.

where $\beta_v > 0$ and $\beta_q > 0$. The solutions of the ARE relevant to the problem at hand are

$$\bar{P} = \alpha_q \begin{bmatrix} 2 & 1 \\ 1 & 1 \end{bmatrix}, \ \bar{\Pi} = \alpha_v \begin{bmatrix} 1 & 1 \\ 1 & 2 \end{bmatrix}$$

where $\alpha_q = 2 + \sqrt{4 + \beta_q}$ and $\alpha_v = 2 + \sqrt{4 + \beta_v}$. If the loop gain of the resulting feedback system is perturbed, i.e., a nondynamic system with gain μ (in nominal conditions $\mu = 1$) is inserted between the controller and Σ, the dynamic matrix of the whole system is

$$F = \begin{bmatrix} 1 & 1 & 0 & 0 \\ 0 & 1 & -\mu\alpha_q & -\mu\alpha_q \\ \alpha_v & 0 & 1 - \alpha_v & 1 \\ \alpha_v & 0 & -(\alpha_q + \alpha_v) & 1 - \alpha_q \end{bmatrix}.$$

It is easy to check that the characteristic polynomial ψ of F is $\psi(s) = s^4 + b_1 s^3 + b_2 s^2 + b_3(\mu)s + b_4(\mu)$ with $b_3(\mu) = \alpha_q + \alpha_v - 4 + 2\alpha_q\alpha_v(\mu - 1)$ and $b_4 = 1 + \alpha_q\alpha_v(1 - \mu)$. From these two relations it follows that, corresponding to arbitrarily small perturbations of the loop gain (that is for values of μ arbitrarily close to 1), it is possible to find sufficiently high values of α_q and α_v (i.e., of β_q and β_v) such that the sign of b_3 differs from the sign of b_4, thus entailing instability of the feedback system.

This unpleasant outcome is caused by the following fact: the transfer function $T_u(s) := -\bar{K}(sI - (A + B\bar{K} + \bar{L}C))^{-1}\bar{L}G(s)$ which results from *cutting* the scheme of Fig. 4.8 at the point P_u does not coincide with the transfer function $T_c(s) := \bar{K}(sI - A)^{-1}B$ which is expedient in proving the robustness of the solution of the LQ problem. Indeed, T_c is the transfer function which results from *cutting* the scheme of Fig. 4.8 at the point P_c. A similar discussion applied to the other side of $G(s)$, with reference to the transfer function $T_y(s) := -G(s)\bar{K}(sI - (A + B\bar{K} + \bar{L}C))^{-1}\bar{L}$ (which results from cutting the scheme in Fig. 4.8 at the point P_y) and to the transfer function $T_f(s) := C(sI - A)^{-1}\bar{L}$ which, in the Kalman filter framework, plays the same role, from the robustness point of view, as T_c does in the optimal regulator setting. It is still easy to check that T_f is the transfer function which results from cutting the above quoted scheme at the point P_f. Therefore, if the four matrices Q, R, V, W are *given* data of the problem and the available knowledge on the controlled process is not accurate, no robustness properties can *a priori* be guaranteed to the control system either on the actuators' or sensors' sides. On the other hand, if, as often is the case, the four matrices above are to be meant as *free* parameters to be selected while carrying over a sequence of trials suggested by a synthesis procedure which exploits the LQG results, then a wise choice of them may again ensure specific robustness properties. In fact, by resorting to reasoning similar to which led to Theorem 3.12 of Subsection 3.4.3 (details can be found in the literature), the following results can be proved. They are stated under the assumptions that the number of control variables u equals the number of output variables y and the matrices B, C are full rank.

Theorem 4.10 *Let the triple (A, B, C) be minimal and $V = \nu BB'$. Then, if no transmission zeros of the triple (A, B, C) has positive real part, the function T_u approaches the function T_c as $\nu \to \infty$.*

Theorem 4.11 *Let the triple (A, B, C) be minimal and $Q = qC'C$. Then, if no transmission zeros of the triple (A, B, C) has positive real part, the function T_y approaches the function T_f as $q \to \infty$.*

Remark 4.12 *(Alternative statement of Theorems 4.10 and 4.11)* By recalling the role played by the matrices Q and R in specifying the meaning of the performance index and by matrices V and W in defining the noises characteristics, it should be fairly obvious that instead of letting matrices Q and V go to infinity, we could let matrices R and W go to zero.

Example 4.12 Consider the LQG problem defined by the matrices

$$A = \begin{bmatrix} 0 & 1 & -1 \\ 0 & 0 & 0 \\ 0 & 0 & 0 \end{bmatrix}, \quad B = \begin{bmatrix} 1 & 0 \\ 1 & 1 \\ 0 & 1 \end{bmatrix},$$

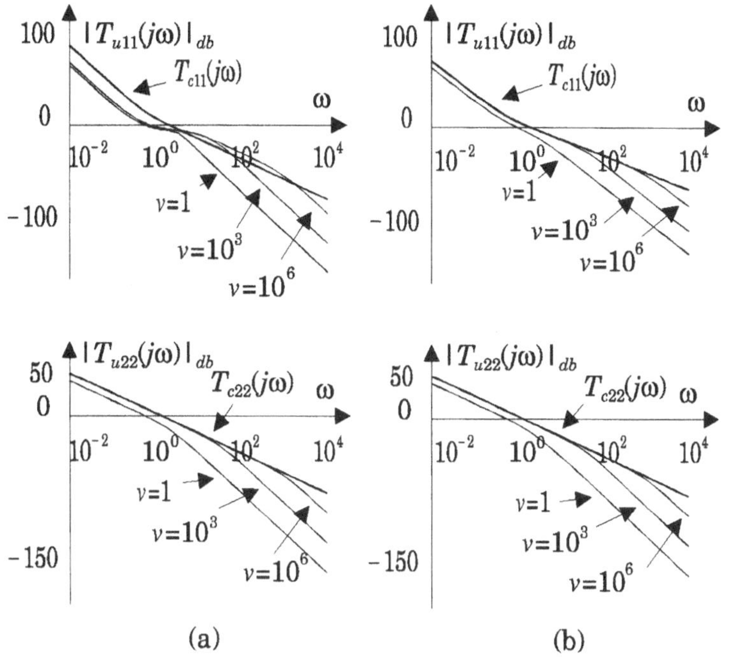

Figure 4.9: Example 4.12: plots of $|T_{u11}(j\omega)|$ and $|T_{u22}(j\omega)|$ when $C = C_d$ (a) and $C = C_s$ (b).

$Q = I_3$, $R = W = I_2$, $V = \nu BB'$ and $C = C_d$ or $C = C_s$, where

$$C_d = \begin{bmatrix} 1 & -2 & 2 \\ 0 & 0 & 1 \end{bmatrix}, \; C_s = \begin{bmatrix} 1 & 0 & 0 \\ 0 & 0 & 1 \end{bmatrix}.$$

When $C = C_d$ the triple (A, B, C) has a transmission zero in the right half-plane ($s = 1$), while this does not happen when $C = C_s$. The plots of the frequency response of $|T_{u11}(j\omega)|$ and $|T_{u22}(j\omega)|$ are shown in Fig. 4.9 for the two cases, corresponding to some values of the parameter ν. Observe that $|T_{u11}|$ does not approach $|T_{c11}|$ when $C = C_d$. However, this circumstance does not prevent some other component of $|T_u|$ from approaching the corresponding component of T_c as ν increases (see the plots of $|T_{u22}|$ in Fig. 4.9).

4.4 Problems

Problem 4.4.1 Find the Kalman filter for the system defined by the matrices

$$A = \begin{bmatrix} 0 & 1 & 0 \\ 1 & 0 & -1 \\ 0 & 0 & 1 \end{bmatrix}, \quad C = \begin{bmatrix} 1 & 0 & 0 \end{bmatrix},$$

$\Xi = I$ when the observation interval is unbounded. Is the filter stable? What can be said of its stability properties if the optimal gain $\bar{L}$ is replaced by $L := 2\bar{L}$?

Problem 4.4.2 Discuss the noisy tracking problem

$$\begin{aligned}
\dot{x} &= Ax + Bu + v, \\
y &= Cx + w, \\
\dot{\vartheta} &= F\vartheta + \zeta, \\
\mu &= H\vartheta + \nu, \\
J &= E\left[\frac{1}{t_f - t_0}\int_{t_0}^{t_f} [(y-\mu)'Q(y-\mu) + u'Ru]\, dt\right]
\end{aligned}$$

where $-\infty < t_0 < t_f < \infty$. Generalize the result to the case of an infinite control interval, that is when the performance index is

$$J = E\left[\lim_{\substack{t_0 \to -\infty \\ t_f \to \infty}} \frac{1}{t_f - t_0}\int_{t_0}^{t_f} [(y-\mu)'Q(y-\mu) + u'Ru]\, dt\right].$$

Problem 4.4.3 Set the complete picture of existence and stability results for the Kalman filter in the time-invariant case when the observation interval is unbounded.

Problem 4.4.4 Find the solution of the optimal filtering problem on the interval $[t_0, t_f]$ defined by the system (4.1a), (4.1b) and (4.2)–(4.6) with $x_0 = x(\tau)$, where τ is a given *interior* point of the interval $[t_0, t_f]$.

Problem 4.4.5 Discuss the optimal filtering problem when the input noise v is the output of a known time-invariant stable system driven by a white noise.

Problem 4.4.6 For $-\infty < \alpha < \infty$ and $-\infty < \beta < \infty$ discuss the existence and stability of the Kalman filter which supplies the optimal estimate of the state of the system

$$\begin{aligned}
\dot{x} &= \begin{bmatrix} 0 & 0 & 0 \\ 1 & 0 & 0 \\ 0 & 1 & \alpha \end{bmatrix} x + v, \\
y &= \begin{bmatrix} 0 & 1 & 1 \end{bmatrix} x + w
\end{aligned}$$

when the intensity of the zero-mean white noise $\begin{bmatrix} v & w \end{bmatrix}'$ is

$$\Xi = \begin{bmatrix} 1 & \beta & 0 & 0 \\ \beta & \beta^2 & 0 & 0 \\ 0 & 0 & 0 & 0 \\ 0 & 0 & 0 & 1 \end{bmatrix}$$

and the observation interval is unbounded.

Problem 4.4.7 Discuss the state estimation problem relative to the system $\dot{x} = Ax + v$, $y = Cx + w$ assuming that a noise-free signal which is the integral of y is also available. The zero-mean white noises v and w are uncorrelated and have intensity V and $W > 0$, respectively.

Problem 4.4.8 Consider the filtering problem over an unbounded observation interval relative to the system $\dot{x} = Ax + v$, $y = Cx + w$ where the intensities of the zero-mean uncorrelated white noises v and w are V and W, respectively, and

$$A = \begin{bmatrix} 0 & 1 & 0 \\ 0 & 0 & 1 \\ 0 & 0 & 0 \end{bmatrix},\ V = \begin{bmatrix} 0 & 0 & 0 \\ 0 & \alpha^2 & \alpha \\ 0 & \alpha & 1 \end{bmatrix},\ C = \begin{bmatrix} 1 & 0 & 0 \end{bmatrix},\ W = \rho.$$

For $-\infty < \alpha < \infty$, what can be said of the variance of the optimal state estimation error as $\rho \to 0^+$?

Problem 4.4.9 Consider the triple of matrices

$$A = \begin{bmatrix} 0 & 0 & 0 \\ 1 & 0 & 0 \\ 0 & 1 & 0 \end{bmatrix},\ C = \begin{bmatrix} 1 & 0 & 0 \end{bmatrix},\ L = \begin{bmatrix} -1 \\ -2 \\ -2 \end{bmatrix}.$$

Might the equation $\dot{\hat{x}} = (A + LC)\hat{x} - Ly$ define the Kalman filter for the system $\dot{x} = Ax + v$, $y = Cx + w$ when the observation interval is unbounded?

Problem 4.4.10 Consider the system $\dot{x}(t) = Ax(t) + Bv(t)$, $y(t) = Cx(t) + Dw(t)$ where the pair (A, C) of constant matrices is observable and the zero-mean uncorrelated white noises v and w have intensities V and $W > 0$, respectively. Select matrices B and D in such a way that the eigenvalues of the resulting Kalman filter corresponding to an unbounded observation interval have real part less than $-\alpha$.

Chapter 5

The Riccati equations

5.1 Introduction

The Riccati equations in the differential (DRE) as well as in the algebraic form (ARE) deserve particular attention since, in a sense, they constitute the keystone of the solution of a number of significant problems. Indeed in the preceding chapters it has been shown how fundamental control and filtering problems can easily be tackled and solved once the main properties of such equations have been well understood and issues such as existence, uniqueness and so on clarified. It is therefore advisable to devote this chapter to the illustration of some basic facts concerning the DRE's and ARE's involved in the LQ and LQG problems. The discussion has been occasionally generalized to encompass also forms of the ARE which are of interest in related frameworks (e.g., H_∞ control) whenever the required extra effort is kept within reasonable limits.

5.2 The differential equation

The differential equation

$$\dot{P}(t) = -P(t)A(t) - A'(t)P(t) + P(t)B(t)R^{-1}(t)B'(t) - Q(t), \quad (5.1a)$$

$$P(t_f) = S, \quad (5.1b)$$

where A, B, Q, R are continuously differentiable functions, is considered over the interval $[t_0, t_f]$. Moreover, it is assumed that matrix A is $n \times n$ and $S = S' \geq 0$, $Q(t) = Q'(t) \geq 0$, $R(t) = R'(t) > 0$, $t \in [t_0, t_f]$.

It will be shown that the solution of eqs. (5.1) can be expressed as a function of the transition matrix of a suitable linear system. In fact, consider the *hamiltonian matrix*

$$Z(t) := \begin{bmatrix} A(t) & -B(t)R^{-1}(t)B'(t) \\ -Q(t) & -A'(t) \end{bmatrix} \tag{5.2}$$

and the system of linear differential equations

$$\begin{bmatrix} \dot{x}(t) \\ \dot{\lambda}(t) \end{bmatrix} = Z(t) \begin{bmatrix} x(t) \\ \lambda(t) \end{bmatrix}. \tag{5.3}$$

Then the following result holds.

Theorem 5.1 *Matrix $P(t)$ is a solution of eq. (5.1a) if and only if eq. (5.3) is solved by $\begin{bmatrix} x'(t) & x'(t)P(t) \end{bmatrix}'$ for all $x(t_0)$.*

Proof. Necessity. If $\lambda := Px$ has to solve eq. (5.3), then, in view of eq. (5.2), it follows that

$$\dot{x} = (A - BR^{-1}B'P)x$$

and

$$\frac{d(Px)}{dt} = \dot{P}x + P\dot{x} = -Qx - A'Px.$$

If P solves eq. (5.1a), these equations are satisfied for all $x(t_0)$. *Sufficiency.* If $\begin{bmatrix} x' & x'P \end{bmatrix}'$ solves eq. (5.3) for all $x(t_0)$, then $\dot{x} = (A - BR^{-1}B'P)x$, so that

$$\dot{\lambda} = \frac{d(Px)}{dt} = \dot{P}x + P\dot{x} = (\dot{P} + PA - PBR^{-1}B')x.$$

On the other side, from eq. (5.3) it follows that

$$\dot{\lambda} = -Q'x - A'\lambda = -Q'x - A'Px.$$

If these two expressions for $\dot{\lambda}$ have to simultaneously hold for all $x(t_0)$, then P must be a solution of eq. (5.2).

The above theorem allows us to express the solution of eqs. (5.1) as a function of the transition matrix Φ associated to Z. As a matter of fact

$$\begin{bmatrix} I \\ S \end{bmatrix} x(t_f) = \Phi(t_f, t) \begin{bmatrix} I \\ P(t) \end{bmatrix} x(t),$$

so that if

$$\Phi(t, t_f) = (\Phi(t_f, t))^{-1} := \begin{bmatrix} \Psi_{11}(t, t_f) & \Psi_{12}(t, t_f) \\ \Psi_{21}(t, t_f) & \Psi_{22}(t, t_f) \end{bmatrix},$$

then

$$x(t) = [\Psi_{11}(t, t_f) + \Psi_{12}(t, t_f)S]x(t_f) \tag{5.4a}$$

$$P(t)x(t) = [\Psi_{21}(t, t_f) + \Psi_{22}(t, t_f)S]x(t_f). \tag{5.4b}$$

The matrix on the right-hand side of eq. (5.4a) is the *inverse* of the transition matrix associated to $A - BR^{-1}B'P$ (recall that $\dot{x} = (A - BR^{-1}B'P)x$) and therefore it is nonsingular, so that from eqs. (5.4) (which hold for all $x(t_0)$ and hence for all $x(t_f)$) it follows that

$$P(t) = [\Psi_{21}(t,t_f) + \Psi_{22}(t,t_f)S][\Psi_{11}(t,t_f) + \Psi_{12}(t,t_f)S]^{-1}. \tag{5.5}$$

A simple example clarifies the discussion above.

Example 5.1 Consider the DRE defined by $A = 0,\ B = Q = R = 1$ and $S = \sigma \geq 0$, so that

$$Z = \begin{bmatrix} 0 & -1 \\ -1 & 0 \end{bmatrix},\ \Phi(t,\tau) = \frac{1}{2}\begin{bmatrix} e^{t-\tau} + e^{\tau-t} & e^{\tau-t} - e^{t-\tau} \\ e^{\tau-t} - e^{t-\tau} & e^{t-\tau} + e^{\tau-t} \end{bmatrix}.$$

By exploiting eq. (5.5) we get

$$P(t) = \frac{1+\sigma-(1-\sigma)e^{2(t-t_f)}}{1+\sigma+(1-\sigma)e^{2(t-t_f)}}.$$

5.3 The algebraic equation

The remaining part of this chapter is devoted to the presentation of some significant results concerning the algebraic equation

$$0 = PA + A'P - PBR^{-1}B'P + C'C \tag{5.6}$$

where matrices A, B, C and R are constant, matrix A is square of dimension n, matrix B has dimensions $n \times m$, $m \leq n$, $\text{rank}(B) = m$ and matrix R is nonsingular and symmetric. For the time being no further assumptions are made on R. Associated to the ARE (5.6) consider the hamiltonian matrix (similar to the one in eq. (5.2))

$$Z := \begin{bmatrix} A & -BR^{-1}B' \\ -C'C & -A' \end{bmatrix}. \tag{5.7}$$

The eigenvalues of this matrix enjoy the peculiar property stated in the following theorem.

Theorem 5.2 *The eigenvalues of matrix Z are symmetric with respect to the origin.*

Proof. Consider the matrix

$$J := \begin{bmatrix} 0 & I_n \\ -I_n & 0 \end{bmatrix}$$

which is such that $J^{-1} = J'$. It is easy to verify that $JZJ^{-1} = -Z'$, from which the theorem follows (recall that Z is a real matrix).

The eigenvalues of Z can be related meaningfully to the eigenvalues of the unreachable and/or unobservable parts of system $\Sigma(A, B, C, 0)$, as specified in the following lemma.

Lemma 5.1 *Let $\Sigma(A, B, C, 0)$ be an unobservable and/or unreachable system and $R = R'$ an arbitrary nonsingular matrix. Then it follows that:*

(a) *λ is an eigenvalue of the unobservable part of Σ if and only if there exists a nonzero vector $w \in C^n$ such that $\begin{bmatrix} w' & 0_{1\times n} \end{bmatrix}'$ is an eigenvector of Z relative to the eigenvalue λ*

(b) *λ is an eigenvalue of the unreachable part of Σ if and only if there exists a nonzero vector $w \in C^n$ such that $\begin{bmatrix} 0_{1\times n} & w' \end{bmatrix}'$ is an eigenvector of Z relative to the eigenvalue λ.*

Proof. *Point (a)* If there exists $w \neq 0$ such that

$$Z \begin{bmatrix} w \\ 0 \end{bmatrix} = \lambda \begin{bmatrix} w \\ 0 \end{bmatrix} \tag{5.8}$$

then, in view of eq. (5.7), $Aw = \lambda w$ and $C'Cw = 0$. This last relation is equivalent to $Cw = 0$, so that from the *PBH* test (see Theorem A.1 of Appendix A), we can conclude that λ is an eigenvalue of the unobservable part of Σ. Vice versa, if λ and $w \neq 0$ are such that $Aw = \lambda w$ and $Cw = 0$, then eq. (5.8) holds and λ, which is an eigenvalue of the unobservable part of Σ because of the *PBH* test, is also an eigenvalue of Z. *Point (b)* If there exists $w \neq 0$ such that

$$Z \begin{bmatrix} 0 \\ w \end{bmatrix} = \lambda \begin{bmatrix} 0 \\ w \end{bmatrix}, \tag{5.9}$$

then, by taking into account eq. (5.7), it is $-A'w = \lambda w$ and $-BR^{-1}B'w = 0$. Since R is nonsingular and B is full rank, this last equation implies $B'w = 0$, so that, in view of the *PBH* test, we can conclude that λ is an eigenvalue of the unreachable part of Σ. Vice versa, if λ and $w \neq 0$ are such that $A'w = \lambda w$ and $B'w = 0$, then eq. (5.9) holds and λ, which is an eigenvalue of the unreachable part of Σ because of the *PBH* test, is also an eigenvalue of Z.

Lemma 5.1 states that *all* the eigenvalues of A which are not eigenvalues of the *jointly* reachable and observable part of $\Sigma(A, B, C, 0)$ are also eigenvalues of Z.

Example 5.2 Let

$$A = \begin{bmatrix} 1 & 1 \\ 0 & 0 \end{bmatrix}, \quad B = \begin{bmatrix} 0 \\ 1 \end{bmatrix}, \quad C = \begin{bmatrix} 0 & 1 \end{bmatrix}$$

and $R \neq 0$ arbitrary. System $\Sigma(A, B, C, 0)$ is not observable and $\lambda = 1$ is an eigenvalue both of the unobservable part and of the corresponding matrix Z, for all R.

A tight relation can be established between the solutions of eq. (5.6) and particular Z-invariant subspaces of C^{2n}. Recall that a subspace

$$\mathcal{S} := Im(\begin{bmatrix} X \\ Y \end{bmatrix}) \subseteq C^{2n}$$

is Z-invariant if and only if there exists a square matrix V such that

$$Z \begin{bmatrix} X \\ Y \end{bmatrix} = \begin{bmatrix} X \\ Y \end{bmatrix} V.$$

If, as will be assumed from now on, the matrix whose columns span $\mathcal{S}$ has full rank, matrix V has dimension n if $\mathcal{S}$ is n-dimensional. Let further

$$\mathcal{W} := Im(\begin{bmatrix} 0_{n \times n} \\ I_n \end{bmatrix})$$

and recall that two subspaces $\mathcal{S}$ and $\mathcal{V}$ of C^{2n} are *complements* each of the other if they are *disjoint* ($\mathcal{S} \cap \mathcal{V} = 0$) and if $\mathcal{S} + \mathcal{V} = C^{2n}$. Moreover, the dimension of $\mathcal{S}$ equals the difference between $2n$ and the dimension of $\mathcal{V}$.

It is now possible to state the following fundamental result which provides the desired relation between the solutions of the ARE and Z-invariant subspaces of C^{2n}.

Theorem 5.3 *There exists a one-to-one correspondence between the solutions of the ARE and the Z-invariant subspaces of C^{2n} which are complement of $\mathcal{W}$.*

Proof. If

$$\mathcal{S} := Im(\begin{bmatrix} X \\ Y \end{bmatrix})$$

is a complement of $\mathcal{W}$, then matrix X ($n \times n$) is nonsingular. Moreover, if $\mathcal{S}$ is Z-invariant, then

$$Z \begin{bmatrix} X \\ Y \end{bmatrix} = \begin{bmatrix} X \\ Y \end{bmatrix} V \Leftrightarrow \begin{cases} AX - BR^{-1}B'Y &= XV \\ -QX - A'Y &= YV \end{cases}$$

which implies $V = X^{-1}(AX - BR^{-1}B'Y)$ and hence $-QX - A'Y = YX^{-1}(AX - BR^{-1}B'Y)$. This equation becomes, after postmultiplication of both sides by X^{-1}, $-Q - A'YX^{-1} = YX^{-1}A - YX^{-1}BR^{-1}B'YX^{-1}$. This relation shows that YX^{-1} solves the ARE.

Note that if $\mathcal{S}$ is specified by a different couple $(\hat{X}, \hat{Y})$ of matrices, then the corresponding solutions coincide. Indeed, if

$$Im(\begin{bmatrix} X \\ Y \end{bmatrix}) = Im(\begin{bmatrix} \hat{X} \\ \hat{Y} \end{bmatrix})$$

then, necessarily,

$$\begin{bmatrix} X \\ Y \end{bmatrix} = \begin{bmatrix} \hat{X} \\ \hat{Y} \end{bmatrix} T$$

for a suitable nonsingular T. Therefore,

$$\hat{Y}\hat{X}^{-1} = \hat{Y}TT^{-1}\hat{X}^{-1} = \hat{Y}T(\hat{X}T)^{-1} = YX^{-1}.$$

Vice versa, let P be a solution of the ARE and

$$\mathcal{S} := Im(\begin{bmatrix} I \\ P \end{bmatrix}).$$

The subspace $\mathcal{S}$ is obviously a complement of $\mathcal{W}$. Furthermore, by recalling that P solves the ARE,

$$\begin{aligned} Z\begin{bmatrix} I \\ P \end{bmatrix} &= \begin{bmatrix} A - BR^{-1}B'P \\ -Q - A'P \end{bmatrix} \\ &= \begin{bmatrix} A - BR^{-1}B'P \\ PA - PBR^{-1}B'P \end{bmatrix} = \begin{bmatrix} I \\ P \end{bmatrix}(A - BR^{-1}B'P) \end{aligned}$$

which proves that $\mathcal{S}$ is a Z-invariant subspace.

This theorem allows some conclusions about the number n_{ARE} of solutions of the ARE. More precisely, by recalling that each n-dimensional, Z-invariant subspace is spanned by n (generalized) linearly independent eigenvectors of Z, it can be stated that:

(*i*) If Z has $2n$ distinct eigenvalues, then

$$n_{ARE} \leq \binom{2n}{n} = \frac{2n(2n-1)\cdots(n+1)}{n!}.$$

(*ii*) If Z has multiple eigenvalues and is cyclic (its characteristic and minimal polynomials coincide), then

$$n_{ARE} < \binom{2n}{n}.$$

(*iii*) If Z has multiple eigenvalues and is not cyclic, then n_{ARE} may be not finite.

Simple examples clarify the above discussion.

Example 5.3 Let

$$A = \begin{bmatrix} 0 & 1 \\ 0 & 0 \end{bmatrix}, \quad B = \begin{bmatrix} 0 \\ 1 \end{bmatrix}, \quad C = \begin{bmatrix} 1 & 0 \end{bmatrix}, \quad R = 1.$$

solution #	p_{11}	p_{12}	p_{21}	p_{22}	columns of V
1	$-\alpha$	1	1	$-\alpha$	1, 2
2	0	j	$-j$	0	1, 3
3	$j\alpha$	-1	-1	$-j\alpha$	1, 4
4	$-j\alpha$	-1	-1	$j\alpha$	2, 3
5	0	$-j$	j	0	2, 4
6	α	1	1	α	3, 4

Table 5.1: Example 5.3: solutions of the ARE.

The eigenvalues of Z are distinct as they are: $\lambda_{1,2} = (1 \pm j)/\sqrt{2}$ and $\lambda_{3,4} = -\lambda_{1,2}$ (recall Theorem 5.2). Consistent with this fact, there exist six 2-dimensional, Z-invariant subspaces which are the span of the columns of the six matrices built up with any pair of columns of

$$V := \frac{1}{\sqrt{2}} \begin{bmatrix} \sqrt{2} & \sqrt{2} & \sqrt{2} & \sqrt{2} \\ 1+j & 1-j & -1-j & -1+j \\ -1+j & -1-j & 1-j & 1+j \\ -j\sqrt{2} & j\sqrt{2} & -j\sqrt{2} & j\sqrt{2} \end{bmatrix}$$

which is the matrix resulting from gathering together four eigenvectors relative to the four above mentioned eigenvalues. Each of these subspaces is a complement of $\mathcal{W}$, so that there exist six solutions of the ARE which can be computed as shown in the proof of Theorem 5.3. They are reported in Table 5.1, where p_{ij} denotes the (i, j) element of matrix P, solution of the ARE and $\alpha := \sqrt{2}$. Note that only the solutions #1 and #6 are real; furthermore, the solution #1 is negative definite, while the solution #6 is positive definite. As for the remaining solutions, the second and the fifth ones are hermitian.

Example 5.4 Let

$$A = \begin{bmatrix} 0 & 0 \\ 0 & 1 \end{bmatrix}, \quad B = \begin{bmatrix} 1 \\ 0 \end{bmatrix}, \quad C = \begin{bmatrix} 2 & 0 \\ 0 & \sqrt{2} \end{bmatrix}, \quad R = 1.$$

The eigenvalues of Z are $\lambda_{1,2} = \pm 1$ and $\lambda_{3,4} = \pm 2$ (recall Theorem 5.2). Observe that the couple (A, B) is not reachable and that the eigenvalue of the unreachable part is 1. Such an eigenvalue appears in the spectrum of Z (Lemma 5.1). The eigenvalues of Z are distinct: consistent with this fact, there exist six 2-dimensional, Z-invariant subspaces. They are the span of the columns of the six matrices built up with any pair of columns of

$$V := \begin{bmatrix} 0 & 0 & 1 & 1 \\ 1 & 0 & 0 & 0 \\ 0 & 0 & -2 & 2 \\ -1 & 1 & 0 & 0 \end{bmatrix}$$

which is the matrix resulting from gathering together four eigenvectors relative to the four above mentioned eigenvalues. Only two of these subspaces are complements of $\mathcal{W}$, namely those relative to columns (1,3) and (1,4) of matrix V, so that there exist only two solutions of the ARE. They are

$$P_1 = \begin{bmatrix} -2 & 0 \\ 0 & -1 \end{bmatrix}, \quad P_2 = \begin{bmatrix} 2 & 0 \\ 0 & -1 \end{bmatrix}.$$

Example 5.5 Let

$$A = \begin{bmatrix} -1 & 1 \\ 0 & 0 \end{bmatrix}, \quad B = \begin{bmatrix} 0 \\ 1 \end{bmatrix}, \quad C'C = \begin{bmatrix} 2 & -1 \\ -1 & 1 \end{bmatrix}, \quad R = 1.$$

The eigenvalues of Z are $\lambda_{1,2} = 1$ and $\lambda_{3,4} = -1$ (recall Theorem 5.2). It is simple to verify that Z is cyclic, so that the number of solutions of the ARE must be less than six. As a matter of fact only three solutions exist, namely

$$P_1 = \begin{bmatrix} 1 & 0 \\ 0 & 1 \end{bmatrix}, \quad P_2 = \begin{bmatrix} 1 & 0 \\ 0 & -1 \end{bmatrix}, \quad P_3 = \begin{bmatrix} -7 & 4 \\ 4 & -3 \end{bmatrix}.$$

Example 5.6 Let $A = 0_{2\times 2}$, $B = C = R = I_2$. The eigenvalues of Z are $\lambda_{1,2} = 1$ and $\lambda_{3,4} = -1$ (recall Theorem 5.2). It is easy to check that the minimal polynomial of Z is $\varphi_Z(s) = (s-1)(s+1)$ so that Z is not cyclic. Hence the number of solutions of the ARE can be not finite. Indeed, there is an infinite number of solutions which are reported in Table 5.2, where p_{ij} denotes the (i,j) element of matrix P, solution of the ARE, $\gamma := \sqrt{1-\alpha\beta}$ and α, β are arbitrary complex numbers.

Example 5.7 Consider the ARE defined by $A = -2I_2$, $B = C = I_2$ and $R = -I_2$. the minimal polynomial of Z is $\varphi_Z(s) = (s-\sqrt{3})(s+\sqrt{3})$, so that Z is not cyclic but n_{ARE} is finite. In fact, there exist only four solutions of the ARE, namely: $P_1 = (2-\sqrt{3})I_2$, $P_2 = (2+\sqrt{3})I_2$ and

$$P_3 = \begin{bmatrix} 2-\sqrt{3} & 0 \\ 0 & 2+\sqrt{3} \end{bmatrix}, \quad P_4 = \begin{bmatrix} 2+\sqrt{3} & 0 \\ 0 & 2-\sqrt{3} \end{bmatrix}.$$

The eigenvalues of Z are meaningfully related to those of $A - BR^{-1}B'P$, the dynamic matrix relative to the feedback connection of $\Sigma(A, B, C, 0)$ with the controller $u = -R^{-1}B'Px$, P being any solution of the ARE. In fact the following theorem holds.

Theorem 5.4 *Let P be a solution of the ARE. Then any eigenvalue of $A_c := A - BR^{-1}B'P$ is also an eigenvalue of Z.*

Proof. Consider the matrix

$$T := \begin{bmatrix} I & 0 \\ -P & I \end{bmatrix}.$$

By exploiting the fact that P solves the ARE it follows that

$$TZT^{-1} = \begin{bmatrix} A_c & -BR^{-1}B' \\ 0 & -A_c' \end{bmatrix}$$

and the theorem is proved.

p_{11}	p_{12}	p_{21}	p_{22}
1	0	0	1
-1	0	0	1
1	0	0	-1
-1	0	0	-1
1	0	α	-1
-1	0	α	1
1	α	0	-1
-1	α	0	1
γ	α	β	$-\gamma$
$-\gamma$	α	β	γ

Table 5.2: Example 5.6: solutions of the ARE.

This result raises the question whether one or more among the solutions of the ARE are *stabilizing*, i.e., whether there exists a solution P_S of the ARE such that all the eigenvalues of $A - BR^{-1}B'P_S$ have negative real parts. In view of Theorem 5.4, this can happen provided only that no eigenvalue of Z has zero real part and hence (recall Lemma 5.1) Σ must not have eigenvalues with zero real part in its unobservable and/or unreachable parts. Furthermore, the couple (A, B) has obviously to be a stabilizable one. *Existence* results for P_S will be given after the presentation of some features of P_S (*uniqueness, symmetry*, etc.). For the sake of convenience, the term *stable subspace* of matrix T denotes the subspace spanned by the *stable* (generalized) eigenvectors of T, namely the eigenvectors associated to eigenvalues with negative real parts (stable eigenvalues) of T. Moreover, in the remaining part of this chapter a matrix is stable if all its eigenvalues are stable.

Theorem 5.5 *Let P_S be a stabilizing solution of the ARE. Then it follows that:*

(a) $P_S = YX^{-1}$, *where* $\mathcal{S}_S := Im(\begin{bmatrix} X \\ Y \end{bmatrix}) := Im(\Gamma)$ *is the stable subspace of* Z*;*

(b) P_S *is unique;*

(c) P_S *is real;*

(d) P_S *is symmetric.*

Proof. Point (a) From Theorem 5.3, the subspace associated to the solution P_S of the ARE is $Im([\ I \quad P_S'\ ]')$ and

$$Z\begin{bmatrix} I \\ P_S \end{bmatrix} = \begin{bmatrix} I \\ P_S \end{bmatrix}(A - BR^{-1}B'P_S).$$

Matrix $A - BR^{-1}B'P_S$ is stable and hence the subspace associated to P_S must be the one spanned by the stable eigenvectors of Z.

Point (b) The uniqueness of the stabilizing solution follows from the uniqueness of the stable subspace of Z. Alternatively, assume, by contradiction, that there exist two stabilizing solutions of the ARE, P_1 and P_2, $P_1 \neq P_2$. Then

$$0 = P_1 A + A' P_1 - P_1 B R^{-1} B' P_1 + C'C,$$
$$0 = P_2 A + A' P_2 - P_2 B R^{-1} B' P_2 + C'C.$$

By subtracting the second equation from the first one we get

$$0 = (P_1 - P_2)(A - BR^{-1}B'P_1) + (A - BR^{-1}B'P_2)'(P_1 - P_2).$$

This equation (interpreted as an equation in the unknown $(P_1 - P_2)$) admits a unique solution, namely $P_1 - P_2 = 0$, since $(A - BR^{-1}B'P_i)$, $i = 1, 2$ are stable matrices. *Point (c)* The claim is easily proved since for an $n \times n$ real matrix it is always possible either to select n real eigenvectors when all the eigenvalues are real, or to make the 2-dimensional subspace spanned by a pair of complex conjugate eigenvectors (corresponding to a pair of complex conjugate eigenvalues) be spanned by their sum and difference, the latter multiplied by j. Therefore the matrices X and Y which specify $\mathcal{S}_S$ can be thought of as real without loss of generality and P_S is real too. *Point (d)* Since $\mathcal{S}_S$ is the Z-invariant subspace corresponding to the stabilizing solution, it follows that $Z\Gamma = \Gamma V$, where V is stable. Letting $W := X^{\sim}Y - Y^{\sim}X$, we obtain $W = \Gamma^{\sim}J\Gamma$, J being the matrix defined in the proof of Theorem 5.2. On the other hand, $JZJ' = -Z'$, so that $JZ + Z'J = 0$ and $0 = \Gamma^{\sim}(JZ + Z'J)\Gamma = \Gamma^{\sim}J\Gamma V + V^{\sim}\Gamma^{\sim}J\Gamma$. The first and last terms of this relation can be rewritten as $0 = WV + V^{\sim}W$ which, in view of the stability of V, implies $0 = W$ and, from the definition of W, $YX^{-1} = (X^{\sim})^{-1}Y^{\sim}$. Therefore, $P_S = YX^{-1} = (YX^{-1})^{\sim} = P_S^{\sim}$. This last relation implies the symmetry of P_S since it is real.

The above results are clarified by the following simple examples.

Example 5.8 Consider the ARE defined in Example 5.3. There exists a unique, real and symmetric stabilizing solution (solution # 6). Indeed, the eigenvalues of the corresponding A_c matrix are $\lambda_{1,2} = -(1 \pm j)/(\sqrt{2})$ (recall also Theorem 5.4). Further, note that this solution corresponds to the stable subspace of Z. On the other hand, the eigenvalues of the A_c matrix corresponding to solution #1 are $\lambda_{1,2} = (1 \pm j)/(\sqrt{2})$ (recall again Theorem 5.4).

Example 5.9 Consider the ARE defined by $A = 0_{2\times 2}$, $B = C = R = I_2$. The eigenvalues of Z are $\lambda_{1,2} = 1$ and $\lambda_{3,4} = -1$. The stable subspace of Z can be specified by the matrices $X = Y = I_2$, so that $P_S = I_2$.

Example 5.10 This example shows that the stable subspace of Z may be not a complement of the subspace $\mathcal{W}$, so that the ARE does not admit a stabilizing solution. Consider the ARE defined in Example 5.4. The stable subspace of Z can be specified by the matrices

$$X = \begin{bmatrix} 0 & 1 \\ 0 & 0 \end{bmatrix}, \quad Y = \begin{bmatrix} 0 & 2 \\ 1 & 0 \end{bmatrix}.$$

Since X is singular, the stabilizing solution does not exist.

Example 5.11 Consider the ARE defined in Example 5.7. The stable subspace of Z can be specified by the matrices $X = I_2$ and $Y = (2-\sqrt{3})I_2$, so that $P_S = Y = P_1$. The remaining three solutions of the ARE which are reported in Example 5.7 (P_2, P_3, P_4) are not stabilizing.

An existence result for the stabilizing solution of the ARE can easily be given when the matrix R is either positive or negative definite as the following theorem states, by showing that two *necessary* conditions become, if taken together, also *sufficient.*

Theorem 5.6 *Let $R > 0$ or $R < 0$. Then the stabilizing solution of the ARE exists if and only if the following conditions both hold:*

(a) *The pair (A, B) is stabilizable;*

(b) *No eigenvalue of matrix Z has zero real part.*

Proof. Necessity is obvious. For sufficiency it will be shown that $\mathcal{S}_S$, the stable subspace of Z (n-dimensional thanks to assumption (b)), is a complement of the subspace $\mathcal{W}$. In other words, it will be shown that matrix X is nonsingular if $\mathcal{S}_S = Im([\ X' \quad Y'\ ]')$.

The Z-invariance of $\mathcal{S}_S$ implies that

$$AX - BR^{-1}B'Y = XV, \tag{5.10a}$$
$$-QX - A'Y = YV \tag{5.10b}$$

where V is stable. From eqs. (5.10) we obtain, by premultiplying the first one of them by $Y^\sim$ and taking into account what has been shown in the proof of Theorem 5.5, point (d) ($W = X^\sim Y - Y^\sim X = 0$),

$$Y^\sim(AX - BR^{-1}B'Y) = Y^\sim XV = X^\sim YV = -X^\sim QX - X^\sim A'Y. \tag{5.11}$$

Now assume, by contradiction, that X is singular, i.e. that there exists a vector $\vartheta \in \ker(X)$, $\vartheta \neq 0$. If eq. (5.11) is premultiplied by $\vartheta^\sim$ and postmultiplied by ϑ we obtain

$$\vartheta^\sim Y^\sim(AX - BR^{-1}B'Y)\vartheta = -\vartheta^\sim X^\sim QX\vartheta - \vartheta^\sim X^\sim A'Y\vartheta$$

which implies that $B'Y\vartheta = 0$ since $X\vartheta = 0$ and because of the sign definition of R. By exploiting these two last relations in eq. (5.10a) after its postmultiplication by ϑ, we can conclude that $V\vartheta \in \ker(X)$. By iterating these considerations starting from the vector $\vartheta^{(1)} := V\vartheta$ we find $V\vartheta^{(1)} = V^2\vartheta \in \ker(X)$. Further, if eq. (5.10b) is postmultiplied by ϑ, we obtain $-A'Y\vartheta = YV\vartheta$.

In conclusion, we have proved that, if $\vartheta \in \ker(X)$, then

$$V^k\vartheta \in \ker(X),\ k \geq 0, \tag{5.12}$$
$$-A'Y\vartheta = YV\vartheta, \tag{5.13}$$
$$B'Y\vartheta = 0. \tag{5.14}$$

Let φ be the least degree monic polynomial such that $\varphi(V)\vartheta = 0$. Such a polynomial certainly exists since the characteristic polynomial ψ_V of V satisfies the equation $\psi_V(V) = 0$. Further let γ be a root of φ. Then it follows that

$$0 = (V - \gamma I)\varphi^\star(V)\vartheta := (V - \gamma I)\xi.$$

Note that $\xi \neq 0$ because of the assumed minimality of the degree of φ. The last equation implies that γ is an eigenvalue of V and therefore $\mathrm{Re}[\gamma] < 0$. On the other hand eq. (5.12) entails that $\xi \in \ker(X)$, so that eq. (5.13) holds also for $\vartheta = \xi$ and it follows that

$$-A'Y\xi = YV\xi = \gamma Y\xi. \tag{5.15}$$

This relation calls for $-\gamma$ to be an eigenvalue of A'. Indeed, $Y\xi \neq 0$ since, otherwise, $X\xi = 0$ and $Y\xi = 0$ with $\xi \neq 0$: as a consequence, $\mathcal{S}_S$ should not be an n-dimensional subspace. Equation (5.14) implies $B'Y\xi = 0$ which, together with eq. (5.15), leads to the conclusion, thanks to the *PBH* test (see Theorem A.1 of Appendix A), that the unreachable part of the pair (A, B) possesses an eigenvalue $(-\gamma)$ with positive real part, thus violating assumption (b). Therefore, X is nonsingular and the stabilizing solution exists.

Example 5.12 Consider the ARE defined in Example 5.3. The pair (A, B) is reachable and no eigenvalue of matrix Z has zero real part. Indeed the stabilizing solution exists.

If, on the other hand, the ARE defined in Example 5.4 is taken into consideration, we can check that the pair (A, B) is not stabilizable and that no eigenvalue of matrix Z has zero real part. Indeed the stabilizing solution does not exist.

Finally, consider the ARE defined in Example 5.7. The pair (A, B) is reachable and all eigenvalues of matrix Z have real parts different from zero. In fact the stabilizing solution (P_1) exists.

Example 5.13 Consider the ARE defined by

$$A = \begin{bmatrix} 0 & 1 \\ 0 & 0 \end{bmatrix}, \quad B = \begin{bmatrix} 0 \\ 1 \end{bmatrix}, \quad C = \begin{bmatrix} 0 & 1 \end{bmatrix}, \quad R = 1.$$

The pair (A, B) is reachable, but the characteristic polynomial of Z is $\psi_Z(s) = s^2(s^2 - 1)$. The solutions of the ARE are

$$P_{1,2} = \begin{bmatrix} 0 & 0 \\ 0 & \pm 1 \end{bmatrix}, \quad P_3 = \begin{bmatrix} 0 & -1 \\ 0 & 0 \end{bmatrix}, \quad P_4 = \begin{bmatrix} 0 & 0 \\ -1 & 0 \end{bmatrix},$$

no one of which is stabilizing.

Example 5.14 Consider the ARE defined by $A = -1$, $B = C = 1$, $R = -1$. The pair (A, B) is reachable, but the characteristic polynomial of Z is $\psi_Z(s) = s^2$. In fact, the only solution of the ARE is $P_1 = 1$ which is not stabilizing.

Conditions for the existence and uniqueness of sign defined solutions of the ARE can be given by further restricting the considered class of equations.

Theorem 5.7 *Let $R > 0$. Then a solution $P = P' \geq 0$ of the ARE exists if and only if the unreachable but observable part of system $\Sigma(A, B, C, 0)$ is stable.*

Proof. Sufficiency. See the results in Section 3.4, specifically Theorem 3.5. *Necessity.* First observe that the eigenvalues of the unreachable but observable part of $\Sigma(A, B, C, 0)$ are also eigenvalues of matrix $A + BK$, whatever K is. Further, it is immediate to ascertain that if P^* satisfies the ARE, then it satisfies the equation

$$0 = P(A - BR^{-1}B'P) + (A - BR^{-1}B'P)'P + C'C + PBR^{-1}B'P \tag{5.16}$$

as well. Assume now that there exists a vector $\xi \neq 0$ such that $(A - BR^{-1}B'P^*)\xi = \lambda\xi$, $\text{Re}[\lambda] \geq 0$. Should this not be true, then the theorem would be proved since all the eigenvalues of the unreachable but observable part are also eigenvalues of $A - BR^{-1}B'P^*$. Thus, if eq. (5.16) (written for P^*) is postmultiplied by such a ξ and premultiplied by $\xi^\sim$, we get

$$\begin{aligned} 0 &= \xi^\sim P^*(A - BR^{-1}B'P^*)\xi + \xi^\sim(A - BR^{-1}B'P^*)'P^*\xi + \xi^\sim C'C\xi \\ &\quad + \xi^\sim P^*BR^{-1}B'P^*\xi \\ &= \lambda\xi^\sim P^*\xi + \lambda^\sim\xi^\sim P^*\xi + \xi^\sim C'C\xi + \xi^\sim P^*BR^{-1}B'P^*\xi. \end{aligned}$$

The sum of the first two terms of the right-hand side of this equation is $2\text{Re}[\lambda]\xi^\sim P^*\xi$ and is nonnegative because $P^* \geq 0$ by assumption. Also the remaining two terms are nonnegative (recall the assumption $R > 0$): thus $C\xi = 0$ and $B'P^*\xi = 0$. This last relation implies that $\lambda\xi = (A - BR^{-1}B'P^*)\xi = A\xi$. In view of the *PBH* test (Theorem A.1 of Appendix A) it is possible to conclude that λ is an eigenvalue of the unobservable part of $\Sigma(A, B, C, 0)$ and the proof is complete.

Example 5.15 Consider the ARE defined in Example 5.3. The pair (A, B) is reachable: consistent with this, there exists a symmetric positive semidefinite solution, the sixth one. If, on the other hand, reference is made to the ARE defined in Example 5.4, it is possible to check that: *i*) the pair (A, B) is not reachable, *ii*) the pair (A, C) is observable, *iii*) the eigenvalues of A are nonnegative. Thus the unreachable but observable part is certainly not stable. Indeed no one of the solutions of the ARE is positive semidefinite.

Theorem 5.8 *Let $R > 0$. Then it follows that:*

(a) *If the pair (A, C) is detectable there exists at most one symmetric and positive semidefinite solution of the ARE. It is also a stabilizing solution;*

(b) *If the pair (A, C) is observable, the positive semidefinite solution (if it exists) of the previous point is positive definite.*

Proof. Point (a) Let $P = P' \geq 0$ be a solution of the ARE. From the proof of Theorem 5.7 it follows that if $\xi \neq 0$ is such that $(A - BR^{-1}B'P)\xi = \lambda\xi$ with $\text{Re}[\lambda] \geq 0$ then λ is an eigenvalue of the unobservable part of $\Sigma(A, B, C, 0)$, against the detectability assumption. Therefore such a solution is a stabilizing one and hence unique thanks to Theorem 5.5.

Point (b) By assumption there exists a solution $P = P' \geq 0$ of the ARE. If it is not positive definite, there exists $\xi \neq 0$ such that $P\xi = 0$. Since P solves the ARE it follows that $\xi^\sim(PA + A'P - PBR^{-1}B'P + C'C)\xi = 0$ which, in turn, implies that $\xi^\sim C'C\xi = 0$, i.e., $C\xi = 0$. In this case $(PA + A'P - PBR^{-1}B'P)\xi = 0$ and hence also $PA\xi = 0$. Again it is $\xi^\sim A'(PA + A'P - PBR^{-1}B'P + C'C)A\xi = 0$ which, in turn, entails that $\xi^\sim A'C'CA\xi = 0$, namely $CA\xi = 0$. By iterating this procedure we find that $CA^i\xi = 0,\ i = 0, 1, \ldots, n-1$, against the observability assumption.

Example 5.16 Consider the ARE defined by

$$A = \begin{bmatrix} 0 & 0 \\ 0 & -1 \end{bmatrix},\ B = \begin{bmatrix} 1 \\ 1 \end{bmatrix},\ C = \begin{bmatrix} 2 & 0 \end{bmatrix},\ R = 1.$$

The relevant solutions are

$$P_1 = \begin{bmatrix} -2 & 0 \\ 0 & 0 \end{bmatrix},\ P_2 = -P_1,\ P_3 = \begin{bmatrix} -4 & 0 \\ 3 & 0 \end{bmatrix},\ P_4 = P_3'$$

and

$$P_5 = \begin{bmatrix} -10 & 12 \\ 12 & -18 \end{bmatrix},\ P_6 = \begin{bmatrix} -6 & 4 \\ 4 & -2 \end{bmatrix}.$$

It is easy to verify that only P_2 is symmetric and positive semidefinite, consistently with detectability of the pair (A, C). The solution P_2 is also stabilizing.

Example 5.17 Consider the ARE defined by $A = B = R = 1$ and $C = 0$. Two symmetric and positive semidefinite solutions exist: $P_1 = 0$ and $P_2 = 2$. Indeed, (A, C) is not a detectable pair.

Example 5.18 Consider the ARE defined in Example 5.13: there exists a unique positive semidefinite solution though (A, C) is not a detectable pair: as a matter of fact, the result in Theorem 5.8 holds in one sense only. Finally, make reference to the equations defined in Examples 5.3, 5.5, 5.6: in all cases (A, C) is an observable pair and, consistently, the relevant ARE admits a unique symmetric, positive semidefinite solution which is positive definite and stabilizing.

When $R > 0$ a further characterization of the stabilizing solution of the ARE can be given.

Theorem 5.9 *Let $R > 0$. If the stabilizing solution P_S exists, then it follows that:*

(a) *It is symmetric and positive semidefinite;*

(b) *It is maximal, that is $P_S - \Pi \geq 0$, $\forall \Pi = \Pi'$ is a solution of the ARE.*

Proof. *Point (a)* The symmetry of P_S follows from Theorem 5.5. If P_S is the stabilizing solution of the ARE then, letting $A_c := A - BR^{-1}B'P_S$ and $T_1 := P_SBR^{-1}B'P_S + C'C$, it follows that

$$0 = P_SA_c + A_c'P_S + T_1.$$

This equation is a Lyapunov equation in the unknown P_S: thus it admits a unique solution which is also positive semidefinite since A_c is stable and $T_1 \geq 0$. *Point (b)* Let P_S be the stabilizing solution and Π any other symmetric solution of the ARE, so that

$$0 = P_S A + A' P_S - P_S B R^{-1} B' P_S + C'C,$$
$$0 = \Pi A + A'\Pi - \Pi B R^{-1} B' \Pi + C'C.$$

By subtracting the second equation from the first one and letting $T_2 := (P_S - \Pi) B R^{-1} B' (P_S - \Pi)$, it follows that

$$0 = (P_S - \Pi) A_c + A_c'(P_S - \Pi) + T_2.$$

This equation is a Lyapunov equation in the unknown $(P_S - \Pi)$: thus it admits a unique solution which is also positive semidefinite, since A_c is stable and $T_2 \geq 0$.

Example 5.19 Consider the ARE defined in Example 5.17: it admits two positive semidefinite solutions: it is easy to check that the stabilizing solution exists, namely $P_S = P_2$ and $P_S - P_1 \geq 0$. Theorem 5.9 can be tested also by making reference to the equations defined in Examples 5.3, 5.5 and 5.6.

Example 5.20 In general, Theorem 5.9 does not hold if R is not positive definite. In fact consider the ARE defined in Example 5.7. It admits four positive semidefinite solutions one of which, P_1, is also stabilizing but not maximal. Indeed such a solution is minimal ($P_i - P_1 \geq 0,\ i = 2, 3, 4$). Instead, the ARE defined by $A = B = 1$, $C = 0$ and $R = -1$ admits two solutions $P_1 = 0$ (which is not stabilizing) and $P_2 = -2$ which is stabilizing although negative definite.

A further useful characterization of the positive semidefinite solution of the ARE can be given after the result in the following lemma has been established.

Lemma 5.2 *Let $R > 0$. The hamiltonian matrix Z has eigenvalues with zero real parts if and only if they are also eigenvalues of the unobservable and/or unreachable part of system $\Sigma(A, B, C, 0)$.*

Proof. The proof will be given for the *only if* claim, the other claim being a direct consequence of Lemma 5.1. Let ω be such that

$$Z \begin{bmatrix} \xi \\ \eta \end{bmatrix} = j\omega \begin{bmatrix} \xi \\ \eta \end{bmatrix}, \begin{bmatrix} \xi \\ \eta \end{bmatrix} \neq 0.$$

From this equation it follows that

$$\begin{bmatrix} \eta^\sim & \xi^\sim \end{bmatrix} Z \begin{bmatrix} \xi \\ \eta \end{bmatrix} = \eta^\sim A\xi - \eta^\sim B R^{-1} B' \eta - \xi^\sim C'C\xi - \xi^\sim A'\eta$$
$$= j\omega(\eta^\sim \xi + \xi^\sim \eta).$$

Observe that the last term of this equation and $\vartheta := \eta^\sim A\xi - \xi^\sim A'\eta$ are imaginary numbers. Thus $-\xi^\sim C'C\xi$ and $-\eta^\sim BR^{-1}B'\eta$ (which, on the contrary, are real and nonpositive) must both be zero, so that $C\xi = 0$ and $B'\eta = 0$. This fact implies that

$$Z\begin{bmatrix}\xi\\ \eta\end{bmatrix} = \begin{bmatrix}A\xi\\ -A'\eta\end{bmatrix} = j\omega\begin{bmatrix}\xi\\ \eta\end{bmatrix}.$$

If $\xi \neq 0$ then $A\xi = j\omega\xi$ and $C\xi = 0$ which in turn entails, in view of the *PBH* test, that $j\omega$ is an eigenvalue of the unobservable part of $\Sigma(A, B, C, 0)$. Instead, if $\eta \neq 0$, then $-A'\eta = j\omega\eta$ and $B'\eta = 0$ which, again in view of the *PBH* test, implies that $-j\omega$ (and hence also $j\omega$) is an eigenvalue of the unreachable part of $\Sigma(A, B, C, 0)$.

Theorem 5.10 *Let $R > 0$. Then there exists a unique symmetric, positive semidefinite solution of the ARE if and only if system $\Sigma(A, B, C, 0)$ is stabilizable and detectable. Moreover, such a solution is also stabilizing.*

Proof. Sufficiency. If $\Sigma(A, B, C, 0)$ is stabilizable and detectable, from Lemma 5.2 it follows that the eigenvalues of the hamiltonian matrix Z have nonzero real part. This fact, together with stabilizability of the pair (A, B), implies the existence of the stabilizing solution of the ARE (Theorem 5.6). Thanks to Theorem 5.9, this solution is symmetric and positive semidefinite, while, from Theorem 5.8, at most one of such solutions can exist. *Necessity.* The existence of a stabilizing solution entails stabilizability of the pair (A, B) and hence also stability of the unreachable but observable part of $\Sigma(A, B, C, 0)$. Let T be a nonsingular matrix which performs the decomposition of system $\Sigma(A, B, C, 0)$ into the observable and unobservable parts, that is let

$$A^* := TAT^{-1} := \begin{bmatrix}A_o & 0\\ A_1 & A_2\end{bmatrix},\quad B^* := TB := \begin{bmatrix}B_o\\ B_1\end{bmatrix},$$

$$C^* := CT^{-1} := \begin{bmatrix}C_o & 0\end{bmatrix},$$

where the pair (A_o, C_o) is observable and the pair (A_o, B_o) is stabilizable. In view of Lemma 5.2 and Theorems 5.6, 5.9 there exists a symmetric, positive semidefinite and stabilizing solution P_o of the Riccati equation

$$0 = \Pi A_o + A_o'\Pi - \Pi B_o R^{-1} B_o'\Pi + C_o'C_o.$$

Letting

$$P^* := \begin{bmatrix}P_o & 0\\ 0 & 0\end{bmatrix},$$

it is easy to check that P^* is: (i) symmetric and positive semidefinite; (ii) such that

$$0 = P^*A^* + A^{*\prime}P^* - P^*B^*R^{-1}B^{*\prime}P^* + C^{*\prime}C^*.$$

By premultiplying both sides of this equation by T', postmultiplying them by T and recalling the form of A^*, B^* and C^*, it is easy to verify that $P := T'P^*T$ is symmetric, positive semidefinite and such that

$$0 = PA + A'P - PBR^{-1}B'P + C'C,$$

namely that it solves the ARE. By assumption, it is the only solution possessing these properties and furthermore it is also stabilizing, so that A_c given by

$$\begin{aligned} A_c &:= A - BR^{-1}B'P = T^{-1}(A^* - B^*R^{-1}B^{*\prime}P^*)T \\ &= T^{-1}(A^* - B^*R^{-1}B^{*\prime}\begin{bmatrix} P_o & 0 \\ 0 & 0 \end{bmatrix})T \\ &= T^{-1}\begin{bmatrix} A_o - B_oR^{-1}B_o'P_o & 0 \\ A_1 - B_1R^{-1}B_o'P_o & A_2 \end{bmatrix}T \end{aligned}$$

is a stable matrix. This implies stability of A_2 and hence detectability of the pair (A, C).

Example 5.21 Consider the ARE defined in Examples 5.3, 5.5 and 5.6. The underlying systems Σ are stabilizable and detectable: consistent with this, in all cases there exists a unique symmetric, positive semidefinite solution which is also stabilizing. If, on the contrary, reference is made to Example 5.13, it can be seen that system Σ is stabilizable but not detectable: consistent with this, there exists a unique symmetric and positive semidefinite solution which however is not a stabilizing one. Finally, if Example 5.17 is again taken into consideration, stabilizability is checked while detectability does not hold. Indeed two symmetric and positive semidefinite solutions exist, one of which is also stabilizing.

For the $R > 0$ case a summary of the most significant results is given in Fig. 5.1.

Finally, a numerical method for the computation of the stabilizing solution (if it exists) of the ARE is now briefly sketched. It is considered to be one of the most efficient among those presently available and is based on Theorem 5.5, point (a). More precisely, assume that a $2n$-dimensional nonsingular matrix T has been determined such that

$$Z^* := TZT^{-1} = \begin{bmatrix} Z_{11} & Z_{12} \\ 0 & Z_{22} \end{bmatrix}$$

where the square n-dimensional submatrix Z_{11} is stable. If

$$T^{-1} := \begin{bmatrix} T_{11} & T_{12} \\ T_{21} & T_{22} \end{bmatrix},$$

then the stable subspace of Z is specified by the first n columns of T^{-1} and hence the stabilizing solution of the ARE exists if and only if T_{11} is nonsingular and is given by $T_{21}T_{11}^{-1}$. The numerical method is therefore efficient if the computation of T is such. This happens when T is the *unitary* transformation which supplies the Schur canonical form.

If, as often is the case, the stabilizing solution has also to be symmetric and positive semidefinite (properties which may not hold when R is not positive definite), then it is necessary to check these features in the computed stabilizing solution.

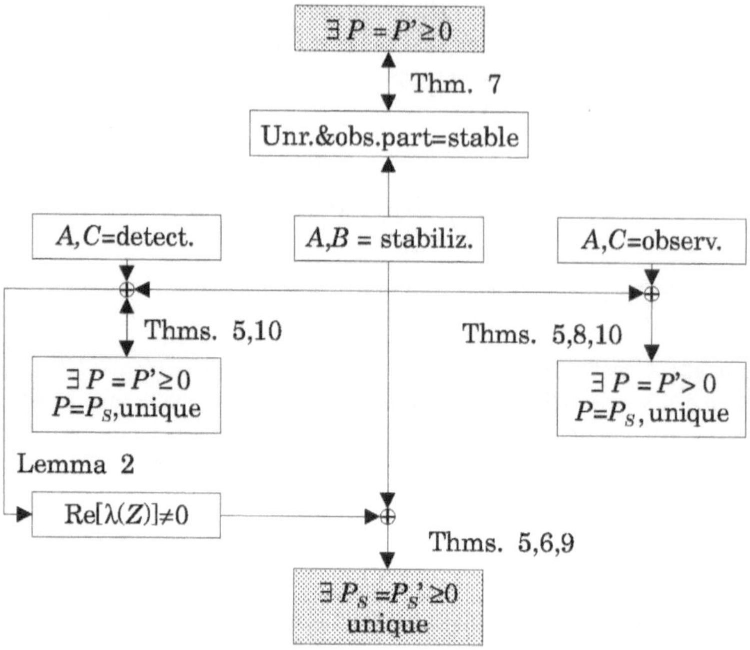

Figure 5.1: A summary of the results when $R > 0$.

Example 5.22 Consider the ARE defined in Example 5.5. Letting

$$T^{-1} = \begin{bmatrix} -0.7071 & 0 & 0.5011 & 0.4989 \\ 0 & -0.7071 & 0.4989 & -0.5011 \\ -0.7071 & 0 & -0.5011 & -0.4989 \\ 0 & -0.7071 & -0.4989 & 0.5011 \end{bmatrix}$$

we get

$$Z^* = TZT^{-1} = \begin{bmatrix} -1 & 1 & 0.7119 & 2.1197 \\ 0 & -1 & -0.7087 & -0.7055 \\ 0 & 0 & 0.5000 & -0.4978 \\ 0 & 0 & 0.5022 & 1.500 \end{bmatrix}$$

and

$$P_S = \begin{bmatrix} -0.7071 & 0 \\ 0 & -0.7071 \end{bmatrix} \begin{bmatrix} -0.7071 & 0 \\ 0 & -0.7071 \end{bmatrix}^{-1} = \begin{bmatrix} 1 & 0 \\ 0 & 1 \end{bmatrix}.$$

Example 5.23 Consider the ARE defined by $A = 2I_2$, $B = C = I_2$, $R = -I_2$. Letting

$$T^{-1} = \begin{bmatrix} -0.2588 & 0 & 0.9659 & 0 \\ 0 & -0.2588 & 0 & 0.9659 \\ 0.9659 & 0 & 0.2588 & 0 \\ 0 & 0.9659 & 0 & 0.2588 \end{bmatrix}$$

we get

$$Z^* = TZT^{-1} = \begin{bmatrix} -\sqrt{3} & 0 & -2 & 0 \\ 0 & -\sqrt{3} & 0 & -2 \\ 0 & 0 & \sqrt{3} & 0 \\ 0 & 0 & 0 & \sqrt{3} \end{bmatrix}$$

and

$$P_S = \begin{bmatrix} 0.9659 & 0 \\ 0 & 0.9659 \end{bmatrix} \begin{bmatrix} -0.2588 & 0 \\ 0 & -0.2588 \end{bmatrix}^{-1} = -(2+\sqrt{3})I_2$$

which is not positive semidefinite.

5.4 Problems

Problem 5.4.1 Consider the equation $0 = PA + A'P - PBB'P + Q$ where

$$A = \begin{bmatrix} 0 & 1 \\ 0 & 0 \end{bmatrix}, \quad B = \begin{bmatrix} 0 \\ 1 \end{bmatrix}, \quad Q = \begin{bmatrix} 0 & 0 \\ 0 & 1 \end{bmatrix}.$$

Does there exists a stabilizing solution of the ARE?

Problem 5.4.2 Consider the equation $0 = PA + A'P + PBSB'P + Q$. Does there exists a quadruple $(A, B, S = S', Q = Q')$ such that the ARE above admits two solutions P_1 and P_2 with $A + BSB'P_1$ and $A + BSB'P_2$ both stable?

Problem 5.4.3 Is it possible to find matrices A and C such that the equation $0 = PA + A'P - P^2 + C'C$ admits only the following three solutions

$$P_1 = \begin{bmatrix} 1 & 0 \\ 0 & 1 \end{bmatrix}, \quad P_2 = \begin{bmatrix} 0.2 & 1 \\ 1 & 5 \end{bmatrix}, \quad P_3 = \begin{bmatrix} 1 & 1 \\ 1 & 1 \end{bmatrix} ?$$

Problem 5.4.4 Consider the equation $0 = PA + A'P - PBB'P + Q$ where

$$A = \begin{bmatrix} 1 & 1 \\ 0 & 0 \end{bmatrix}, \quad B = \begin{bmatrix} 1 \\ \beta \end{bmatrix}, \quad C = \begin{bmatrix} 1 & \alpha \end{bmatrix}.$$

Find the values of the parameters α and β corresponding to which there exists a stabilizing solution of the ARE.

Problem 5.4.5 Consider the equation $0 = PA + A'P - PBR^{-1}B'P + Q$ where

$$A = \begin{bmatrix} 0 & 1 & 0 \\ 0 & 1 & 0 \\ 0 & 1 & -1 \end{bmatrix}, \; B = \begin{bmatrix} 0 & 1 \\ 1 & 0 \\ 0 & 0 \end{bmatrix}, \; Q = \begin{bmatrix} 0 & 0 & 0 \\ 0 & 1 & 0 \\ 0 & 0 & 0 \end{bmatrix}, \; R = \begin{bmatrix} 1 & 0 \\ 0 & -1 \end{bmatrix}.$$

Does there exist a stabilizing solution of the ARE?

Problem 5.4.6 Is it possible that the equation $0 = PA + A'P - PBB'P + Q$ with

$$A = \begin{bmatrix} 0 & 1 \\ 0 & 0 \end{bmatrix}, \; B = \begin{bmatrix} 0 \\ 1 \end{bmatrix}, \; Q = \begin{bmatrix} 1 & 0 \\ 0 & 0 \end{bmatrix}$$

admits a solution of the form

$$P = \begin{bmatrix} \alpha & 2 \\ 2 & 3 \end{bmatrix}?$$

Problem 5.4.7 Consider the equation $0 = PA + A'P - PBB'P + Q$ with

$$A = \begin{bmatrix} 0 & 0 \\ 0 & 1 \end{bmatrix}, \; B = \begin{bmatrix} 1 \\ 1 \end{bmatrix}, \; Q = \begin{bmatrix} 1 & 0 \\ 0 & 0 \end{bmatrix}.$$

Does there exist a stabilizing solution of the ARE?

Problem 5.4.8 By exploiting the content of Section 5.2 find the solution of the equation $\dot{P} = -PA - A'P + PBB'P$ with boundary condition $P(1) = I$, where

$$A = \begin{bmatrix} 0 & 1 \\ 0 & 0 \end{bmatrix}, \; B = \begin{bmatrix} 0 \\ 1 \end{bmatrix}.$$

Problem 5.4.9 Consider the equation $0 = PA + A'P + PBR^{-1}B'P + C'C$ where $R \neq 0$ and

$$A = \begin{bmatrix} 1 & 1 \\ 0 & 1 \end{bmatrix}, \; B = \begin{bmatrix} 1 \\ 0 \end{bmatrix}, \; C = \begin{bmatrix} 0 & 1 \end{bmatrix}.$$

Give the tightest possible upper bound on the number of its solutions.

Problem 5.4.10 Consider the equation $0 = PA + A'P + PBB'P + C'C$ where

$$A = \begin{bmatrix} 0 & 1 & 0 & 0 \\ 0 & 0 & 1 & 0 \\ 0 & 0 & -1 & 0 \\ 0 & 0 & 1 & 0 \end{bmatrix}, \; B = \begin{bmatrix} 0 \\ 1 \\ 0 \\ 0 \end{bmatrix}, \; C = \begin{bmatrix} 1 & 0 & 1 & 0 \end{bmatrix}.$$

Which of the following statements are false?

(a) There exists a stabilizing solution.

(b) An infinite number of solutions might exist.

(c) No solution is positive semidefinite.

(d) There exists a solution P_1 such that the spectrum of $(A - BB'P_1)$ is $\{1, 2, 3, 4\}$.

Part II

Variational methods

Chapter 6

The Maximum Principle

6.1 Introduction

This chapter is devoted to the presentation of necessary optimality conditions which rely on a first variation analysis and apply to control problems specified by the following six items.

(a) The controlled system which is a continuous-time, finite dimensional dynamic system;

(b) The set $\bar{S}_0$ to which the initial state x_0 must belong;

(c) The set $\bar{S}_f$ to which the final state x_f must belong;

(d) The set of functions which can be selected as inputs to the system;

(e) The performance index;

(f) A set of constraints on the state and/or control variables which can be given many different forms and have both an integral or pointwise nature.

More in detail, the system to be controlled is defined, as in Part I, by the equation

$$\dot{x}(t) = f(x(t), u(t), t) \tag{6.1}$$

where $x \in R^n$, $u \in R^m$ and the functions f, $\partial f/\partial x$, $\partial f/\partial t$ are continuous with respect to all their arguments. Together with this equation the boundary conditions

$$x(t_0) = x_0 \in \bar{S}_0, \tag{6.2a}$$

$$x(t_f) = x_f \in \bar{S}_f \tag{6.2b}$$

are also present and express the need that the state at the initial and final time belongs to given sets. Both the initial time t_0 and the final time t_f may or may not be specified. The set of functions which can be selected as control inputs is denoted by $\bar{\Omega}$, which is the subset of the set Ω^m constituted by the m-tuples of piecewise continuous functions taking values in a given closed subset U of R^m: thus $u(\cdot) \in \bar{\Omega}$. The performance index J (to be minimized) is the sum of an integral term with a term which is a function of the final event, namely

$$J = \int_{t_0}^{t_f} l(x(t), u(t), t)dt + m(x(t_f), t_f) \tag{6.3}$$

where the functions l and m possess the same continuity properties as the function f. Finally, various kinds of specific limitations on both the state and control variables can be settled by means of the constraints mentioned at point (f) above. The detailed description of such constraints, which, for the sake of convenience, are referred to as *complex* constraints, is postponed to Section 6.3.

As before, letting $\varphi(\cdot; t_0, x_0, u(\cdot))$ be the system motion, the optimal control problem to be considered in this second part is, in essence, of the same kind as the one tackled in Part I and can be stated in the following way.

Problem 6.1 (Optimal control problem) *Determine a vector of functions $u^o(\cdot) \in \bar{\Omega}$ defined on the interval $[t_0^o, t_f^o]$ in such a way that eqs. (6.2) and, if it is the case, other complex constraints are satisfied and the performance index (6.3), when evaluated in correspondence of $\varphi(\cdot, t_0^o, x^o(t_0^o), u^o(\cdot))$ and $u^o(\cdot)$ over the interval $[t_0^o, t_f^o]$, takes on the least possible value.*

Indeed a little more general statement could have been adopted where the eqs. (6.2) are replaced by

$$(x(t_0), t_0) \in S_0, \tag{6.4a}$$

$$(x(t_f), t_f) \in S_f, \tag{6.4b}$$

which apparently constrain the initial and final events. Handling this new framework is however an easy job, as will be shown in the sequel: thus the simpler setting of Problem 6.1 has been chosen for the sake of simplicity.

The so-called Maximum Principle will be presented in the forthcoming sections, supplying a set of necessary optimality conditions. First, problems with *simple* constraints are dealt with in Section 6.2, i.e., problems which require that only the constraints (6.2) and $u(\cdot) \in \bar{\Omega}$ are satisfied, while in the subsequent Section 6.3 problems characterized by the presence of complex constraints are tackled. Finally, particular problems are discussed in Sections 6.4 and 6.5: those characterized by the presence of *singular arcs* and the *minimum time* problems.

A couple of definitions are expedient in the forthcoming discussion: the first of them concerns the *hamiltonian function* which slightly differs from the one introduced in Part I, while the second one specifies a particular dynamical system.

Definition 6.1 (Hamiltonian function) *The function*

$$H(x,u,t,\lambda_0,\lambda) := \lambda_0 l(x,u,t) + \lambda' f(x,u,t) \tag{6.5}$$

where $\lambda_0 \in R$ *and* $\lambda \in R^n$, *is the hamiltonian function (relative to the system (6.1) and the performance index (6.3)).*

Definition 6.2 (Auxiliary system) *With reference to the hamiltonian function (6.5) and a pair of functions* $(x(\cdot), u(\cdot))$ *satisfying eq. (6.1), the system of linear differential equations*

$$\dot{\lambda}(t) = - \left. \frac{\partial H(x,u(t),t,\lambda_0(t),\lambda(t))}{\partial x} \right|'_{x=x(t)}, \tag{6.6a}$$

$$\dot{\lambda}_0(t) = 0 \tag{6.6b}$$

is the auxiliary system for Problem 6.1.

Equation (6.6b) simply amounts to stating that λ_0 is a constant: for this reason, only eq. (6.4a) is often referred to as the auxiliary system and eq. (6.4b) is taken into account by just requiring the constancy of λ_0. Finally, note that eq. (6.1) can be replaced by

$$\dot{x}(t) = \left(\frac{\partial H(x(t),u(t),t,\lambda_0,\lambda)}{\partial \lambda} \right)', \tag{6.7}$$

This last equation, together with eq. (6.6a), constitutes the so-called *hamiltonian system.*

6.2 Simple constraints

The optimal control problems dealt with in this section require that no complex constraints are present on the state and/or control variables and that both the set $\bar{S}_0$ and the set $\bar{S}_f$ are regular varieties, that is a set of the kind

$$\bar{S} := \{x |\ x \in R^n,\ \alpha_i(x) = 0,\ i = 1,2,\dots,q \leq n\}$$

where the α_i's are continuously differentiable functions such that $\text{rank}(\Sigma(x)) = q$, $\forall x \in \bar{S}$, with

$$\Sigma(x) := \begin{bmatrix} \dfrac{d\alpha_1(x)}{dx} \\ \vdots \\ \dfrac{d\alpha_q(x)}{dx} \end{bmatrix}$$

The nature of a regular variety is clarified by the following example.

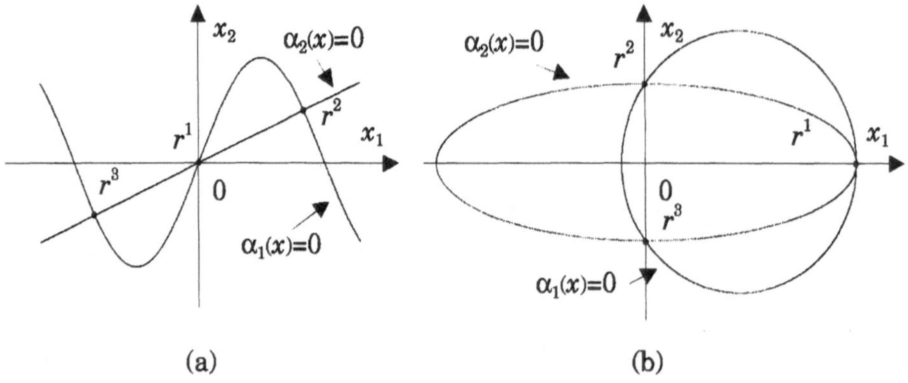

Figure 6.1: The set constituted by the vectors r^i, $i = 1, 2, 3$ is (a) or is not (b) a regular variety.

Example 6.1 Consider the set $\bar{S} \subset R^2$ defined by the two functions

$$\alpha_1(x) = x_2 - \sin(x_1),$$
$$\alpha_2(x) = x_2 - \frac{3}{5\pi}x_1.$$

This set is constituted by the three vectors $r^1 = \begin{bmatrix} 0 & 0 \end{bmatrix}'$, $r^2 = \begin{bmatrix} 5\pi/6 & 1/2 \end{bmatrix}'$ and $r^3 = -r^2$ (see Fig. 6.1 (a)) and is a regular variety since the matrix

$$\Sigma(x) = \begin{bmatrix} -\cos(x_1) & 1 \\ -\frac{3}{5\pi} & 1 \end{bmatrix}$$

has rank equal to 2 in all points of $\bar{S}$.

On the other hand, the set $\bar{S} \subset R^2$ defined by the two functions

$$\alpha_1(x) = x_2^2 + (x_1 - 4)^2 - 25,$$
$$\alpha_2(x) = 9x_2^2 + x_1^2 - 81,$$

which is constituted by the three vectors $r^1 = \begin{bmatrix} 9 & 0 \end{bmatrix}'$, $r^2 = \begin{bmatrix} 0 & 3 \end{bmatrix}'$ and $r^3 = -r^2$ (see Fig. 6.1 (b)), is not a regular variety since the matrix

$$\Sigma(x) = \begin{bmatrix} 2(x_1 - 4) & 2x_2 \\ 2x_1 & 18x_2 \end{bmatrix}$$

has rank less than 2 when $x = r^1$.

The problems in this section are defined by:

(a) The model of the controlled system

$$\dot{x}(t) = f(x(t), u(t), t);$$

(b) The constraint on the state at the initial time t_0

$$x(t_0) \in \bar{S}_0 = \{x|\ x \in R^n,\ \alpha_{0i}(x) = 0,\ i = 1, 2, \ldots, q_0 \leq n\}$$

where $\bar{S}_0$ is a regular variety;

(c) The constraint on the state at the final time t_f

$$x(t_f) \in \bar{S}_f = \{x|\ x \in R^n,\ \alpha_{fi}(x) = 0,\ i = 1, 2, \ldots, q_f \leq n\}$$

where $\bar{S}_f$ is a regular variety;

(d) The constraint on the control functions

$$u(\cdot) \in \bar{\Omega};$$

(e) The performance index

$$J = \int_{t_0}^{t_f} l(x(t), u(t), t)dt + m(x(t_f), x(t_f))$$

where the functions f, l, m possess the continuity properties previously mentioned.

Problems where the performance index is purely integral are first considered in Subsection 6.2.1, while this restriction is removed in the subsequent Subsection 6.2.2.

6.2.1 Integral performance index

The following result is available for optimal control problems where only simple constraints are imposed to the state and/or control variables and the performance index simply consists of an integral term.

Theorem 6.1 *Let $u^o(\cdot) \in \bar{\Omega}$ be a control which is defined over the interval $[t_0^o, t_f^o]$, $t_0^o < t_f^o$, and transfers the state of the system (6.1) from a suitable point $x^o(t_0^o) \in \bar{S}_0$ to a suitable point $x^o(t_f^o) \in \bar{S}_f$ and $x^o(\cdot) = (\varphi(\cdot; t_0^o, x^o(t_0^o), u^o(\cdot))$. A necessary condition for the quadruple $(x^o(\cdot), u^o(\cdot), t_0^o, t_f^o)$ to be optimal is the existence of a solution $\lambda^o(\cdot)$, λ_0^o of the auxiliary system (6.6) corresponding to $(x^o(\cdot), u^o(\cdot))$ such that*

(i) $$H(x^o(t), u^o(t), t, \lambda_0^o, \lambda^o(t)) \leq H(x^o(t), u, t, \lambda_0^o, \lambda^o(t)) \qquad t \in [t_0^o, t_f^o], \text{ a.e.}; \ \forall u \in U;$$

(ii) $$H(x^o(t_f^o), u^o(t_f^o), t_f^o, \lambda_0^o, \lambda^o(t_f^o)) = 0;$$

(iii) $$H(x^o(t_0^o), u^o(t_0^o), t_0^o, \lambda_0^o, \lambda^o(t_0^o)) = 0;$$

(iv) $$\lambda_0^o \geq 0;$$

(v) $$\begin{bmatrix} \lambda_0^o \\ \lambda^o(\cdot) \end{bmatrix} \neq 0;$$

(vi) $$\lambda^o(t_f^o) = \sum_{i=1}^{q_f} \vartheta_{fi} \left. \frac{d\alpha_{fi}(x)}{dx} \right|'_{x=x^o(t_f^o)}, \quad \vartheta_{fi} \in R;$$

(vii) $$\lambda^o(t_0^o) = \sum_{i=1}^{q_0} \vartheta_{0i} \left. \frac{d\alpha_{0i}(x)}{dx} \right|'_{x=x^o(t_0^o)}, \quad \vartheta_{0i} \in R.$$

Some comments are in order. The first one refers to condition (i) which amounts to saying that a control must minimize the hamiltonian function in order to be optimal. It is quite obvious that a control which differs from an optimal one over a set of zero measure is still an optimal control. This fact motivates the insertion of *a.e.* (which stands for *almost everywhere*) into condition (i). The second remark concerns conditions (ii) and (iii), usually referred to as *transversality* conditions at the final and initial time. They hold only if the final time or the initial time are not given, respectively. Thus, if the final (initial) time is specified in the optimal control problem at hand, any optimal solution might not comply with the relevant transversality condition. A third comment refers to conditions (vi) and (vii) which are often mentioned as transversality conditions as well: there ϑ_{fi}, $i = 1, 2, \ldots, q_f$ and ϑ_{0i}, $i = 1, 2, \ldots, q_0$ are suitable constants. In order to avoid misunderstandings, they are here referred to as *orthogonality* conditions at the final and initial time, respectively. Indeed, a geometric interpretation of these conditions can be given, since they simply require that the n-dimensional vector $\lambda^o(t_f^o)$ ($\lambda^o(t_0^o)$) be orthogonal to the hyperplane $T_f(x^o(t_f^o))$ ($T_0(x^o(t_0^o))$) which is tangent to $\bar{S}_f$ at $x^o(t_f^o)$ ($\bar{S}_0$ at $x^o(t_0^o)$). Note that when the state at the final and/or initial time is completely specified, the relevant orthogonality condition is anyhow satisfied.

Example 6.2 This very simple problem has a trivial solution: thus it is expedient in checking Theorem 6.1. Consider a cart moving without friction on a straight rail under the effect of a unique force u. Denoting by x_1 and x_2 the cart position and velocity, respectively, the equations for this system are

$$\dot{x}_1 = x_2,$$
$$\dot{x}_2 = u.$$

Let the position and velocity at the initial time 0 be given and assume that the problem is to bring the cart to a specified position in the shortest possible time. Obviously, this new position has to be different from the initial one. The intensity of the force which can be applied ranges between $u_{\max} > 0$ and $u_{\min} < 0$. Therefore the problem is completely specified by the above equations and $x(0) \in \bar{S}_0$, $x(t_f) \in \bar{S}_f$, $u(t) \in U$, where $\bar{S}_0 = \{x| \ \alpha_{0i}(x) = 0, \ i = 1, 2\}$, $\bar{S}_f = \{x| \ \alpha_{f1}(x) = 0\}$, $U = \{u| \ u_{\min} \leq u \leq u_{\max}\}$ with $\alpha_{01}(x) = x_1 - x_{10}$, $\alpha_{02}(x) = x_2 - x_{20}$, $\alpha_{f1}(x) = x_1 - x_{1f}$, $x_{10} \neq x_{1f}$, as well as by the performance index

$$J = \int_0^{t_f} dt.$$

It is quite obvious that the solution of the problem is $u^o(t) = \mu$, $t \geq 0$, where $\mu = u_{\max}$ if $x_{10} < x_{1f}$ and $\mu = u_{\min}$ if $x_{10} > x_{1f}$. Corresponding to this choice we obtain

$$x_1^o(t) = x_{10} + x_{20}t + \frac{\mu}{2}t^2,$$
$$x_2^o(t) = x_{20} + \mu t$$

and

$$t_f^o = -\frac{x_{20}}{\mu} + \frac{\sqrt{x_{20}^2 - 2(x_{10} - x_{1f})\mu}}{|\mu|}.$$

It is now shown how λ^o and λ_0^o can be determined so as to satisfy the conditions (i), (ii), (iv)–(vi) of Theorem 6.1 (recall that the initial time and state are given). As a preliminary, note that the sets $\bar{S}_0$ and $\bar{S}_f$ are regular varieties, so that the quoted theorem applies and the hamiltonian function $H = \lambda_0 + \lambda_1 x_2 + \lambda_2 u$ can be defined. Consistent with this, $\dot{\lambda}_1 = 0$, $\dot{\lambda}_2 = -\lambda_1$, $\dot{\lambda}_0 = 0$ are the equations of the system (6.6), and

$$\lambda_1^o = -\frac{1}{x_2^o(t_f^o)},$$
$$\lambda_2^o = -\frac{1}{x_2^o(t_f^o)}(t_f^o - t),$$
$$\lambda_0^o = 1$$

is a solution of them, since $x_2^o(t_f^o) \neq 0$. These functions apparently verify conditions (ii), (iv) and (v). The orthogonality condition (vi) calls for the second component of $\lambda^o(t_f^o)$ to be zero and this is the case. Finally, observe that the sign of $\lambda_2^o(t)$, $0 \leq t < t_f^o$ does not change and is different from the sign of $x_2^o(t_f^o)$, the latter being equal to the sign of μ. Therefore, also condition (i) is satisfied.

Remark 6.1 *(Limits of Theorem 6.1)* It should be emphasized that Theorem 6.1 supplies conditions which are only *necessary* (NC). Therefore, if a quadruple $(x(\cdot), u(\cdot), t_0, t_f)$ complies with the problem constraints and allows us to determine a solution (λ, λ_0) of the auxiliary system which verifies the theorem conditions, the only conclusion to be drawn is that such a quadruple satisfies the NC and *might* be an optimal solution. In short, one often says that the control u at hand satisfies the NC and is a *candidate* optimal control.

Remark 6.2 *(Pathological problems)* For a given problem, assume that the pair (λ^o, λ_0^o) satisfies the conditions of Theorem 6.1: then, it is not difficult to check that they are also satisfied by the pair $(\hat{\lambda}^o, \hat{\lambda}_0^o)$ with $\hat{\lambda}^o = k\lambda^o$, $\hat{\lambda}_0^o = k\lambda_0^o$, $k > 0$. Therefore, it is always possible to set $\lambda_0^o = 1$ apart from those cases where it is mandatory to set $\lambda_0^o = 0$. Such cases are referred to as *pathological* cases.

Remark 6.3 *(Constructive use of the necessary conditions)* In principle, the necessary conditions of Theorem 6.1 allow us to determine a quadruple $(x(\cdot), u(\cdot), t_0, t_f)$ which satisfies them. Indeed, let $\lambda_0 = 1$ and enforce condition (i): a function $u_h(x, t, \lambda)$ results which, when substituted for $u(t)$ in eqs. (6.6a), (6.7), defines a hamiltonian system of $2n$ differential equations for the $2n$ unknown functions x and λ. This system is endowed with $2n$ boundary conditions deriving from: *a*) the constraint $x(t_0) \in \bar{S}_0$ (q_0 conditions), *b*) the constraint $x(t_f) \in \bar{S}_f$ (q_f conditions), *c*) the orthogonality condition at the initial time ($n - q_0$ conditions) and *d*) the orthogonality condition at the final time ($n - q_f$ conditions). Thus a solution can be found which is parameterized by the yet unknown values of t_0 and t_f: this solution can be specified fully by exploiting the two transversality conditions. A control which satisfies the NC results from the substitution of the available expressions for x, λ into the function u_h. Finally, observe that the number of conditions to be imposed always equals the number of unknown parameters, even when the initial (final) state and/or time is specified.

If in so doing no control can be found which satisfies the NC, one can set $\lambda_0 = 0$, i.e., explore the possibility that the problem at hand is pathological. If also this second attempt is unsuccessful, the conclusion should be drawn that no solution exists for the optimal control problem under consideration.

Example 6.3 The problem in Example 6.2 is again considered by letting $x_{10} = 0$, $x_{20} = 0$, $x_{1f} = 5$, $u_{\max} = 1$, $u_{\min} = -1$: thus conditions (i), (ii), (iv)–(vi) of Theorem 6.1 must be imposed. Choosing $\lambda_0 = 1$, the minimization of the hamiltonian function with respect to u leads to defining the function $u_h(\lambda) = -\text{sign}(\lambda_2)$ which fully specifies u but in the case $\lambda_2 = 0$. This circumstance does not entail any significant difficulty since, as will later be shown, it can only occur at an isolated time instant. The solution of the auxiliary system is $\lambda_1 = \lambda_1(0)$, $\lambda_2 = \lambda_2(0) - \lambda_1(0)t$ and the orthogonality condition (vi) requires $\lambda_2(t_f) = 0$: thus $\lambda_2(t) = \lambda_1(0)(t_f - t)$ and λ_2 does not change sign and the control is constant and equal to ± 1. As a consequence, it follows that $x_1 = \pm t^2/2$, $x_2 = \pm t$. The transversality condition (ii) imposes $\lambda_1(0) = \mp 1/t_f$, while complying with the constraint on the final state entails $5 = \pm (t_f)^2/2$. Therefore, it is $t_f = \sqrt{10}$, $\lambda_1(0) = -1/\sqrt{10}$ and the control $u(t) = 1$ satisfies the NC.

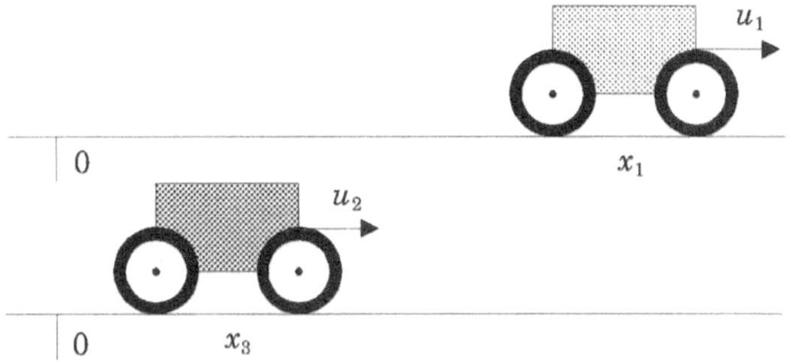

Figure 6.2: The two carts considered in Example 6.4.

Example 6.4 Consider the system in Fig. 6.2: it is constituted by two carts with unitary mass which move without friction over two parallel rails. Two forces u_1 and u_2 act on the carts. By denoting with x_1 and x_3 their positions and with x_2 and x_4 their velocities, the equations for the system are $\dot{x}_1 = x_2$, $\dot{x}_2 = u_1$, $\dot{x}_3 = x_4$, $\dot{x}_4 = u_2$. If the initial state and time are given, namely, $x_1(0) = 1$, $x_3(0) = -1$, $x_2(0) = x_4(0) = 0$, the control actions must be determined which make, at a given time T, $x_1(T) = x_3(T)$, $x_2(T) = x_4(T)$ and minimize the performance index

$$J = \frac{1}{2}\int_0^T (u_1^2 + u_2^2)dt.$$

The sets $\bar{S}_0$ and $\bar{S}_f$ are regular varieties, thus conditions (i), (iv)–(vi) of Theorem 6.1 must be satisfied. The minimization of the hamiltonian function $H = \lambda_0(u_1^2+u_2^2)/2+\lambda_1 x_2 + \lambda_2 u_1 + \lambda_3 x_4 + \lambda_4 u_2$ with respect to u is possible only if the problem is not pathological: therefore, by choosing $\lambda_0 = 1$, it leads to defining the function $u_h(\lambda)$ specified by $u_{h1}(\lambda) = -\lambda_2$ and $u_{h2}(\lambda) = -\lambda_4$. The solution of the auxiliary system is $\lambda_1(t) = \lambda_1(0)$, $\lambda_2(t) = \lambda_2(0) - \lambda_1(0)t$, $\lambda_3(t) = \lambda_3(0)$, $\lambda_4(t) = \lambda_4(0) - \lambda_3(0)t$, while the orthogonality condition requires that $\lambda_1(T) = -\lambda_3(T)$ and $\lambda_2(T) = -\lambda_4(T)$. In view of these results, complying with the constraints on the final state allows us to determine the values of the parameters $\lambda_i(0)$, yielding $\lambda_1(0) = -\lambda_3(0) = 12/T^3$ and $\lambda_2(0) = -\lambda_4(0) = 6/T^2$, from which the expressions for the control variables u_1 and u_2 which satisfy the NC can easily be derived.

Now consider a slight modification of this problem where the final time is not specified and the term 2 is added inside the integral in the performance index in order to emphasize the need for the control interval to be reasonably short. In this case the transversality condition (ii) must also be imposed. Once the previously found (and still valid) relations among the parameters $\lambda_i(0)$ are taken into account, we obtain $T = \sqrt{6}$.

Example 6.5 Once more consider the system described in Example 6.2, but assume that only the velocity is given at the known initial time, while the state at the known final time is required to belong to a specified set. More precisely, let $x(0) \in \bar{S}_0$, $x(1) \in \bar{S}_f$ with $\bar{S}_0 = \{x | \; x_2 = 0\}$, $\bar{S}_f = \{x | \; x_1^2 + (x_2 - 2)^2 - 1 = 0\}$ and the performance index be

$$J = \frac{1}{2} \int_0^1 u^2 dt.$$

Both $\bar{S}_0$ and $\bar{S}_f$ are regular varieties: thus conditions (i), (iv)–(vii) of Theorem 6.1 must be satisfied. The problem is not pathological since otherwise the hamiltonian function $H = \lambda_0 u^2/2 + \lambda_1 x_2 + \lambda_2 u$ could not be minimized with respect to u. By choosing $\lambda_0 = 1$ we obtain $u_h(\lambda) = -\lambda_2$, while the equations of the auxiliary system together with condition (vii), which requires $\lambda_1(0) = 0$, imply $\lambda_2(t) = \lambda_2(0)$, $\lambda_1(t) = 0$. Therefore it follows that $x_1(t) = x_1(0) - \lambda_2(0)t^2/2$, $x_2(t) = -\lambda_2(0)t$. The orthogonality condition (vi) is

$$\begin{bmatrix} \lambda_1(1) \\ \lambda_2(1) \end{bmatrix} = \vartheta_f \begin{bmatrix} 2x_1(1) \\ 2(x_2(1) - 2) \end{bmatrix}$$

so that either $\vartheta_f = 0$ or $x_1(1) = 0$. The first occurrence must not be considered since it would also entail $\lambda_2(0) = 0$, i.e., $u(\cdot) = 0$ which does not comply with the constraint at the final time. By imposing the control feasibility (i.e., that the constraint on the final state is satisfied) we get $\lambda_2(0) = -1$ or $\lambda_2(0) = -3$ with $x_1(0) = -1/2$ and $x_1(0) = -3/2$, respectively. Therefore two distinct controls exist which satisfy the NC: the form of the performance index points out that $u(t) = 3$ is certainly not optimal.

Example 6.6 Consider the first order system $\dot{x} = u$ with $x(t_0) = x_0$, $x(t_f) = x_f$, where x_0 and x_f are given and different from each other, while t_0 and t_f must be determined, together with u, so as to minimize the performance index

$$J = \frac{1}{2} \int_{t_0}^{t_f} (t^4 + u^2) dt.$$

This problem can be handled by resorting to Theorem 6.1 and imposing that conditions (i)–(v) be satisfied. The nature of the hamiltonian function ($H = \lambda_0(t^4 + u^2)/2 + \lambda u$) prevents the problem from being pathological: thus, letting $\lambda_0 = 1$, we get $u_h(\lambda) = -\lambda = -\lambda(0)$. From the two transversality conditions it follows that $t_0 = -\sqrt{|\lambda(0)|}$ and $t_f = -t_0$. Finally, by requiring that the control be admissible, i.e., imposing that $x_f = x_0 - \lambda(0)(t_f - t_0)$, we obtain $\lambda(0) = ((x_0 - x_f)/2)^{2/3} \text{sign}(x_0 - x_f)$.

Remark 6.4 *(Free final and/or initial state)* A number of optimal control problems do not require that the final or initial state satisfy any specific constraint (see, for instance, the LQ problem dealt with in Part I). Such a particular situation can not be tackled directly by making reference to the notion of regular variety which was expedient in stating Theorem 6.1. On the other hand, when one of the state components x_i can be freely selected at the final and/or initial time, the relevant multiplier λ_i is zero at that instant (see Examples 6.2, 6.5): thus, it should not be surprising that the orthogonality condition (vi) ((vii)) becomes $\lambda^o(t_f^o) = 0$ ($\lambda^o(t_0^o) = 0$) if the final (initial) state is free.

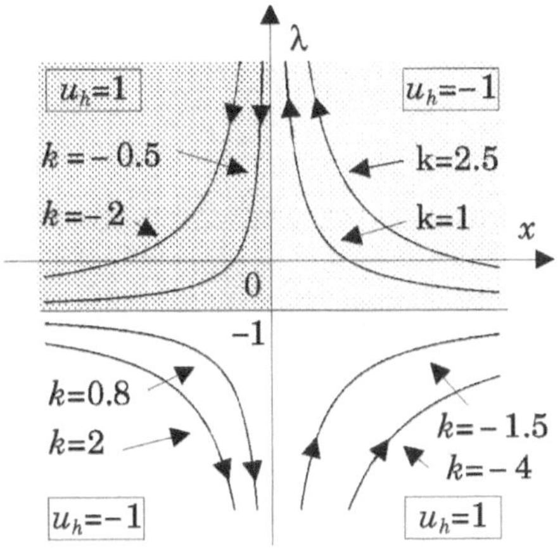

Figure 6.3: Example 6.8: trajectories of the hamiltonian system.

Example 6.7 The solution of the LQ problem which was found in Chapter 3.1 by resorting to the Hamilton-Jacobi theory is now shown to satisfy the NC. The system is $\dot{x}(t) = Ax(t) + Bu(t)$ with given initial state $x(0) = x_0$ and free final state $x(T)$, T being specified. The performance index is

$$J = \frac{1}{2}\int_0^T \left[x'Qx + u'Ru\right] dt$$

with $R = R' > 0$ and $Q = Q' \geq 0$. The hamiltonian function is $H = \lambda_0(u'Ru + x'Qx)/2 + \lambda'(Ax + Bu)$ and, for $\lambda_0 = 1$, is minimized by $u_h = -R^{-1}B'\lambda$, so that the equations of the hamiltonian system are

$$\dot{x} = Ax - BR^{-1}B'\lambda,$$
$$\dot{\lambda} = -Qx - A'\lambda.$$

In view of the material in Chapter 3.1, the pair (x^o, λ^o), where x^o is the solution of the equation $\dot{x} = (A - BR^{-1}B'P)x$, $x^o(0) = x_0$ and $\lambda^o(t) = P(t)x^o(t)$ with P a solution of the DRE $\dot{P} = -PA - A'P + PBR^{-1}B'P - Q$, $P(T) = 0$, must satisfy the hamiltonian system and the relevant boundary conditions for λ, specifically $\lambda^o(T) = 0$. This last request is apparently verified, while it is straightforward to check that also the first request is fulfilled by inserting λ^o and x^o into the hamiltonian system.

Example 6.8 Consider the optimal control problem defined by the first order system $\dot{x} = xu$ with given initial state $x(0) = x_0 \neq 0$. The final state is free, while the final time is 1. The control variable is constrained by $u(t) \in U = \{u| \ -1 \leq u \leq 1\}$ and the performance index is

$$J = \int_0^1 xudt.$$

The hamiltonian function is $H = (\lambda_0 + \lambda)xu$ and its minimization with respect to u yields $u_h(x, \lambda_0, \lambda) = -\text{sign}((\lambda_0 + \lambda)x)$. The hamiltonian system is

$$\dot{x} = xu_h(x, \lambda_0, \lambda),$$
$$\dot{\lambda} = -(\lambda_0 + \lambda)u_h(x, \lambda_0, \lambda),$$

with boundary conditions $x(0) = x_0$ and $\lambda(1) = 0$ (recall that the final state is free). Note that this problem can not be pathological since, otherwise, $\lambda(\cdot) = 0$ and condition (v) of Theorem 6.1 would be violated. Therefore $\lambda_0 = 1$. In a similar way observe that the control which minimizes H is always well defined because neither x nor $1 + \lambda(\bar{t})$ can be 0. Indeed the first circumstance is inconsistent with the nature of the set U and $x_0 \neq 0$, while the second one would entail, if verified, $\lambda(t) = -1$, $t \geq \bar{t}$, thus preventing the fulfillment of the orthogonality condition. The trajectories of the hamiltonian system are the solution of the equation

$$\frac{d\lambda}{dx} = -\frac{1+\lambda}{x},$$

i.e., are the curves $(1 + \lambda)x = k$ which are shown in Fig. 6.3 corresponding to some values of k. This figure, where also the values taken on by the function u_h are recorded, clearly points out that the control which satisfies the NC is $u = -\text{sign}(x_0)$.

Remark 6.5 *(Time-invariant problems)* If the functions f and l do not explicitly depend on time, i.e., when the optimal control problem is time-invariant, an interesting property of the hamiltonian function can be proved. More precisely, if the problem is time-invariant, then

$$H(x^o(t), u^o(t), \lambda_0^o, \lambda^o(t)) = \text{const.}$$

provided that the quadruple $(x^o, u^o, \lambda_0^o, \lambda^o)$ verifies condition (i) of Theorem 6.1, (λ_0^o, λ^o) being a solution of the auxiliary system corresponding to (x^o, u^o). This property allows us to check/impose (if it is the case) the transversality condition at any time instant within the interval $[0, t_f^o]$ (recall that the initial time can be thought of as given and set equal to 0, without any loss of generality, if the problem is time-invariant).

This claim can easily be tested by making reference to the time-invariant LQ problem. Indeed in this case

$$\begin{aligned}\frac{dH(x^o(t), u^o(t), 1, \lambda^o(t))}{dt} = & \left.\frac{\partial H(x, u^o(t), 1, \lambda^o(t))}{\partial x}\right|_{x=x^o(t)} \dot{x}^o(t) \\ & + \left.\frac{\partial H(x^o(t), u, 1, \lambda^o(t))}{\partial u}\right|_{u=u^o(t)} \dot{u}^o(t) \\ & + \left.\frac{\partial H(x^o(t), u^o(t), 1, \lambda)}{\partial \lambda}\right|_{\lambda=\lambda^o(t)} \dot{\lambda}^o(t).\end{aligned}$$

This relation is correct since $u^o = -R^{-1}B'Px^o$ is differentiable in view of the equations which are solved by x^o and P. The first and third term of the right-hand side

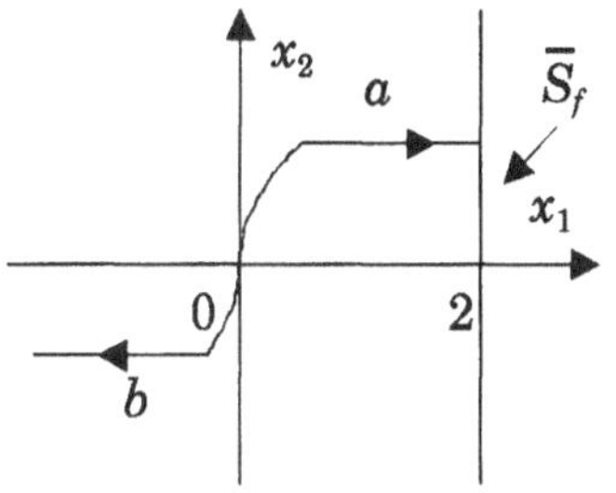

Figure 6.4: Example 6.9: the feasibility issue for the controls a and b.

of the above equation cancel out since

$$\dot{\lambda}^o(t) = -\left.\frac{\partial H(x, u^o(t), 1, \lambda^o(t))}{\partial x}\right|'_{x=x^o(t)},$$

$$\dot{x}^o(t) = \left.\frac{\partial H(x^o(t), u^o(t), 1, \lambda)}{\partial \lambda}\right|'_{\lambda=\lambda^o(t)}$$

Finally, no constraints are imposed on the current value of u: thus condition (i) of Theorem 6.1 is satisfied only if

$$\left.\frac{\partial H(x^o(t), u, 1, \lambda^o(t))}{\partial u}\right|_{u=u^o(t)} = 0$$

Example 6.9 Consider the optimal control problem defined by the system $\dot{x}_1 = x_2$, $\dot{x}_2 = u$ with initial state $x(0) = 0$ and final state constrained to belong to the regular variety $\bar{S}_f = \{x|\ x_1 = 2\}$. The control variable must be such that $-1 \leq u(t) \leq 1$. Finally, the performance index to be minimized is

$$J = \int_0^{t_f} [1 + |u|]dt.$$

Thus the hamiltonian function is $H = \lambda_0(1 + |u|) + \lambda_1 x_2 + \lambda_2 u$, so that $\dot{\lambda}_1 = 0$, $\dot{\lambda}_2 = -\lambda_1$, $\lambda_1(t) = \lambda_1(0)$ and $\lambda_2(t) = \lambda_2(0) - \lambda_1(0)t$. The problem cannot be pathological since otherwise it would follow that $u_h(\lambda) = \text{sign}(\lambda_2)$ and the transversality condition, set at $t = 0$, would require $\lambda_2(0) = 0$. In turn, the orthogonality condition implies that $\lambda_2(t_f) = 0$, so that it should also be $\lambda_1(0) = 0$ and condition (v) of Theorem 6.1 would be violated. Therefore, one can set $\lambda_0 = 1$. Note that $u(0) \neq 0$, since, otherwise, the transversality condition could not be verified. In view of the above discussion the conclusion can be drawn that λ_2 is a linear function of time which vanishes for $t = t_f$, has always the same sign and $|\lambda_2(0)| \geq 1$. The minimization of the hamiltonian function yields

$$u_h(\lambda) = \begin{cases} 1, & \lambda_2 < -1 \\ 0, & -1 < \lambda_2 < 1 \\ -1, & 1 < \lambda_2. \end{cases}$$

When $\lambda_2 = -1$ ($\lambda_2 = 1$) any value of $u \in [0,1]$ ($u \in [-1,0]$) is minimizing. However, $\lambda_2(t)$ can equal ± 1 only at isolated time instants, since otherwise it would follow that $\lambda_2(\cdot) = \pm 1$ and the request $\lambda_2(t_f) = 0$ would be violated.
Therefore, the conclusion can be drawn that only two are the allowable u:

(a)
$$u(t) = \begin{cases} 1, & 0 \le t < \tau \\ 0, & \tau < t \le t_f, \end{cases}$$

(b)
$$u(t) = \begin{cases} -1, & 0 \le t < \tau \\ 0, & \tau < t \le t_f. \end{cases}$$

The second control turns out to be not feasible (see also Fig. 6.4) and it remains to specify the parameters of the only control which could satisfy the NC. From the transversality condition we get $\lambda_2(0) = -2$ so that, by imposing $\lambda_2(\tau) = -1$ and $\lambda_2(t_f) = 0$ it follows that $\tau = t_f/2$, $\lambda_1(0) = -2/t_f$. Finally we find $x_2(\tau) = x_2(t_f) = \tau$ and $x_1(t_f) = \tau t_f - \tau^2/2$: thus, by imposing feasibility, i.e., $x_1(t_f) = 2$, it results that $t_f = 4\sqrt{3}$. All elements necessary to define a control which satisfies the NC have been determined.

Example 6.10 Consider three optimal control problems defined by the same system $\dot{x}_1 = x_2$, $\dot{x}_2 = u$, $x(0) = 0$, $x_1(t_f) = x_{1f} > 0$, $x_2(t_f) = 0$. The performance index for the first two problems are

$$J_1 = \frac{1}{2}\int_0^{t_f} (2 + u^2)dt,$$
$$J_2 = \int_0^{t_f} (1 + |u|)dt,$$

respectively. As for the third problem, the performance index J_3 is equal to J_1. For all problems the final time is free, while the control variable must comply with the constraint $-1 \le u(t) \le 1$, $\forall t$ when dealing with the first two problems.

The first problem is now considered. By choosing $\lambda_0 = 1$, the hamiltonian function is $H = 1 + u^2/2 + \lambda_1 x_2 + \lambda_2 u$ which yields

$$u_h(\lambda) = \begin{cases} 1, & \lambda_2 \le -1 \\ -\lambda_2, & -1 \le \lambda_2 \le 1 \\ -1, & \lambda_2 \ge 1. \end{cases}$$

The transversality condition at the initial time requires $u(0) = \pm 1$. By noticing that the auxiliary system implies that λ_2 is a linear function of time, the conclusion can be drawn that the control is a function of time which is either constant or first constant and then linear or first constant, then linear and finally constant again. A little thought reveals that only the third alternative may lead to an admissible control (recall the values of the initial and final states) and $u(0) = 1$. Therefore $\lambda_2(0) = -3/2$ and a control which satisfies the NC must be of the form

$$u(t) = \begin{cases} 1, & 0 \le t \le \tau_1 \\ -\lambda_2(t), & \tau_1 \le t \le \tau_2 \\ -1, & \tau_2 \le t \le t_f \end{cases}$$

where τ_1 and τ_2 are the times where λ_2 equals -1 and 1 respectively. These requirements, together with control feasibility, allow us to determine all the parameters which specify a solution satisfying the NC, precisely: $\tau_1 = t_{f1}/6$, $\tau_2 = 5t_{f1}/6$, $\lambda_1(0) = -3/t_{f1}$, $t_{f1} = \sqrt{108x_{1f}/23}$. As a consequence we find

$$x(t) = \begin{bmatrix} \dfrac{t^2}{2} \\ t \end{bmatrix}, \quad 0 \le t \le \tau_1,$$

$$x(t) = \begin{bmatrix} \dfrac{t_{f1}^2}{432} - \dfrac{t_{f1}}{24}t + \dfrac{3}{4}t^2 - \dfrac{1}{2t_{f1}}t^3 \\ -\dfrac{t_{f1}}{24} + \dfrac{3}{2}t - \dfrac{3}{2t_{f1}}t^2 \end{bmatrix}, \quad \tau_1 \le t \le \tau_2,$$

$$x(t) = \begin{bmatrix} -\dfrac{31}{108}t_{f1}^2 + t_{f1}t - \dfrac{1}{2}t^2 \\ t_{f1} - t \end{bmatrix}, \quad \tau_2 \le t \le t_{f1}.$$

The second problem is now taken into consideration. A similar discussion leads to the conclusion that a control which satisfies the NC is a piecewise constant function of time which first equals 1, then equals 0 and finally equals -1. The parameters which identify the corresponding solution are $\lambda_2(0) = -2$, $t_{f2} = \sqrt{16x_{1f}/3}$, $t_1 = t_{f2}/4$, $t_2 = 3t_{f2}/4$, $\lambda_1(0) = -4/t_{f2}$. As a consequence, we get

$$x(t) = \begin{bmatrix} \dfrac{t^2}{2} \\ t \end{bmatrix}, \quad 0 \le t \le t_1, \qquad x(t) = \begin{bmatrix} -\dfrac{t_{f2}^2}{32} + \dfrac{t_{f2}}{4}t \\ \dfrac{t_{f2}}{4} \end{bmatrix}, \quad t_1 \le t \le t_2,$$

$$x(t) = \begin{bmatrix} -\dfrac{5}{16}t_{f2}^2 + t_{f2}t - \dfrac{1}{2}t^2 \\ t_{f2} - t \end{bmatrix}, \quad t_2 \le t \le t_{f2}.$$

Finally the third problem is tackled. It is easy to check that a control which satisfies the NC is a linear function of time, precisely $u(t) = -\lambda_2(0) + \lambda_1(0)t$ which yields $x_2(t) = -\lambda_2(0)t + \lambda_1(0)t^2/2$, $x_1(t) = -\lambda_2(0)t^2/2 + \lambda_1(0)t^3/6$ and $t_{f3} = \sqrt{-6x_{1f}/\lambda_2(0)}$, where $\lambda_2(0) = -\sqrt{2}$, $\lambda_1(0) = -\sqrt{-4\lambda_2(0)/(3x_{1f})}$. The solutions of the three problems are compared in Fig. 6.5.

Remark 6.6 *(Alternative approach to time-varying problems)* It is interesting to note that a time-varying optimal control problem might be restated in terms of an equivalent *time-invariant* problem. In fact, add $\dot{z} = 1$, $z(t_0) = t_0$, to the system equations, so that $z(t) = t$. By letting $\xi := \begin{bmatrix} x' & z \end{bmatrix}'$, a new problem is obtained which is defined by $\dot{\xi} = f^*(\xi, u)$, $\xi(0) \in \bar{S}_0^*$, $\xi(t_f^*) \in \bar{S}_f^*$ and the performance index

$$J = \int_0^{t_f^*} l^*(\xi, u)dt$$

where the functions f^* and l^* are related to the functions f and l in an obvious way, while $\bar{S}_0^* = \{\xi | \ \xi = \begin{bmatrix} x' & z \end{bmatrix}', \ x \in \bar{S}_0\}$, $\bar{S}_f^* = \{\xi | \ \xi = \begin{bmatrix} x' & z \end{bmatrix}', \ x \in \bar{S}_f\}$. The

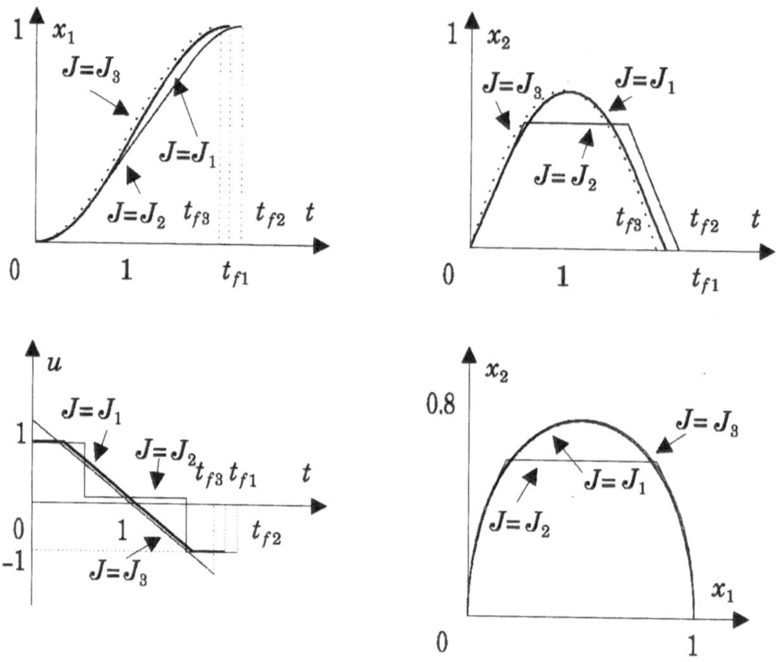

Figure 6.5: Example 6.10: state and control responses, state trajectories when $x_{1f} = 1$.

solution of this new problem supplies $\xi(0)$, the last component of which ($\xi_{n+1}(0)$) is the initial time for the given problem, while the final time is $\xi_{n+1}(0) + t_f^*$, t_f^* being the final time for the restated problem. This approach can be checked by applying it to Example 6.6.

Remark 6.7 *(Constraints on the final and/or initial event)* The discussion in Remark 6.6 is expedient in dealing with optimal control problems where the *final and/or initial event* rather than the final and/or initial state is constrained, i.e., where eqs. (6.4) enter into the statement, as done in Part I. In fact, first suppose that the final event is constrained (see eq.(6.4b)), namely that

$$(x(t_f), t_f) \in S_f = \{(x,t)|\ \alpha_i(x,t) = 0,\ i = 1, 2, \ldots, q \le n+1\}$$

where the α_i's are continuously differentiable functions and the set S_f is a regular variety, i.e., is such that $\text{rank}(\Sigma(x,t)) = q$, $\forall (x,t) \in S_f$ with

$$\Sigma(x,t) := \begin{bmatrix} \dfrac{\partial \alpha_1(x,t)}{\partial x} & \dfrac{\partial \alpha_1(x,t)}{\partial t} \\ \vdots & \vdots \\ \dfrac{\partial \alpha_q(x,t)}{\partial x} & \dfrac{\partial \alpha_q(x,t)}{\partial t} \end{bmatrix}.$$

Then a customary problem with constrained final state can be obtained by simply adding the state variable z defined by $\dot{z} = 1$. Note that in so doing the hamiltonian function becomes

$$\begin{aligned} H^*(x, u, z, \lambda_0, \lambda, \mu) &:= H(x, u, t, \lambda_0, \lambda) + \mu \\ &= \lambda_0 l(x, u, z) + \lambda' f(x, u, z) + \mu \end{aligned}$$

while the equation

$$\dot{\mu}(t) = - \left. \frac{\partial H^*(x(t), u(t), z, \lambda_0, \lambda(t), \mu(t))}{\partial z} \right|_{z=z(t)}$$

has to be added to the auxiliary system. The orthogonality condition can be expressed as

$$\begin{aligned} \lambda^o(t_f^o) &= \sum_{i=1}^{q} \vartheta_i \left. \frac{\partial \alpha_i(x, t_f^o)}{\partial x} \right|'_{x=x^o(t_f^o)}, \\ \mu(t_f^o) &= \sum_{i=1}^{q} \vartheta_i \left. \frac{\partial \alpha_i(x^o(t_f^o), z)}{\partial z} \right|_{z=t_f^o}, \end{aligned}$$

where ϑ_i are suitable scalars. These equations show that a problem restatement is actually not needed when handling constrained final events: indeed it suffices to modify the transversality condition which now is

$$H(x^o(t_f^o), u(t_f^o), t_f^o, \lambda_0^o, \lambda^o(t_f^o)) + \sum_{i=1}^{q} \vartheta_i \left. \frac{\partial \alpha_i(x^o(t_f^o), t)}{\partial t} \right|_{t=t_f^o} = 0.$$

A similar discussion can be carried on when the initial event rather than the state is constrained (see eq.(6.4a)).

Example 6.11 Consider the optimal control problem defined by the first order system $\dot{x} = u$ with initial state $x(0) = 0$ and final event belonging to the regular variety $S_f = \{(x,t)|\ (x-2)^2 + (t-2)^2 - 1 = 0\}$. The performance index to be minimized is

$$J = \int_0^{t_f} \frac{1}{2} u^2 dt.$$

Therefore the hamiltonian function is $H = \lambda_0 u^2/2 + \lambda u$ and $\lambda(t) = \lambda(0)$. The problem can not be pathological and $u_h = -\lambda(0)$, $x(t) = -\lambda(0)t$. The transversality and orthogonality conditions together with the constraint on the final event imply that

$$\begin{aligned} 0 &= -\frac{1}{2}\lambda^2(0) + 2\vartheta(t_f - 2), \\ 0 &= \lambda(0) - 2\vartheta(x(t_f) - 2), \\ 1 &= (x(t_f) - 2)^2 + (t_f - 2)^2. \end{aligned}$$

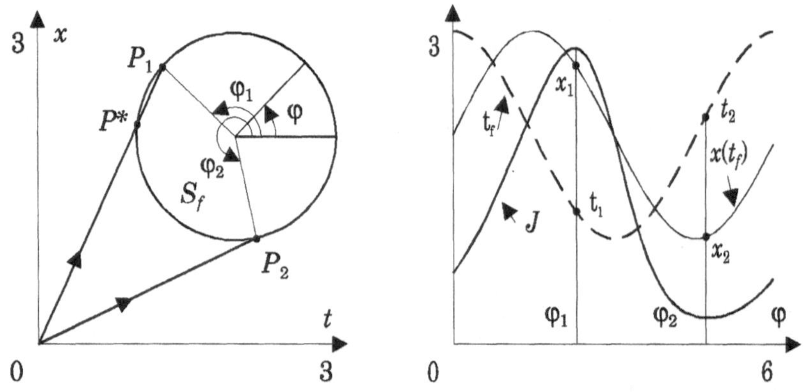

Figure 6.6: Example 6.11: state motions which satisfy the NC and optimal solutions for given φ.

By taking into account that $x(t_f) = -\lambda(0)t_f$, these equations yield

$$0 = 3\lambda^4(0) + 8\lambda^3(0) + 16\lambda^2(0) + 32\lambda(0) + 12,$$

$$t_f = \frac{4 - 2\lambda(0)}{\lambda^2(0) + 2},$$

$$\vartheta = -\frac{\lambda(0)}{2(\lambda(0)t_f + 2)}$$

The only two real solutions of the first equation are $\lambda_1 = -2.11$, $\lambda_2 = -0.46$ and the relevant values for the final time, the final state and performance index are: $t_1 = 1.27$, $t_2 = 2.23$, $x_1 = 2.69$, $x_2 = 1.03$, $J_1 = 2.83$, $J_2 = 0.24$. Figure 6.6 (a) shows what has been obtained: point P_1 is the event corresponding to the first value of $\lambda(0)$, while point P_2 is the event relevant to the second value. Note that point P^*, though corresponding to a final event which can be attained with a better control than the one yielding point P_1 (recall the performance index), has not been found by imposing the fulfillment of the NC. This can easily be understood by looking at Fig. 6.6 (b) where t_f, $x(t_f)$ and J are plotted against the angle φ which specifies a particular element of S_f. These quantities are obtained by exploiting the Hamilton-Jacobi theory with reference to the problem where the final state (and hence also the final time) is given.

6.2.2 Performance index function of the final event

Optimal control problems where the state and control variables undergo simple constraints and a function of the final event is included in the performance index are now taken into consideration. Therefore, unlike before (Section 6.2.1), the structure of the performance index is

$$J = \int_{t_0}^{t_f} l(x(t), u(t), t)dt + m(x(t_f), t_f). \tag{6.8}$$

Moreover, the initial state x_0 and time t_0 are given while the final state is either free or constrained to be an element of a regular variety $\bar{S}_f$. The following result applies to problems of this kind.

Theorem 6.2 *Let $\bar{S}_f = R^n$ and/or $l(\cdot,\cdot,\cdot) \neq 0$. Further let $u^o(\cdot) \in \bar{\Omega}$ be a control defined over the interval $[t_0, t_f^o]$, $t_0 < t_f^o$, which transfers the state of the system (6.1) from x_0 to $x^o(t_f^o) \in \bar{S}_f$ and $x^o(\cdot) = (\varphi(\cdot; t_0^o, x^o(t_0^o), u^o(\cdot))$. A necessary condition for the triple $(x^o(\cdot), u^o(\cdot), t_f^o)$ to be optimal is the existence of a solution $\lambda^o(\cdot)$, λ_0^o of the auxiliary system (6.6) corresponding to $(x^o(\cdot), u^o(\cdot))$ such that*

(i) $$H(x^o(t), u^o(t), t, \lambda_0^o, \lambda^o(t)) \leq H(x^o(t), u, t, \lambda_0^o, \lambda^o(t)) \quad t \in [t_0^o, t_f^o], \text{ a.e.}; \ \forall u \in U,$$

(ii) $$H(x^o(t_f^o), u^o(t_f^o), t_f^o, \lambda_0^o, \lambda^o(t_f^o)) + \lambda_0^o \left. \frac{\partial m(x^o(t_f^o), t)}{\partial t} \right|_{t=t_f^o} = 0,$$

(iii) $$\lambda_0^o \geq 0,$$

(iv) $$\begin{bmatrix} \lambda_0^o \\ \lambda^o(\cdot) \end{bmatrix} \neq 0,$$

(v) $$\lambda^o(t_f^o) - \lambda_0^o \left. \frac{\partial m(x, t_f^o)}{\partial x} \right|_{x=x^o(t_f^o)}' = \sum_{i=1}^{q} \vartheta_i \left. \frac{d\alpha_i(x)}{dx} \right|_{x=x^o(t_f^o)}', \quad \vartheta_i \in R.$$

Example 6.12 A simplified model of a d.c. motor with independent and constant excitation and purely inertial load is

$$\begin{aligned} \dot{x}_1 &= x_2, \\ \dot{x}_2 &= k_1 x_3, \\ \dot{x}_3 &= k_2 u + k_3 x_2 + k_4 x_3, \end{aligned}$$

where x_1, x_2 are the shaft position and velocity, respectively, while x_3 and u are the armature current and the applied voltage. The constants $k_i, i = 1, 2, 3, 4$, depend on the motor electrical parameters, the field current and the applied load. The description of the optimal control problem is complete once the initial state $x(0) = 0$ and the performance index

$$J = \frac{1}{2} \left[\int_0^1 u^2 dt + \sigma(x_1(1) - 1)^2 \right], \quad \sigma > 0$$

are given. The condition (i) of Theorem 6.2 prevents the problem from being pathological, while from condition (v) and the equations of the auxiliary system it follows that $\lambda_1(t) = \lambda_1(0)$ and λ_2, λ_3, x, u are all *linear* functions of $\lambda_1(0)$, precisely of the form $f_i(t)\lambda_1(0)$. Therefore, by recalling the orthogonality condition we get

$$x_1(1) = K\lambda_1(0) = 1 + \frac{\lambda_1(0)}{\sigma}.$$

In this relation K is a constant: thus, in particular, $\lambda_1^{-1}(0) = K - 1/\sigma$, so that when σ increases $\lambda_1(0)$ approaches $1/K$ and $x_1(1)$ tends to 1, as should have been expected.

Example 6.13 Consider the optimal control problem concerning the system $\dot{x}_1 = x_2$, $\dot{x}_2 = u$ with $x(0) = 0$ and $x(t_f)$ free. The performance index is

$$J = \frac{1}{2}\int_0^{t_f} u^2 dt + t_f - x_1(t_f).$$

By choosing $\lambda_0 = 1$ (the problem cannot be pathological), the orthogonality condition together with the equations of the auxiliary system imply $\lambda_1(t) = -1$, $\lambda_2(t) = t - t_f$, so that from the transversality condition it follows that $x_2(t_f) = 1$. Since $u_h = -\lambda_2$ we get $t_f = \sqrt{2}$.

Remark 6.8 *(A sketch of the derivation of the Maximum Principle)* A simple derivation of the Maximum Principle conditions can be obtained by making reference to a particular optimal control problem. As a preliminary, note that, in view of Remark 6.6, it is possible to assume the problem to be stationary, without loss of generality, while, in order to simplify the discussion, the initial state is given and the final state is free. Therefore the controlled system is described by the equation $\dot{x}(t) = f(x(t), u(t))$ and must comply with the simple constraints $x(0) = x_0$ and $u(\cdot) \in \bar{\Omega}$. Since the initial state is specified, the performance index can be given the (only seemingly) particular form $J = c'x(t_f)$. Indeed, observe that

$$\begin{aligned} m(x(t_f)) &= m(x(0)) + \int_0^{t_f} \frac{dm(x(t))}{dt} dt \\ &= m(x(0)) + \int_0^{t_f} \left.\frac{dm(x)}{dx}\right|_{x=x(t)} f(x(t), u(t)) dt \end{aligned}$$

so that, by neglecting the constant term $m(x(0))$, the given performance index becomes

$$J = \int_0^{t_f} [l(x(t), u(t)) + \left.\frac{dm(x)}{dx}\right|_{x=x(t)} f(x(t), u(t))] dt := \int_0^{t_f} l^*(x(t), u(t)) dt.$$

Now introduce the new state variable z by means of the equation $\dot{z}(t) = l^*(x(t), u(t))$ and the boundary condition $z(0) = 0$: in so doing we obtain $J = z(t_f)$.

The forthcoming discussion relies on the (trivial) idea that, given an optimal triple $(x^o(\cdot), u^o(\cdot), t_f^o)$, a perturbation of the control $u^o(\cdot)$ and final time t_f^o should not cause an improvement of the performance index. Obviously, the more general is

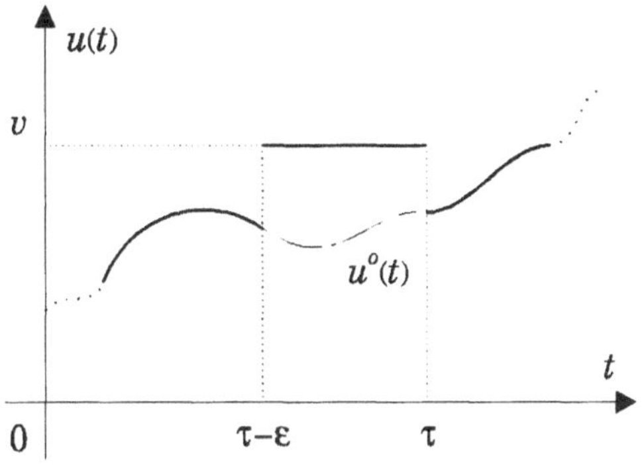

Figure 6.7: The perturbation for $u^o(\cdot)$.

the set of considered perturbations, the more significant are the resulting necessary conditions. In the specific case the final time is allowed to change to $t_f = t_f^o + \eta$, where $\eta \in R$. The control variable is perturbed in a neighbourhood of a time instant τ, $0 < \tau < t_f^o$, where it is continuous, by allowing that $u(t) = u^o(t)$, $0 \leq t \leq \tau - \varepsilon$, $u(t) = v$, $\tau - \varepsilon < t \leq \tau$, $u(t) = u^o(t)$, $\tau < t \leq t_f^o$, with $v \in U$, $\varepsilon > 0$ (see also Fig. 6.7).

The perturbation δJ_f of the performance index caused by a change in the final time is given by

$$\delta J_f = c'(x(t_f) - x^o(t_f^o)) = c'f(x^o(t_f^o), u^o(t_f^o))\eta + o(\eta)$$

if $u(t) = u^o(t_f^o)$, $t_f^o \leq t \leq t_f^o + \eta$ when $\eta > 0$. Here $o(\cdot)$ denotes a generic quantity which is infinitesimal of higher order than the argument. In order that $\delta J_f \geq 0$ it is necessary that

$$c'f(x^o(t_f^o), u^o(t_f^o)) = 0 \tag{6.9}$$

since η can be either positive or negative.

The perturbation δJ_u of the performance index caused by a control perturbation of the above described form can be evaluated by first determining the perturbation of the state at time τ. We get

$$\begin{aligned} x(\tau) - x^o(\tau) &= [f(x^o(\tau - \varepsilon), v) - f(x^o(\tau - \varepsilon), u^o(\tau - \varepsilon))]\,\varepsilon + o(\varepsilon) \\ &= [f(x^o(\tau), v) - f(x^o(\tau), u^o(\tau))]\,\varepsilon + o(\varepsilon). \end{aligned}$$

Therefore, by denoting with Φ the transition matrix associated to the matrix $\partial f/\partial x$ evaluated for $x = x^o$ and $u = u^o$, it follows that

$$\begin{aligned} \delta J_u &= c'(x(t_f^o) - x^o(t_f^o)) = c'\Phi(t_f^o, \tau)\,[x(\tau) - x^o(\tau)] + o(\varepsilon) \\ &= c'\Phi(t_f^o, \tau)\,[f(x^o(\tau), v) - f(x^o(\tau), u^o(\tau))]\,\varepsilon + o(\varepsilon). \end{aligned}$$

Hence it must result that $c'\Phi(t_f^o,\tau)\,[f(x^o(\tau),v) - f(x^o(\tau),u^o(\tau))] \geq 0$. In view of the material collected in Appendix A.2, we can set $c'\Phi(t_f^o,\tau) = c'\Psi'(\tau,t_f^o)$ where Ψ is the transition matrix associated to the matrix $-(\partial f/\partial x)'$ evaluated for $x = x^o$ and $u = u^o$, so that the last inequality becomes

$$\lambda^{o\prime}(\tau)\,[f(x^o(\tau),v) - f(x^o(\tau),u^o(\tau))] \geq 0 \tag{6.10}$$

provided that λ^o solves the equations

$$\dot{\lambda}(t) = -\left.\frac{\partial f(x,u^o(t))}{\partial x}\right|'_{x=x^o(t)}, \tag{6.11}$$

$$\lambda(t_f^o) = c. \tag{6.12}$$

The hamiltonian function relevant to the problem at hand is $H(x,u,\lambda) = \lambda' f(x,u)$: hence eq. (6.10) simply states the minimality of such a function, while eq. (6.11) is the auxiliary system and eqs. (6.9), (6.12) are the transversality and orthogonality conditions, respectively.

Remark 6.9 *(Nonpathological problems)* The discussion in Remark 6.8 does not mention the scalar λ_0: this is consistent with the fact (which can be proved) that problems where the final state is free can not be pathological.

Remark 6.10 *(Constancy of the hamiltonian function)* When the problem is stationary the hamiltonian function is proved to be constant if it is evaluated along a solution which satisfies the NC. By letting $M(t) := H(x^o(t),u^o(t),\lambda_0^o,\lambda^o(t))$, it follows that

$$\begin{aligned} H(x^o(t+\delta t),u^o(t),\lambda_0^o,\lambda^o(t+\delta t)) - M(t) &\geq M(t+\delta t) - M(t) \\ &\geq M(t+\delta t) - H(x^o(t),u^o(t+\delta t),\lambda_0^o,\lambda^o(t)). \end{aligned} \tag{6.13}$$

When $\delta t \to 0$ the first element of this chain of inequalities goes to 0 since H, x^o, λ^o are continuous functions. Even if u^o might be a discontinuous function, also the last element of the chain goes to 0 since the two terms appearing there depend in the *same* way on $u^o(t+\delta t)$: thus that element is a continuous function of δt. Therefore, it follows that

$$\lim_{\delta t \to 0} M(t+\delta t) - M(t) = 0.$$

This equation proves the continuity of the hamiltonian function. By letting

$$H_x^o(t) := \left.\frac{\partial H(x,u^o(t),\lambda_0^o,\lambda^o(t))}{\partial x}\right|_{x=x^o(t)},$$

$$H_\lambda^o(t) := \left.\frac{\partial H(x^o(t),u^o(t),\lambda_0^o,\lambda)}{\partial \lambda}\right|_{\lambda=\lambda^o(t)},$$

it follows that for any $t \in [0,t_f^o]$,

$$H(x^o(t+\delta t),u^o(t),\lambda_0^o,\lambda^o(t+\delta t)) = M(t) + [H_x^o(t)\dot{x}^o(t) + H_\lambda^o\dot{\lambda}^o(t)]\delta t + o(\delta t)$$

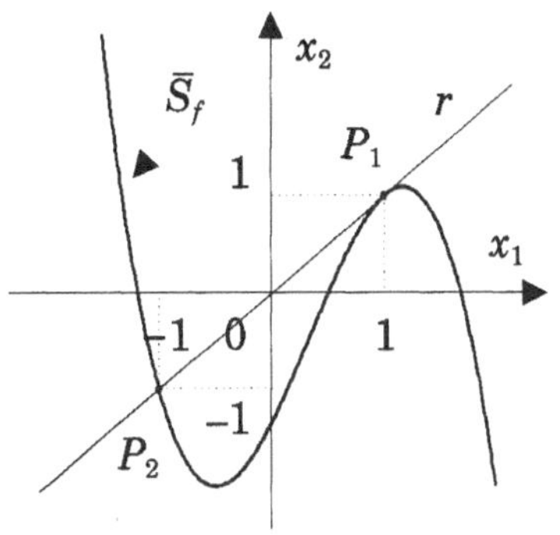

Figure 6.8: The state space for the problem in Example 6.14.

so that

$$\lim_{\delta t \to 0} \frac{H(x^o(t+\delta t), u^o(t), \lambda_0^o, \lambda^o(t+\delta t)) - M(t)}{\delta t}$$
$$= H_x^o(t)\dot{x}^o(t) + H_\lambda^o(t)\dot{\lambda}^o(t) = -\dot{\lambda}^{o\prime}(t)\dot{x}^o(t) + \dot{x}^{o\prime}(t)\dot{\lambda}^o(t) = 0. \tag{6.14}$$

In a similar way, for any $t \in [0, t_f^o]$ where u^o is continuous, it results that

$$\lim_{\delta t \to 0} \frac{M(t+\delta t) - H(x^o(t), u^o(t+\delta t), \lambda_0^o, \lambda^o(t)}{\delta t} = 0. \tag{6.15}$$

In view of eqs. (6.13)–(6.15) we can conclude that

$$\lim_{\delta t \to 0} \frac{M(t+\delta t) - M(t)}{\delta t} = 0,$$

i.e., $\dot{M}(t) = 0$ for each t where u is continuous. Since u is piecewise continuous, this fact, together with the continuity of M, implies the constancy of M.

Example 6.14 Problems where the variable λ_0 can be zero are now presented. Consider the second order system $\dot{x}_1 = u$, $\dot{x}_2 = u$ with initial state $x(0) = 0$, final state $x(1) \in \bar{S}_f$ and performance index

$$J = \int_0^1 \left[\frac{u^2}{2} + \rho u\right] dt$$

where ρ is a given constant and $\bar{S}_f = \{x |\ 4x_1^3 - 4x_1^2 - 7x_1 + 3x_2 - 4 = 0\}$ (see Fig. 6.8). Note that the only possible trajectory for the system is the straight line r described by $x_2 = x_1$ (see Fig. 6.8) which has two points in common with $\bar{S}_f$, precisely the points P_1 and P_2 with coordinates $x_1 = x_2 = 1$ and $x_1 = x_2 = -1$, respectively. Therefore, the problem amounts to steering the state of the first order system $\dot{\xi} = u$ from $\xi(0) = 0$ to $\xi(1) = \pm 1$ while minimizing the performance index above. Three different values for ρ are taken into consideration: (i), $\rho = 2$; (ii), $\rho = 0$; (iii), $\rho = -2$. By resorting to the Hamilton-Jacobi theory it is straightforward to check that $u = -1$ is an optimal control for case (i), $u = 1$ is optimal for case (iii) and both $u = 1$ and $u = -1$ are optimal for case (ii). Going back to the NC we find that the hamiltonian function is $H = \lambda_0(u^2/2 + \rho u) + (\lambda_1 + \lambda_2)u$ so that both λ_1 and λ_2 are constant. When considering case (i) the NC are satisfied by $\lambda_0 = 1$, $\lambda_1 = -13/16$, $\lambda_2 = -3/16$; when considering case (iii) the NC are satisfied only if $\lambda_0 = 0$, $\lambda_1 = -\lambda_2 \neq 0$, so that the problem is pathological; when considering case (ii) the NC are satisfied only by $\lambda_0 = 0$, $\lambda_1 = -\lambda_2 \neq 0$ if $u = 1$ and by $\lambda_0 = 1$, $\lambda_1 = 13/16$, $\lambda_2 = 3/16$ if $u = -1$.

The forthcoming examples are aimed at emphasizing some peculiarities of the NC and shedding light on some intimate limitations of the corresponding results.

Example 6.15 This example shows that many solutions consistent with the NC might exist, only one of them being actually optimal. Consider the system $\dot{x}_1 = x_2$, $\dot{x}_2 = u$ with $x(0) = 0$, $x(1) \in \bar{S}_f$, $\bar{S}_f = \{x | \ 5x_1^2 - x_2 - 5 = 0\}$ and the performance index

$$J = \int_0^1 \frac{u^2}{2} dt.$$

The following equations are obtained by enforcing the orthogonality condition and the feasibility of the adopted control (which is of the form $u(t) = \lambda_1(0)t - \lambda_2(0)$):

$$\begin{aligned} 0 &= \lambda_1(0)(10x_1(1) - 1) - 10\lambda_2(0)x_1(1), \\ x_2(1) &= \frac{1}{2}\lambda_1(0) - \lambda_2(0), \\ x_1(1) &= \frac{1}{6}\lambda_1(0) - \frac{1}{2}\lambda_2(0), \\ 0 &= 5x_1^2(1) - x_2(1) - 5. \end{aligned}$$

These equations admit three solutions (the value of $x_2(1)$ is not reported as of no interest)

$$S_1 : \begin{cases} x_1(1) = 1.15 \\ \lambda_1(0) = -3.98 \\ \lambda_2(0) = -3.63, \end{cases} \quad S_2 : \begin{cases} x_1(1) = 0.1522 \\ \lambda_1(0) = -31.1316 \\ \lambda_2(0) = -10.6816, \end{cases} \quad S_3 : \begin{cases} x_1(1) = -0.85 \\ \lambda_1(0) = 2.18 \\ \lambda_2(0) = 2.44. \end{cases}$$

The significance of these three solutions can be understood by assuming that $x_1(1)$ (and hence $x_2(1)$, too) is given. For each $x_1(1)$ it is easy to compute an optimal solution and the relevant value $J^o(x_1(1))$ of the performance index by resorting to the Hamilton-Jacobi theory. The plot of such a function is reported in Fig. 6.9 which shows that the solution S_2 corresponds to a local *maximum*, while the remaining two correspond to local *minima*. Furthermore, it is obvious that the solution S_3 is *globally* optimal.

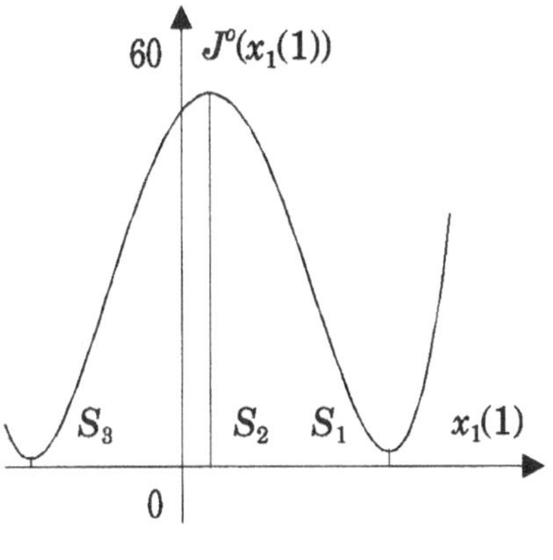

Figure 6.9: Example 6.15: the optimal value of the performance index.

Example 6.16 This example emphasizes the role played in Theorems 6.2 and 6.1 by the request that the final time be *strictly* greater than the initial time. Consider the problem defined by the first order system $\dot{x} = -x + u$, $x(0) = 1$, with free final state and the performance index

$$J = \frac{1}{2}\left\{\int_0^{t_f} [3x^2 + u^2]dt + sx^2(t_f)\right\}, \; s > 0$$

where the final time t_f must be determined. The hamiltonian function is $H = 3x^2/2 + u^2/2 + \lambda(u - x)$, thus, $u_h = -\lambda$ and $\dot{\lambda} = -3x + \lambda$. The solution of the hamiltonian system is of the form

$$x(t) = \alpha e^{2t} + \beta e^{-2t},$$
$$\lambda(t) = -3\alpha e^{2t} + \beta e^{-2t},$$

where α, β and t_f can be computed by enforcing $x(0) = 1$, the transversality and orthogonality conditions, the latter of them being $\lambda(t_f) = sx(t_f)$. From the first two requirements we find the two pairs $\alpha = 0$, $\beta = 1$ and $\alpha = 1$, $\beta = 0$. However, the second pair prevents satisfaction of the orthogonality condition (recall that $s > 0$). On the other hand, this can be done for the first pair, provided only that $s = 1$, but the value for t_f is arbitrary. In conclusion, no solution complying with the NC exists if $s \neq 1$, while an infinite number of solutions exists if $s = 1$.

The reason for this outcome can easily be understood if reference is made to a slightly modified version of the problem, where the final time is given. Thus an LQ problem results, the solution of which is completely specified once the relevant DRE is integrated. We get

$$P(t, t_f, s) = \begin{cases} \dfrac{s + 3 + 3(s-1)e^{4(t-t_f)}}{s + 3 - (s-1)e^{4(t-t_f)}} & , s \neq 1 \\[2ex] 1 & , s = 1 \end{cases}$$

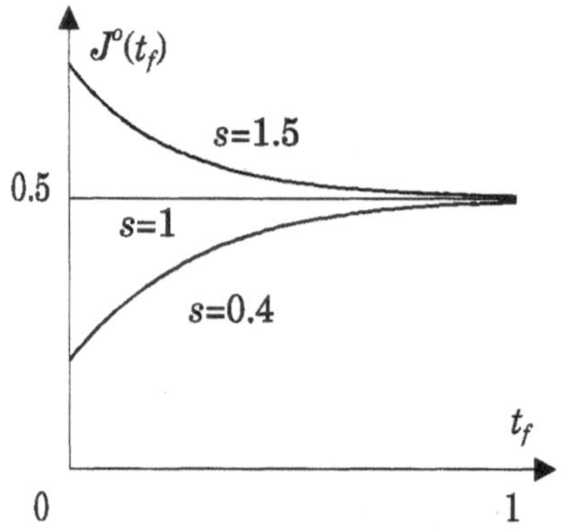

Figure 6.10: Example 6.16: the optimal value of the performance index.

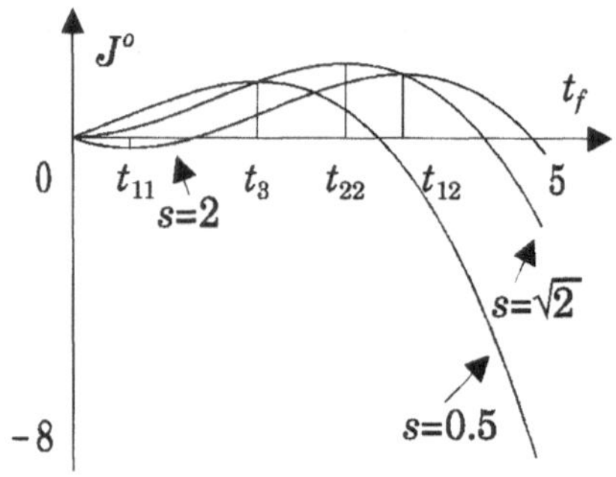

Figure 6.11: Example 6.17: the optimal value of the performance index.

where the dependence of the solution of the DRE on the boundary condition has been put into evidence. The plot of the optimal value of the performance index $J^o(t_f) = P(0, t_f, s)/2$ vs. the final time t_f is reported in Fig. 6.10 corresponding to three values of s. The following conclusions can be drawn: if $s > 1$ no optimal solution of the problem exists, as was pointed out by the NC; if $s = 1$ there exists an infinite number of optimal solutions, once more consistent with the NC; if $s < 1$ there *exists* an optimal solution which is characterized by $t_f = 0$, but the NC are not satisfied just because the final time is *not* strictly greater than the initial time (recall Remark 6.8 and note that the derivative of $J^o(t_f)$ with respect to t_f is not zero for $t_f = 0$).

Example 6.17 This example shows that it might happen that more than one control satisfies the NC even if the problem does *not* admit a solution. Consider the problem defined by the first order system $\dot{x} = u$, $x(0) = 0$, with free final state and the performance index

$$J = \int_0^{t_f} [1 + \frac{u^2}{2} - x]dt + sx(t_f), \; s > 0$$

where, obviously, t_f is not given. The hamiltonian function is $H = 1 + u^2/2 - x + \lambda u$ so that $u_h = -\lambda$, $\lambda(t) = t + \lambda(0)$. By enforcing the transversality and orthogonality conditions we get $\lambda(0) = \pm\sqrt{2}$ and $t_f = s - \lambda(0)$. Therefore, when $s \geq \sqrt{2}$ two values for t_f result, while if $0 < s < \sqrt{2}$ only one value can be deduced. For instance, if $s = s_1 := 2$ we find $t_{11} = 2 - \sqrt{2}$ and $t_{12} = 2 + \sqrt{2}$; if $s = s_2 := \sqrt{2}$ we find $t_{21} = 0$ and $t_{22} = 2\sqrt{2}$; if $s = s_3 := 0.5$ we find $t_3 = 1/2 + \sqrt{2}$. The intimate nature of these outcomes can be explained by making reference to a slight modification of the problem where the final time is assumed to be (arbitrarily) given. By exploiting the Hamilton-Jacobi theory an optimal control can easily be found together with the relevant optimal value of the performance index, which is $J^o(s, t_f) = t_f - (3(\alpha(s, t_f))^2 t_f + 3\alpha(s, t_f)t_f^2 + t_f^3)/6$, where $\alpha(s, t_f) := s - t_f$. The plots reported in Fig. 6.11 are thus obtained: they show that no solution of the given problem exists whatever the value of s is and that the values of t_f resulting from the NC correspond either to local minima or global maxima of the optimal performance index. When $s = \sqrt{2}$, note that the derivative of the optimal value of the performance index with respect to t_f is zero at $t_f = 0$: this is the reason why such a value has been found while enforcing the NC.

Example 6.18 This last example shows that more than one control might satisfy the NC, but only some of them are actually *distinct* solutions of the problem. Consider the problem defined by the system $\dot{x}_1 = x_2$, $\dot{x}_2 = u$ with $x(0) \in \bar{S}_0$, $x(1) = 0$, $\bar{S}_0 = \{x | \; x_1^2/4 + x_2^2 - 1 = 0\}$ and the performance index

$$J = \int_0^1 \frac{u^2}{2} dt.$$

The orthogonality condition at the initial time requires that $\lambda_1(0) = \vartheta x_1(0)/4$ and $\lambda_2(0) = \vartheta x_2(0)$. The minimization of the hamiltonian function and the auxiliary system imply that $u(t) = -\lambda_2(0) + \lambda_1(0)t$: thus it is easy to conclude, by enforcing feasibility, that $x_1(0) \neq 0$ and $x_2(0) \neq 0$. Therefore it is $\lambda_2(0) = 4\lambda_1(0)x_2(0)/x_1(0)$. From the condition $x(1) = 0$ it follows that $x_2(0) = x_1(0)(-11 \pm \sqrt{157})/12$: thus we can conclude that the initial states consistent with the NC are four (see Fig. 6.12). In view of the nature of the problem it is possible to state that the values of the performance index corresponding to the choices $x(0) = P_1$ and $x(0) = P_3$ are equal and similarly equal are the values corresponding to the choices $x(0) = P_2$ and $x(0) = P_4$. The optimal value of the performance index can easily be computed as a function of the angle φ which identifies the initial state (again see Fig. 6.12) by resorting to the Hamilton-Jacobi theory. The simple conclusion is that two distinct optimal solutions exist corresponding to the points P_2 and P_4.

Remark 6.11 *(A computational algorithm)* Many iterative algorithms have been proposed for the numerical computation of a (candidate) optimal control: one of them,

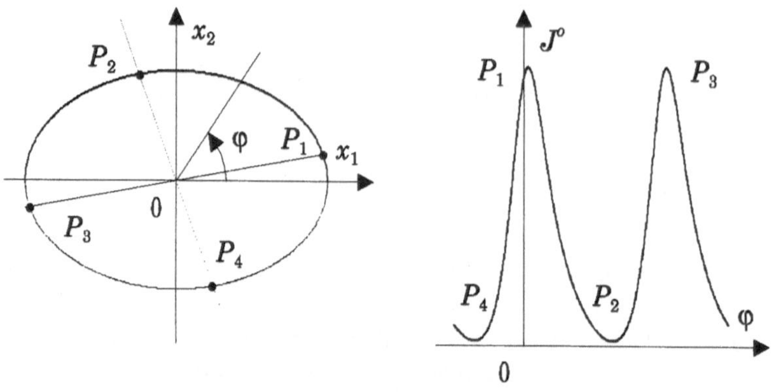

Figure 6.12: Example 6.18: initial states which satisfy the NC and the optimal value of the performance index vs. φ.

particularly simple and easily understandable, is here briefly presented with the aim of showing how the NC can be fruitfully exploited. The (trivial) idea on which it relies is basically the following: assume that the solution at hand is *far* from complying with the NC: then determine a perturbation which (hopefully) makes the new solution *closer* to the satisfaction of the NC and iterate until a solution is found which satisfies the NC up to a specified degree of accuracy. In its present form, the algorithm is suited to tackle the class of problems defined by the dynamical system

$$\dot{x}(t) = f(x(t), u(t), t), \ x(t_0) = x_0 \tag{6.16}$$

the performance index

$$J = \int_{t_0}^{t_f} l(x(t), u(t), t)dt + m(x(t_f), t_f) \tag{6.17}$$

and the constraint on the final events

$$(x(t_f), t_f) \in S_f = \{(x,t)| \ \alpha(x,t) = 0\} \tag{6.18}$$

where the number of components of the vector α is not greater than $n+1$. In these equations t_0 and x_0 are given, the functions f, l, m, α possess the (frequently) mentioned continuity properties and the set S_f, if actually constraining the final event, is a regular variety. No further limitations are imposed on the state and/or control variables and it is assumed that the problem is not pathological. Then it is possible to describe the i-iteration of the algorithm by denoting with the superscript $^{(i)}$ the variable or parameter available at that iteration, while $\mathcal{T}^{(i)}$ is the relevant control interval $[t_0, t_f^{(i)}]$.

Algorithm 6.1

1. *For* $u = u^{(i)}$ *compute* $x^{(i)}$ *by integrating eq. (6.16) over the interval* $\mathcal{T}^{(i)}$.
2. *Compute* $\lambda^{(i)}$ *and* $\Omega^{(i)}$ *by integrating eqs.*

$$\dot{\lambda}(t) = -\left.\frac{\partial H(x, u^{(i)}(t), t, \lambda(t))}{\partial x}\right|'_{x=x^{(i)}(t)},$$
$$\dot{\Omega}(t) = -A^{(i)\prime}(t)\Omega(t)$$

over the interval $\mathcal{T}^{(i)}$ *with*

$$\lambda(t_f^{(i)}) = \left.\frac{\partial m(x, t_f^{(i)})}{\partial x}\right|'_{x=x^{(i)}(t_f^{(i)})},$$
$$\Omega(t_f^{(i)}) = \left.\frac{\partial \alpha(x, t_f^{(i)})}{\partial x}\right|'_{x=x^{(i)}(t_f^{(i)})},$$
$$A^{(i)}(t) := \left.\frac{\partial f(x, u^{(i)}(t), t)}{\partial x}\right|_{x=x^{(i)}(t)}.$$

3. *Compute*

$$G_1^{(i)} := \int_{t_0}^{t_f^{(i)}} \Omega^{(i)\prime}(t)B^{(i)}(t)W^{(i)}(t)B^{(i)\prime}(t)\Omega^{(i)}(t)dt,$$
$$G_2^{(i)} := \int_{t_0}^{t_f^{(i)}} H_u^{(i)}(t)W^{(i)}(t)B^{(i)\prime}(t)\Omega^{(i)}(t)dt,$$
$$G_3^{(i)} := \int_{t_0}^{t_f^{(i)}} H_u^{(i)}(t)W^{(i)}(t)H_u^{(i)\prime}(t)dt,$$

where

$$B^{(i)}(t) := \left.\frac{\partial f(x^{(i)}(t), u, t)}{\partial u}\right|_{u=u^{(i)}(t)},$$
$$H_u^{(i)}(t) := \left.\frac{\partial H(x^{(i)}(t), u, t, \lambda^{(i)}(t))}{\partial u}\right|_{u=u^{(i)}(t)},$$

while $W^{(i)}(t) = W^{(i)\prime}(t) > 0$, $t \in \mathcal{T}^{(i)}$ *is a matrix to be suitably chosen.*

4. *Select* $\delta\alpha^{(i)}$ *in such a way that*

$$||\alpha^{(i)} + \delta\alpha^{(i)}|| < ||\alpha^{(i)}||, \; \alpha^{(i)} := \alpha(x^{(i)}(t_f^{(i)}), t_f^{(i)})$$

where $||\cdot||$ *is the euclidean norm.*

5. *Compute*

$$\mu^{(i)} = -\left\{G_1^{(i)} + \beta^{(i)} \left.\frac{d\alpha(x^{(i)}(t),t)}{dt}\right|_{t=t_f^{(i)}} \left.\frac{d\alpha(x^{(i)}(t),t)}{dt}\right|_{t=t_f^{(i)}}^{\prime}\right\}^{\dagger}$$
$$\cdot\left\{\delta\alpha^{(i)} + G_2^{(i)\prime} + \beta^{(i)}\left[\left.\frac{dm(x^{(i)}(t),t)}{dt}\right|_{t=t_f^{(i)}} + l^{(i)}\right]\right.$$
$$\left.\cdot\left.\frac{d\alpha(x^{(i)}(t),t)}{dt}\right|_{t=t_f^{(i)}}\right\}$$

where $l^{(i)} := l(x^{(i)}(t_f^{(i)}), u^{(i)}(t_f^{(i)}), t_f^{(i)})$*, while* $\beta^{(i)} > 0$ *is a scalar to be suitably selected.*

6. *If*

$$\| \alpha^{(i)} \| < \varepsilon_1, \tag{6.19}$$

$$\left\| \frac{d}{dt}\left[m(x^{(i)}(t),t) + \mu^{(i)\prime}\alpha(x^{(i)}(t),t)\right]\right|_{t=t_f^{(i)}} + l^{(i)}\right\| < \varepsilon_2, \tag{6.20}$$

$$\| G_3^{(i)} - G_2^{(i)} G_1^{(i)\dagger} G_2^{(i)\prime} \| < \varepsilon_3, \tag{6.21}$$

where $\varepsilon_i > 0$*,* $i = 1,2,3$ *are suitable scalars, then the triple* $(x^{(i)}, u^{(i)}, t_f^{(i)})$ *satisfies the NC with an approximation the more accurate the smaller the scalars* ε_i *are. Otherwise, let*

$$u^{(i+1)}(t) := u^{(i)}(t) + \delta u^{(i)}(t),$$
$$t_f^{(i+1)} := t_f^{(i)} + \delta t_f^{(i)},$$

where

$$\delta u^{(i)}(t) := -W^{(i)}(t)\left[H_u^{(i)\prime}(t) + B^{(i)\prime}(t)\Omega^{(i)}(t)\mu^{(i)}\right],$$

$$\delta t_f^{(i)} := -\beta^{(i)}\left\{\frac{d}{dt}\left[m(x^{(i)}(t),t) + \mu^{(i)\prime}\alpha^{(i)}(x^{(i)}(t),t)\right]\right|_{t=t_f^{(i)}} + l^{(i)}\right\}.$$

The i-iteration of this algorithm can be justified as follows. First, it is easy to check that the perturbations $\delta u^{(i)}$ of the control $u^{(i)}$ and $\delta t_f^{(i)}$ of the final time $t_f^{(i)}$ cause *first* variations $[\delta J]_1^{(i)}$ of the performance index and $[\delta\alpha]_1^{(i)}$ of the vector $\alpha^{(i)}$ which are given by

$$[\delta J]_1^{(i)} = \left[\left. \frac{dm(x^{(i)}(t),t)}{dt} \right|_{t=t_f^{(i)}} + l^{(i)} \right] \delta t_f^{(i)} + \int_{t_0}^{t_f^{(i)}} H_u^{(i)}(t)\delta u^{(i)}(t)dt, \quad (6.22)$$

$$[\delta\alpha]_1^{(i)} = \left. \frac{d\alpha(x^{(i)}(t),t)}{dt} \right|_{t=t_f^{(i)}} \delta t_f^{(i)} + \left. \frac{\partial\alpha(x,t_f^{(i)})}{\partial x} \right|_{x=x^{(i)}(t_f^{(i)})} \delta x^{(i)}(t_f^{(i)}) \quad (6.23)$$

if $\lambda^{(i)}$ is chosen according to step 2. In eq. (6.23), $\delta x^{(i)}(t_f^{(i)})$ is the solution of the equation

$$\delta\dot{x}(t) = A^{(i)}(t)\delta x(t) + B^{(i)}(t)\delta u^{(i)}(t),\ \delta x(t_0) = 0 \quad (6.24)$$

evaluated at time $t_f^{(i)}$. Therefore it follows that

$$\delta x^{(i)}(t_f^{(i)}) = \int_{t_0}^{t_f^{(i)}} \Phi^{(i)}(t_f^{(i)},t)B^{(i)}(t)\delta u^{(i)}(t)dt,$$

$\Phi^{(i)}$ being the transition matrix associated to $A^{(i)}$: thus, by exploiting the properties of such matrices (see Appendix A.2, point (iv)), eq. (6.23) becomes

$$[\delta\alpha]_1^{(i)} = \left. \frac{d\alpha(x^{(i)}(t),t)}{dt} \right|_{t=t_f^{(i)}} \delta t_f^{(i)} + \int_{t_0}^{t_f^{(i)}} \Omega^{(i)\prime}(t)B^{(i)}(t)\delta u^{(i)}(t)dt \quad (6.25)$$

with $\Omega^{(i)}$ specified at step 2. Then it is reasonable to require that the perturbations to be given to the control and the final time should minimize $[\delta J]_1$, while complying with the condition

$$[\delta\alpha]_1^{(i)} = \delta\alpha^{(i)} \quad (6.26)$$

which is intended to reduce the amount the constraint (6.18) is violated, if we have chosen

$$\delta\alpha^{(i)} = -\varepsilon_4\alpha^{(i)} =,\ 0 < \varepsilon_4 \le 1 \quad (6.27)$$

at step 4. Since both $[\delta\alpha]_1^{(i)}$ and $[\delta J]_1^{(i)}$ linearly depend on the perturbations $\delta u^{(i)}$ and $\delta t_f^{(i)}$, the choice of these perturbations can be performed by minimizing the objective function

$$\delta\bar{J} = [\delta J]_1^{(i)} + \frac{(\delta t_f^{(i)})^2}{2\beta^{(i)}} + \frac{1}{2}\int_{t_0}^{t_f^{(i)}} \delta u^{(i)\prime}(t)(W^{(i)}(t))^{-1}\delta u^{(i)}(t)dt \quad (6.28)$$

subject to eqs. (6.25)–(6.27). By adding these constraints to $\delta\bar{J}$ through a multiplier $\mu^{(i)}$, it is easy to verify that the perturbations which minimize the new objective function are those shown in step 6 and their substitution into eqs. (6.25), (6.26) yields the value for $\mu^{(i)}$ reported at step 5. By inserting the perturbations of the control and final time given in step 6 into the right side of eqs. (6.22), (6.25) (note that $u^{(i+1)}(t)$ must be defined also for $t \in (t_f^{(i)}, t_f^{(i+1)}]$ if $\delta t_f^{(i)} > 0$, for instance by letting $u^{(i+1)}(t) := u^{(i)}(t_f^{(i)}) + \delta u^{(i)}(t_f^{(i)})$) the first variations of the performance index J and the vector α can be computed, yielding

$$[\delta J]_1^{(i)} = -\left[G_3^{(i)} + G_2^{(i)}\mu^{(i)}\right] - \beta^{(i)}\left[\left.\frac{dm(x^{(i)}(t),t)}{dt}\right|_{t=t_f^{(i)}} + l^{(i)}\right]$$
$$\left[\frac{d}{dt}\left[m(x^{(i)}(t),t) + \mu^{(i)\prime}\alpha(x^{(i)}(t),t)\right]\Big|_{t=t_f^{(i)}} + l^{(i)}\right], \tag{6.29}$$

$$[\delta\alpha]_1^{(i)} = -\left[G_2^{(i)\prime} + G_1^{(i)}\mu^{(i)}\right] - \beta^{(i)}\left.\frac{d\alpha(x^{(i)}(t),t)}{dt}\right|_{t=t_f^{(i)}}$$
$$\left[\frac{d}{dt}\left[m(x^{(i)}(t),t) + \mu^{(i)\prime}\alpha(x^{(i)}(t),t)\right]\Big|_{t=t_f^{(i)}} + l^{(i)}\right] \tag{6.30}$$

The comparison of these values with the *actual* variations, namely those corresponding to $u^{(i+1)}$ and $t_f^{(i+1)}$, provides useful information for the parameters β and W at the subsequent iteration (e.g., large deviations suggest their reduction).

When the inequalities in step 6 are verified (the quantities at the left side can be viewed as zero), the triple at hand complies with the NC and the iterative procedure is stopped. In fact, eqs. (6.19), (6.27), imply that $\delta\alpha^{(i)}$ is zero and, together with eqs. (6.20), (6.26), (6.30), $\mu^{(i)} = -(G_1^{(i)})^{\dagger}G_2^{(i)\prime}$. In view of these results and eqs. (6.20), (6.21), from eq. (6.29) it follows that $[\delta J]_1^{(i)}$ has become *zero*.

The Algorithm 6.1 can be specialized suitably to the case of control problems where the final time is given and/or the final state is not constrained by modifying its steps in a fairly obvious way. For instance, if the final time is known and equal to T and the final state is free, it is necessary:

(a) To set $t_f^{(i)} = T$ everywhere;

(b) Not to introduce the matrix $\Omega^{(i)}$;

(c) Not to introduce the matrices $G_1^{(i)}$, $G_2^{(i)}$;

(d) To discard steps 4 and 5;

(e) To perform the only comparison $\| G_3^{(i)} \| < \varepsilon_3$ at step 6 and, if it is the case, to compute $\delta u^{(i)}(t) = -W^{(i)}(t)H_u^{(i)\prime}(t)$ and set $u^{(i+1)}(t) := u^{(i)}(t) + \delta u^{(i)}(t)$.

Example 6.19 Consider the optimal control problem defined by the first order system $\dot{x} = u$ with $x(0) = 1$, $x(t_f) = 0$ and the performance index

$$J = \int_0^{t_f} [1 + \frac{u^2}{2}]dt$$

i	$u^{(i)}$	$t_f^{(i)}$	$G_1^{(i)}$	$G_2^{(i)}$	$G_3^{(i)}$	$\alpha^{(i)}$	$a_1^{(i)}$	$a_2^{(i)}$
1	−1.000	1.000	1.000	−1.000	1.000	0.000	0.250	0.000
2	−1.250	0.750	0.750	−0.938	1.172	0.063	0.054	0.000
3	−1.382	0.696	0.696	−0.962	1.329	0.038	0.019	0.000
4	−1.413	0.867	0.867	−1.226	1.732	−0.226	0.056	0.000
5	−1.374	0.856	0.856	−1.175	1.616	−0.176	0.061	0.000
6	−1.370	0.794	0.794	−1.088	1.491	−0.088	0.039	0.000
7	−1.387	0.755	0.755	−1.047	1.453	−0.047	0.023	0.000
8	−1.398	0.732	0.732	−1.023	1.431	−0.023	0.012	0.000
9	−1.405	0.720	0.720	−1.012	1.422	−0.012	0.006	0.000
10	−1.410	0.713	0.713	−1.006	1.418	−0.006	0.003	0.000

Table 6.1: Example 6.19: iterations of the Algorithm 6.1.

where t_f is free. The initial guesses for u and t_f are $u^{(1)} = -1$ and $t_f^{(1)} = 1$, while the parameters which characterize the Algorithm 6.1 are: $W^{(i)}(t) = 1$, $\beta^{(i)} = 1$, $\delta\alpha^{(i)} = -\alpha^{(i)}/2$, $\forall i$. Finally, the end condition is specified by $\varepsilon_j = 0.01$, $j = 1, 2, 3$. The algorithm requires ten iterations which are synthesized in Tab. 6.1, where $a_1^{(i)}$ and $a_2^{(i)}$ are the values of the left sides of eqs. (6.20), (6.21). The obtained triple is $x(t) = 1 - 1.41t$, $u(t) = -1.41$, $t_f = 0.713$ which has to be meant as a good approximation of the triple resulting from directly enforcing the NC (i.e., by solving the relevant equations), the latter being $x^o(t) = 1 - \sqrt{2}t$, $u^o(t) = -\sqrt{2}$, $t_f^o = 1/\sqrt{2}$.

Example 6.20 Consider the optimal control problem defined by the first order system $\dot{x} = u$ with $x(0) = 0$ and free final state. The performance index is

$$J = \int_0^1 [x + u^2]dt.$$

The initial guess for the control is $u^{(1)} = -1$ and the end condition is characterized by $\varepsilon_3 = 0.01$. Three distinct and constant values, namely, 0.3, 1, 1.5, have been chosen for the parameter $W^{(i)}$ which is the only parameter to be specified in order to apply the Algorithm 6.1 (in its modified form as given at the end of Remark 6.11). It is easy to check that, no matter of actual value for $W^{(i)}$, it results that $u^{(i)}(t) = c_0^{(i)} + c_1^{(i)}t$, so that the evolution of the algorithm can efficiently be summarized as in Tab. 6.2 which shows that there is not convergence when $W^{(i)} = 1$ or $W^{(i)} = 1.5$, $\forall i$, while, when $W^{(i)} = 0.3$, $\forall i$, the algorithm ends at the fourth iteration by supplying $u(t) = -0.532 + 0.468t$ rather than $u(t) = -0.5 + 0.5t$ which is the control resulting from directly enforcing the NC.

In this particular case the behaviour of the algorithm can easily be understood since the initial guess for $u^{(1)}$ and the constancy of $W^{(i)}$ ($W^{(i)} = \bar{W}$) imply that

$$c_0^{(i+1)} = (1 - 2\bar{W})c_0^{(i)} - \bar{W},$$
$$c_1^{(i+1)} = (1 - 2\bar{W})c_1^{(i)} + \bar{W}.$$

	$W^{(i)} = 0.3$			$W^{(i)} = 1$			$W^{(i)} = 1.5$		
i	$c_0^{(i)}$	$c_1^{(i)}$	$G_3^{(i)}$	$c_0^{(i)}$	$c_1^{(i)}$	$G_3^{(i)}$	$c_0^{(i)}$	$c_1^{(i)}$	$G_3^{(i)}$
1	−1.000	0.000	0.700	−1.0	0.0	0.7	−1.000	0.000	0.7
2	−0.700	0.300	0.112	0.0	1.0	0.7	0.500	1.500	2.8
3	−0.580	0.420	0.018	−1.0	0.0	0.7	−2.500	−1.500	11.2
4	−0.532	0.468	0.003	0.0	1.0	0.7	3.500	4.500	44.8

Table 6.2: Example 6.20: iterations of the Algorithm 6.1.

Therefore the evolution of the algorithm can be described by a linear, discrete-time dynamical system, the unique equilibrium state of which is $\bar{c}_0 = -0.5$, $\bar{c}_1 = 0.5$ and does not depend on $\bar{W}$. The convergence of the algorithm requires asymptotic stability of this system, i.e., that $|1 - 2\bar{W}| < 1$, which in turn calls for $0 < \bar{W} < 1$, since $\bar{W} > 0$.

6.3 Complex constraints

Optimal control problems where the state and/or control variables are constrained in a somehow more complicated way are considered in this section. As a result, necessary optimality conditions will be established for a much wider class of problems at a price of an acceptable increase in the presentation complexity. The problems to be discussed are characterized by the presence, at the same time, of only one of the constraints reported below, the extension to the case where more than one of them is contemporarily active being conceptually trivial. More precisely, the control problems with complex constraints which are taken into consideration are:

(a) Problems with the final state constrained to belong to the set

$$\hat{S}_f = \{x| \ x \in \bar{S}_f, \ a_i \leq x_i \leq b_i, \ i \in \mathcal{I}\}$$

where $\bar{S}_f$ is either a regular variety or the whole state space, a_i and b_i are known constants, $\mathcal{I}$ is a set of $q \leq n$ distinct integers n_j, $1 \leq n_j \leq n$. Thus, the set $\hat{S}_f$ is not a regular variety;

(b) Problems where some functions of the state and/or control variables are subject to integral constraints, i.e., constraints of the form

$$\int_{t_0}^{t_f} w_e(x(t), u(t), t)dt = \bar{w}_e,$$

$$\int_{t_0}^{t_f} w_d(x(t), u(t), t)dt \leq \bar{w}_d;$$

(c) Problems where some functions of the state and/or control variables must satisfy instantaneous equality constraints over the whole control interval, i.e., constraints of the form

$$w(x(t), u(t), t) = 0, \;\; t_0 \leq t \leq t_f;$$

(d) Problems where some functions of the state variables must satisfy instantaneous equality constraints at isolated points, i.e., constraints of the form

$$w^{(i)}(x(t_i), t_i) = 0, \;\; t_0 < t_1 < \cdots < t_i < \cdots < t_s < t_f.$$

The time instants t_i may or may not be specified;

(e) Problems where some functions of the state and/or control variables must satisfy instantaneous inequality constraints over the whole control interval, i.e., constraints of the form

$$w(x(t), u(t), t) \leq 0, \;\; t_0 \leq t \leq t_f.$$

In the preceding equations w_e, w_d, w, $w^{(i)}$ and their partial derivatives with respect to x and t are continuous functions. However, in some cases the function w is required to have higher order derivatives (cases (c) and (e) when w does not explicitly depend on u). Finally, the quantities $\bar{w}_e$, $\bar{w}_d$ are given vectors and the notation $\alpha \leq \beta$ means $\alpha_i \leq \beta_i$, $i = 1, 2, \ldots, r$, if α and β are r-dimensional vectors.

6.3.1 Nonregular final varieties

With reference to the class of problems mentioned in item (a), the case where only one state variable must belong to a given interval is first considered. Thus the set $\mathcal{I}$ is constituted by a single element and a possible way of handling the problem consists in *ignoring*, as a first attempt, the constraint $a_i \leq x_i(t_f) \leq b_i$ and then proceeding with the computation of a control which satisfies the NC. If the resulting state motion is such that $x(t_f) \in \hat{S}_f$, the solution at hand obviously verifies the NC also for the original problem. If this does not happen, a couple of problems must be tackled, namely those where the final state must belong to the set $\bar{S}_{fa} := \{x | \; x \in \bar{S}_f, \; x_i = a_i\}$ or to the set $\bar{S}_{fb} := \{x | \; x \in \bar{S}_f, \; x_i = b_i\}$. These sets, if not empty, are regular varieties (possibly after some suitable and obvious rearrangements of the equations $\alpha_i(x) = 0$ which define $\bar{S}_f$) and it is apparent that the solutions which are dominated by some others should be discarded.

Example 6.21 Consider a cart with unitary mass which moves without friction along a straight rail subject to a force u, the absolute value of which must be less than 1.

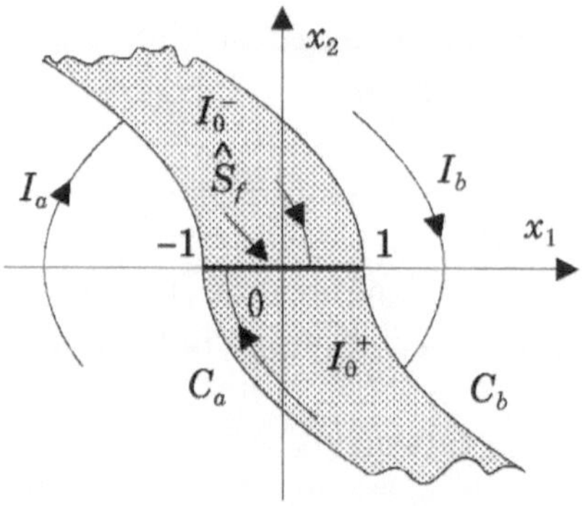

Figure 6.13: Example 6.21: partition of the state space into the three subregions I_0, I_a and I_b.

Therefore the system is described by $\dot{x}_1 = x_2$, $\dot{x}_2 = u$, $|u(t)| \leq 1$, $0 \leq t \leq t_f$, x_1 and x_2 being the cart position and velocity, respectively. The initial state is given, while at the free final time the state must belong to the set $\hat{S}_f = \{x|\ x_2 = 0, -1 \leq x_1 \leq 1\}$. The elapsed time is the performance index and the problem is not trivial only if the initial state does not lie inside $\hat{S}_f$. As a first attempt the constraint on x_1 is ignored, so that the orthogonality condition implies that $\lambda_1(\cdot) = 0$ and $\lambda_2(\cdot) = \lambda_2(0)$, since the solution of the auxiliary system is $\lambda_1 = \lambda_1(0)$ and $\lambda_2(t) = \lambda_2(0) - \lambda_1(0)t$. By requiring that $x_2(t_f) = 0$ we obtain $t_f = |x_2(0)|$, $u(t) = -\text{sign}(x_2(0))$ and $x_1(t_f) = x_1(0) + x_2(0)|x_2(0)| - x_2^2(0)\text{sign}(x_2(0))/2$ (recall that the minimization of the hamiltonian function yields $u_h(\lambda) = -\text{sign}(\lambda_2))$. Therefore the constraint on x_1 can safely be ignored if $-1 \leq x_1(0) + x_2(0)|x_2(0)| - x_2^2(0)\text{sign}(x_2(0))/2 \leq 1$, i.e., if the initial state belongs to the region I_0 of the $x_1 - x_2$ plane which is constituted by the two curves C_a and C_b (defined by the equations $x_1 = -1 - x_2|x_2|/2$ and $x_1 = 1 - x_2|x_2|/2$, respectively) and that part of the plane lying between them (see Fig. 6.13). Such a region is the union of the two subregions I_0^- and I_0^+ shown in the figure: if $x(0) \in I_0^-$, then $u = -1$, otherwise, if $x(0) \in I_0^+$, then $u = 1$. If the initial state does not belong to the region I_0 it is necessary to consider the problem with $x_1(t_f) = 1$ or $x_1(t_f) = -1$. In both cases the control must switch from the value 1 to the value -1 or vice versa. By taking into account the form of the trajectories corresponding to these values of the control (curves of equation $x_1 = \pm x_2^2/2 + k$ according to whether $u = \pm 1$) it is straightforward to conclude that when the initial state belongs to the region I_b located to the right of the curve C_b, the control is given by $u(t) = -1$, $0 \leq t < \tau$, $u(t) = 1$, $\tau < t \leq t_f$, τ being the time instant where the trajectory starting from the initial state and corresponding to $u = -1$ intersects the curve C_b. In fact, in view of the nature of the performance index and by noticing that the time which is required to cover any trajectory (relative to $u = \pm 1$) from a point where $x_2 = \alpha$ to a point where $x_2 = \beta$ is $|\alpha - \beta|$, there is no convenience in switching the control at the intersection with the curve C_a. Vice versa, when the

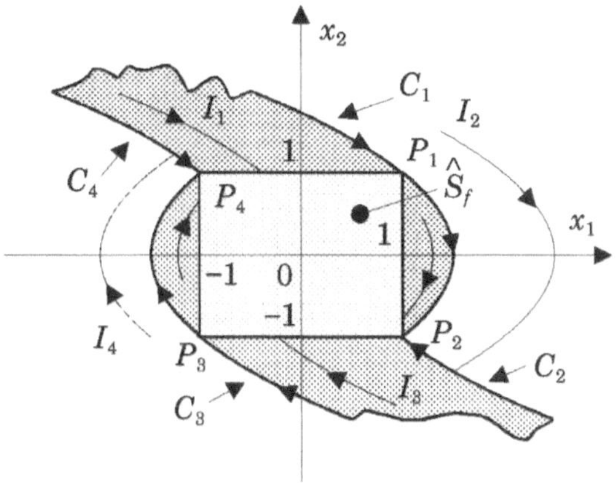

Figure 6.14: Example 6.22: partition of the state space into the four regions $I_i,\ i = 1, 2, 3, 4$.

initial state belongs to the region located to the left of the curve C_a, the control is given by $u(t) = 1,\ 0 \le t < \tau,\ u(t) = -1,\ \tau < t \le t_f$, τ being the time instant where the trajectory starting from the initial state and corresponding to $u = 1$ intersects the curve C_a. As before it can be concluded that there is no convenience in switching the control at the intersection with the curve C_b. Some trajectories consistent with the above discussion are reported in Fig. 6.13.

The procedure illustrated for the case where only one state variable must belong to a given interval can fairly trivially be generalized to the case where more than one component of the state must comply with requirements of such a kind. Indeed a combinatorial-type check of the various possibilities can conveniently be performed, as shown in the following example.

Example 6.22 Again consider the system described in Example 6.21 with the same performance index and final state $x(t_f) \in \hat{S}_f := \{x|\ -1 \le x_i \le 1,\ i = 1, 2\}$. As before, the problem is not trivial only if the given initial state does not lie inside $\hat{S}_f$. It is apparent that the situation where both the requirements on the final state are ignored should not be considered: thus, consistent with the above discussion, the cases to be dealt with refer to the constraints $x_2 = \pm 1,\ -1 \le x_1 \le 1$ and $x_1 = \pm 1,\ -1 \le x_2 \le 1$. It is simple to check that the control must be a piecewise constant function which takes on the values ± 1 and switches at most once between these values. By taking into account the form of the performance index, the shapes of the trajectories corresponding to such values of the control and the whole set of the NC, one can easily conclude that (see Fig. 6.14):

(a) If the initial state belongs to the region I_1 constituted by that part of the $x_1 - x_2$ plane which is delimited by the curves C_1 and C_4 (defined by the equations $x_1 = 1.5 - x_2^2/2,\ -1 \le x_2$ and $x_1 = -0.5 - x_2^2/2,\ 1 \le x_2$, respectively) and the portion $P_4 - P_1 - P_2$ of the boundary of $\hat{S}_f$, then $u = -1$;

(b) If the initial state belongs to the region I_3 constituted by that part of the $x_1 - x_2$ plane which is delimited by the curves C_2 and C_3 (defined by the equations $x_1 = 0.5 + x_2^2/2$, $x_2 \leq -1$ and $x_1 = -1.5 + x_2^2/2$, $x_2 \leq 1$, respectively) and the portion $P_2 - P_3 - P_4$ of the boundary of $\hat{S}_f$, then $u = 1$;

(c) If the initial state belongs to the region I_2 constituted by that part of the $x_1 - x_2$ plane which is located to the right of the curves C_1 and C_2, then $u(t) = -1$, $0 \leq t < \tau$, $u(t) = 1$, $\tau < t \leq t_f$, τ being the time instant where the trajectory starting at the initial state and corresponding to $u = -1$ intersects the curve C_2;

(d) If the initial state belongs to the region I_4 constituted by that part of the $x_1 - x_2$ plane which is located to the left of the curves C_3 and C_4, then $u(t) = 1$, $0 \leq t < \tau$, $u(t) = -1$, $\tau < t \leq t_f$, τ being the time instant where the trajectory starting at the initial state and corresponding to $u = 1$ intersects the curve C_4.

6.3.2 Integral constraints

Problems where integral constraints on functions of the state and/or control variables must be satisfied are now considered. First reference is made to the case of equality constraints which call for the relation

$$\int_{t_0}^{t_f} w_e(x(t), u(t), t)dt = \bar{w}_e \tag{6.31}$$

to be verified. Here w_e is an r_e-dimensional vector of functions which are continuous together with their first derivatives with respect to x and t, while $\bar{w}_e$ is a given vector. This constraint can be handled simply by adding a vector z of r_e new state variables to the original ones. These new variables are defined by the equations

$$\dot{z}(t) = w_e(x(t), u(t), t)$$
$$z(t_0) = 0$$

so that eq. (6.31) holds if $z(t_f) = \bar{w}_e$. Thus the given problem has been restated in terms of a problem relative to which the NC have already been presented.

Example 6.23 Consider the system described by the equations $\dot{x}_1 = x_2$, $\dot{x}_2 = u$ with $x(0) = 0$ and given final state $x_1(t_f) = 1$, $x_2(t_f) = 0$, t_f being free. The performance index is the elapsed time and the integral equality constraint

$$\int_0^{t_f} \frac{u^2}{2} dt = E$$

must be verified. Here $E > 0$ is a given constant representative of the control energy amount to be exploited. By resorting to the discussion above we find that the

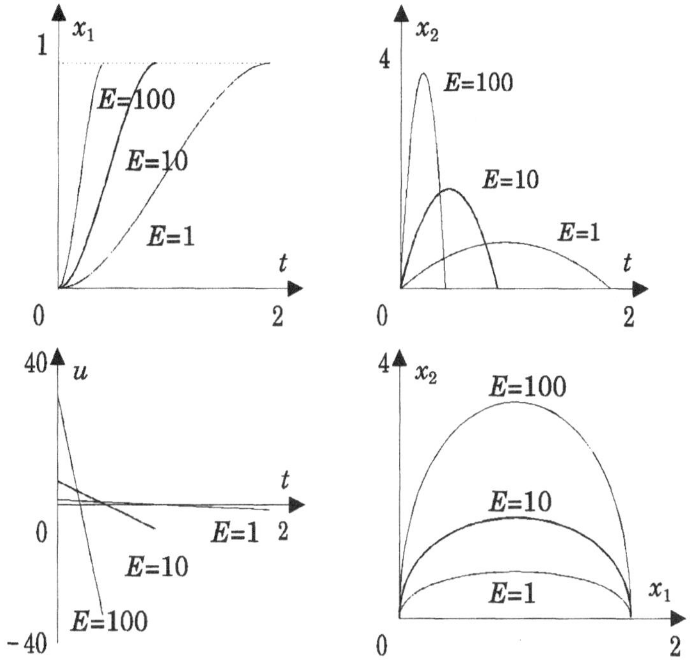

Figure 6.15: Example 6.23: state and control responses and state trajectories corresponding to some values of E.

hamiltonian function is $H = 1 + \lambda_1 x_2 + \lambda_2 u + \mu u^2/2$, once λ_0 has been set to 1. The variable μ is the multiplier associated to the new state variable z which has been introduced in order to account for the integral constraint. The solution of the auxiliary system is $\lambda_1(t) = \lambda_1(0)$, $\lambda_2(t) = \lambda_2(0) - \lambda_1(0)t$ and $\mu(t) = \mu(0)$, while the minimization of the hamiltonian function can be performed provided only that $\mu(0) > 0$, yielding $u_h = -\lambda_2/\mu(0)$. Thus $\mu(0)$ must *a posteriori* be checked to be positive. By enforcing the transversality condition and the control feasibility (in particular, $z(t_f) = E$), we find $t_f = \sqrt[3]{6/E}$, $\lambda_1(0) = -2/3t_f$, $\mu(0) = t_f^4/18$, $\lambda_2(0) = -t_f^2/3$. Note that $\mu(0) > 0$. The state and control responses and the state trajectories are shown in Fig. 6.15 corresponding to three values of E. Notice that faster transients and more demanding control actions result when the constant E is increased.

In a similar way it is possible to handle problems where constraints of the form

$$\int_{t_0}^{t_f} w_d(x(t), u(t), t)dt \leq \bar{w}_d \tag{6.32}$$

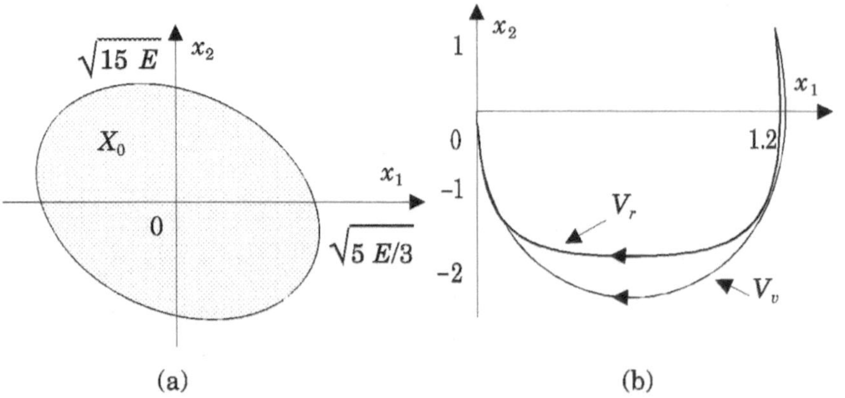

Figure 6.16: Example 6.24: (a) the set X_0; (b) state trajectories when the constraint is satisfied (V_s) and is not satisfied (V_v).

are present, w_d being an r_d-dimensional vector of functions which are continuous together with their first derivatives with respect to x and t, while $\bar{w}_d$ is a given vector. As before this constraint can be taken into account by adding a vector z of r_d new state variables to the original ones. These variables are defined by the equations

$$\dot{z}(t) = w_d(x(t), u(t), t),$$
$$z(t_0) = 0$$

so that eq. (6.32) is verified if $z(t_f) \leq \bar{w}_d$ and the given problem is restated as the problem discussed in Subsection 6.3.1.

Example 6.24 Consider the system presented in Example 6.23 with $x(0) = x_0$ and $x(1) = 0$. The performance index is

$$J = \int_0^1 \frac{u^2}{2} dt$$

and the constraint

$$\int_0^1 \frac{x_2^2}{2} dt \leq E, \; E > 0$$

has to be verified. Thus a unitary-mass-cart moving without friction along a straight rail has to be stopped in a given position by minimizing the involved control energy and requiring that its velocity be kept small. By proceeding as mentioned above, it is easy to conclude that the constraint can be ignored, provided that the initial state belongs to the region X_0 of the $x_1 - x_2$ plane bounded by the ellipse $2x_2^2 +$

$18x_1^2 + 3x_1x_2 = 30E$ (see Fig. 6.16 (a)). If the initial state does not belong to X_0, an extra state variable is introduced through the equations $\dot{z} = x_2^2/2$, $z(0) = 0$, $z(1) = E$. This new problem can be dealt with in the standard way. As an example, if $E = 1$, $x_1(0) = \sqrt{5/3}$, $x_2(0) = 1$, the system trajectories are shown in Fig. 6.16 (b), when the constraint is complied with or violated: correspondingly, the value of the performance index reduces from 88.81 to 19.75 and, for the second case, we find

$$\int_0^1 \frac{x_2^2}{2} dt = 2.39.$$

6.3.3 Global instantaneous equality constraints

Optimal control problems with global instantaneous equality constraints on some functions of the state and control variables are now considered. Thus, the equation

$$w(x(t), u(t), t) = 0, \ t_0 \le t \le t_f$$

comes up together with the usual problem statement: here the (vector-valued) function w and a suitable number (to be later specified) of derivatives with respect to x and t are continuous.

First assume that each component of w explicitly depends on u, so that these constraints may be thought of only as limiting the values which can be taken on by the control while minimizing the hamiltonian function. Consistent with this, these constraints may adequately be taken into account by inserting them into the hamiltonian function through suitable multipliers yielding

$$H^*(x, u, t, \lambda_0, \lambda, \mu) := H(x, u, t, \lambda_0, \lambda) + \mu' w(x, u, t)$$

and all previous results can be exploited with reference to H^*.

On the other hand, if one or more of the components of w do not explicitly depend on u, the situation above can still be recovered by noticing that the condition $w_i(x(t), t) = 0$, $t_0 \le t \le t_f$, implies that all total time derivatives of such a function must be identically zero. Assume that each state variable is either directly or indirectly affected by the control: then it must result, for some $q \ge 1$, that

$$\frac{d^q w_i(x(t), t)}{dt^q} := w_i^*(x(t), u(t), t)$$

since $\dot{x}$ is a function of u. By denoting with ν_i the smallest value of q corresponding to which the q-th derivative explicitly depends on u, the constraint $w_i = 0$ is equivalent to the $\nu_i + 1$ constraints

$$w_i^*(x(t), u(t), t) = 0, \ t_0 \le t \le t_f,$$
$$\left. \frac{d^j w_i(x(t), t)}{dt^j} \right|_{t=\tau} = 0, \ j = 0, 1, 2, \ldots, \nu_i - 1,$$

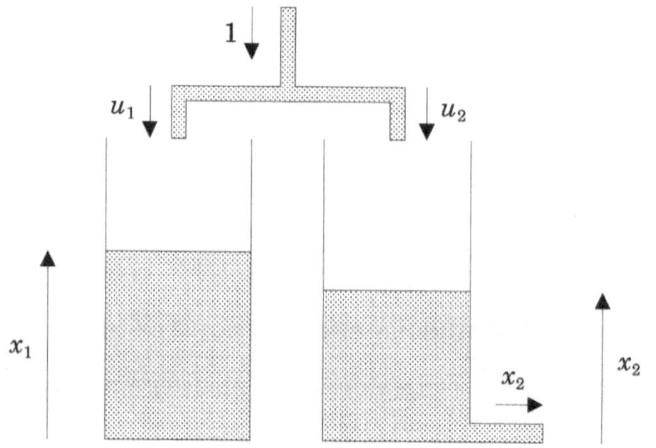

Figure 6.17: The system considered in Example 6.26.

where τ is any time instant belonging to the interval $[t_0, t_f]$. Note that the choice $\tau = t_0$ or $\tau = t_f$ might require us to suitably shrink the set $\bar{S}_0$ or $\bar{S}_f$, or even imply the infeasibility of the problem.

Example 6.25 Consider a system similar to the one described in Example 6.4, i.e., constituted by two carts with mass m_1 and m_2 which can move without friction along two straight and parallel rails. Two forces u_1 and u_2 act on the carts. By denoting with x_1 and x_2 the position and velocity of the first cart and with x_3 and x_4 the position and velocity of the second cart, the equations for this system are $\dot{x}_1 = x_2$, $m_1\dot{x}_2 = u_1$, $\dot{x}_3 = x_4$, $m_2\dot{x}_4 = u_2$. The initial state and time are given, namely $x(0) = \begin{bmatrix} 0 & 1 & 0 & 0 \end{bmatrix}'$, and the objective is to find the control actions which minimize the performance index

$$J = \frac{1}{2}\int_0^1 [u_1^2 + u_2^2]dt$$

while complying with the constraints $x_1(1) = 2$, $x_3(1) = 1$ and $w(x(t), t) = x_1(t) - x_3(t) - t = 0$, $0 \le t \le 1$. The function w does not explicitly depend on the control variable: its differentiation with respect to time leads to the further constraints $x_2(t) - x_4(t) - 1 = 0$ and $w^*(x(t), u(t), t) = a_1u_1 - a_2u_2 = 0$, where $a_i := 1/m_i$, $i = 1, 2$. Note that the first of these two new constraints is verified at the initial time so that no feasibility issues arise. By choosing $\lambda_0 = 1$ the modified hamiltonian function is

$$H^* = \frac{u_1^2 + u_2^2}{2} + \lambda_1 x_2 + \lambda_2 a_1 u_1 + \lambda_3 x_4 + \lambda_4 a_2 u_2 + \mu(a_1u_1 - a_2u_2)$$

and its minimization yields

$$u_h(\lambda) = \begin{bmatrix} -\dfrac{a_1 a_2^2}{a_1^2 + a_2^2}(\lambda_2 + \lambda_4) \\ -\dfrac{a_1^2 a_2}{a_1^2 + a_2^2}(\lambda_2 + \lambda_4) \end{bmatrix}$$

while the equations of the auxiliary system imply $\lambda_2(t)+\lambda_4(t) = (\lambda_2(0)+\lambda_4(0))(1-t)$. Here the orthogonality condition at the final time, $\lambda_2(1)+\lambda_4(1) = 0$, has been taken into account. Finally, by imposing the control feasibility we find $\lambda_2(0) + \lambda_4(0) = -3(a_1^2 + a_2^2)/(a_1 a_2)^2$.

Example 6.26 Consider the system shown in Fig. 6.17 constituted by two cylindrical tanks of equal area. An outgoing flow proportional to the liquid level exists in one of the two tanks. The two tanks are fed through a constant flow which can be subdivided at will between them. By denoting with x_1, x_2, u_1 and u_2 the levels and the flows, the equations for this system are $\dot{x}_1 = u_1$, $\dot{x}_2 = -x_2 + u_2$, $u_1 + u_2 = 1$, provided that the physical parameters have suitably been chosen. The problem is to drive the system in the shortest time from a given initial state $x(0) = x_0$ to a given final state $x(t_f) = x_f$. By choosing $\lambda_0 = 1$, the hamiltonian function is $H = 1+\lambda_1 u_1+\lambda_2(u_2-x_2)$. A solution satisfying the NC can be found by approaching the problem in a way which differs from the one outlined above, namely by noticing that the equality constraint can be taken into account by setting $u_2 = 1 - u_1$ with $0 \leq u_1 \leq 1$, so that

$$u_h = \begin{cases} \begin{bmatrix} 0 \\ 1 \end{bmatrix}, & \lambda_1 - \lambda_2 > 0, \\[2ex] \begin{bmatrix} 1 \\ 0 \end{bmatrix}, & \lambda_1 - \lambda_2 < 0. \end{cases}$$

The solutions of the auxiliary system are $\lambda_1(t) = \lambda_1(0)$, $\lambda_2(t) = \lambda_2(0)e^t$: thus the sign of $\lambda_1 - \lambda_2$ changes at most once (this quantity can not be identically zero since otherwise it would follow that $\lambda = 0$ and $\lambda_0 = 0$ because of the transversality condition). Obviously no feasible solutions exist if $x_{01} > x_{f1}$, while the analysis is trivial if $x_{01} = x_{f1}$. Therefore it is assumed that $\Delta := x_{f1} - x_{01} > 0$: this fact rules out the situation $u_1(\cdot) = 0$. The only controls satisfying the NC have one of the following forms:

(a) $u_1(t) = 1, \ 0 \leq t \leq t_f$,

(b) $u_1(t) = \begin{cases} 1 & 0 \leq t < \tau \\ 0 & \tau < t \leq t_f, \end{cases}$

(c) $u_1(t) = \begin{cases} 0 & 0 \leq t < \tau \\ 1 & \tau < t \leq t_f, \end{cases}$

where the values for τ and t_f are computed by enforcing feasibility. With reference to the three alternatives above we find

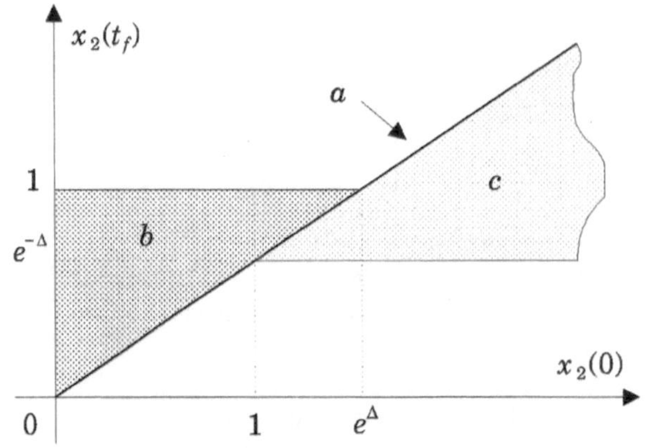

Figure 6.18: Example 6.26: the pairs (x_{02}, x_{f2}) corresponding to which a control satisfying the NC exists for a given Δ.

(a) $t_f = \Delta$,

(b) $t_f = \tau + \ln(\frac{x_{02}e^{-\Delta} - 1}{x_{f2} - 1})$, $\tau = \Delta$,

(c) $t_f = \tau + \Delta$, $\tau = \ln(\frac{x_{02} - 1}{x_{f2}e^{\Delta} - 1})$.

Since $0 \leq \tau \leq t_f$ the conclusion can be drawn that for a given Δ the only pairs (x_{02}, x_{f2}) which are consistent with the NC are those belonging to the shadowed regions of Fig. 6.18 (notice that the pairs $(\alpha, 1)$, $0 \leq \alpha < 1$ are not consistent). The comparison of the required final times allows us to specify the best form of the control for each region. Finally, the actual values can be determined by enforcing the transversality condition and the requirement $\lambda_2(\tau) = 0$.

6.3.4 Isolated equality constraints

The optimal control problems to be considered are characterized by instantaneous equality constraints on functions of the state variables which must be satisfied at isolated time instants. Thus, the customary problem statement is completed by expressions of the kind

$$w^{(i)}(x(t_i), t_i) = 0, \; t_0 < t_1 < \cdots < t_i < \cdots < t_s < t_f$$

where the s instants t_i may or may not be specified and the vector valued functions $w^{(i)}$ are continuous together with their first derivatives.

When these constraints are present the hamiltonian function as well as the functions λ may be discontinuous at the times t_i. More precisely, if x^o, u^o

are optimal and λ_0^o, λ^o are the corresponding solutions of the auxiliary system relative to which the NC are satisfied, then, for $1 \leq i \leq s$, it must happen that

$$\lim_{t \to t_i^-} \lambda^o(t) = \lim_{t \to t_i^+} \lambda^o(t) + \left. \frac{\partial w^{(i)}(x, t_i)}{\partial x} \right|'_{x = x^o(t_i)} \mu^{(i)}$$

where $\mu^{(i)}$ is a suitable vector. Moreover, if the time t_i is not given, it should also be

$$\lim_{t \to t_i^-} H(x^o(t), u^o(t), t, \lambda_0^o, \lambda^o(t)) = \lim_{t \to t_i^+} H(x^o(t), u^o(t), t, \lambda_0^o, \lambda^o(t)) - \left. \frac{\partial w^{(i)}(x^o(t_i), t)}{\partial t} \right|'_{t = t_i} \mu^{(i)}$$

for $1 \leq i \leq s$. Note that, according to the nature of the functions $w^{(i)}$ at hand, H and λ may or may not actually be discontinuous.

Example 6.27 Consider a *rendez-vous* problem which amounts to placing side-by-side two carts which move along two straight parallel rails. The motion of both carts is frictionless and the time of the *rendez-vous* is unspecified. A force u of arbitrary intensity acts on the first cart, while the other one is subject to a constant unitary force. At the initial time the first cart is in the reference position with zero velocity: the same situation has to be achieved at the unspecified final time. The second cart has initial velocity equal to 1 and its position is 1. The whole operation has to be accomplished in a reasonably short time while requiring control actions of limited amplitude. By denoting with x_1 and x_2 the position and velocity of the first cart, the problem is to find the control which steers the system $\dot{x}_1 = x_2$, $\dot{x}_2 = u$ from the initial state $x_1(0) = x_2(0) = 0$ to the final state $x_1(t_f) = x_2(t_f) = 0$ while complying with the constraint

$$w(x(t_1), t_1) = \begin{bmatrix} x_1(t_1) - 1 - t_1 - \dfrac{t_1^2}{2} \\ x_2(t_1) - 1 - t_1 \end{bmatrix} = 0$$

and minimizing the performance index

$$J = \int_0^{t_f} (1 + \frac{u^2}{2}) dt.$$

The hamiltonian function for the problem at hand is $H = 1 + u^2/2 + \lambda_1 x_2 + \lambda_2 u$ so that $u_h = -\lambda_2$ with $\dot{\lambda}_1 = 0$, $\dot{\lambda}_2 = -\lambda_1$. Due to the presence of the constraint, λ and H might be discontinuous at time t_1. If λ_1^+ denotes the constant value of λ_1 for $t > t_1$ and λ_2^- (λ_2^+) is the value approached by λ_2 when t tends to t_1 from the left (right), the equations which complete the set of the NC are

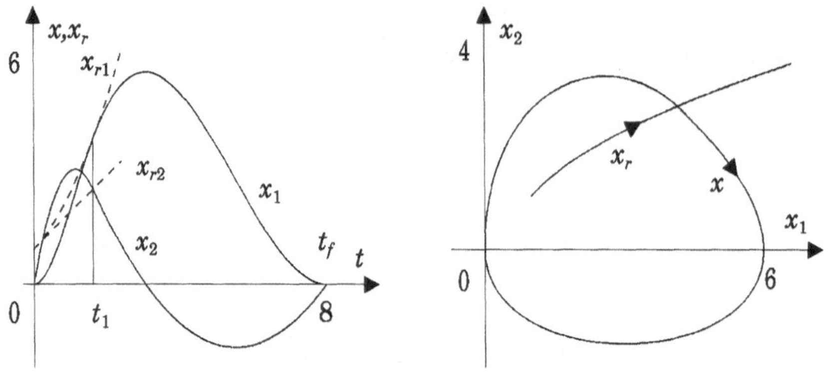

Figure 6.19: Example 6.27: state responses and trajectories.

$$\lambda_1(0) = \lambda_1^+ + \mu_1,$$
$$\lambda_2^- = \lambda_2^+ + \mu_2,$$
$$1 - \frac{(\lambda_2^-)^2}{2} + \lambda_1(0)x_2(t_1) = 1 - \frac{(\lambda_2^+)^2}{2} + \lambda_1^+ x_2(t_1) + (1+t_1)\mu_1 + \mu_2,$$
$$x_1(t_1) = 1 + t_1 + \frac{t_1^2}{2},$$
$$x_2(t_1) = 1 + t_1,$$
$$x_1(t_f) = 0,$$
$$x_2(t_f) = 0,$$
$$1 - \frac{\lambda_2^2(t_f)}{2} = 0.$$

This is a set of eight equations for the eight unknowns $\lambda_1(0)$, $\lambda_2(0)$, λ_1^+, λ_2^+, μ_1, μ_2, t_1, t_f (recall that $\lambda_2^- = \lambda_2(0) - \lambda_1(0)t_1$). The fourth equation together with the fifth one enforce the constraint, while the last equation is the transversality condition. We find $t_1 = 1.70$, $t_f = 8.23$, $\lambda_1(0) = -4.50$, $\lambda_2(0) = -5.42$, $\lambda_1^+ = 0.56$, $\lambda_2^+ = 2.24$. The time responses of x_1 and x_2 together with the position x_{r1} and velocity x_{r2} of the cart to be approached and the corresponding trajectories in the state space are shown in Fig. 6.19.

Example 6.28 Consider the cart described in Example 6.27 with the same initial and final conditions but let the control interval be given and equal to 2. The performance index is

$$J = \int_0^2 \frac{u^2}{2} dt$$

which has to be minimized by complying, every time, with one of the following constraints: (a) $x_1(1) = x_2(1) = 1$, (b) $x_1(1) = 1$, (c) $x_2(1) = 1$, (d) $x_1(t_1) = x_2(t_1) = 1$, (e) $x_1(t_1) = 1$, (f) $x_2(t_1) = 1$. In all cases the hamiltonian function is

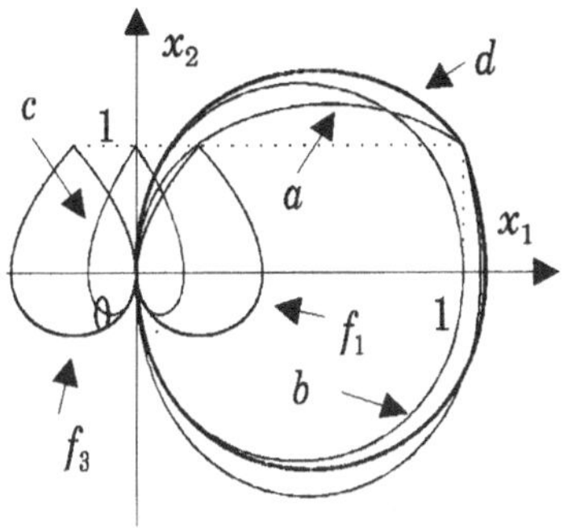

Figure 6.20: Example 6.28: state trajectories.

$H = u^2/2 + \lambda_1 x_2 + \lambda_2 u$ so that $u_h = -\lambda_2$, $\dot{\lambda}_1 = 0$ and $\dot{\lambda}_2 = -\lambda_1$. If α^+ (α^-) denotes the right (left) limit of the generic function α at a discontinuity point, the picture of the equations to be taken into consideration and the relevant conclusions are as follows.

Case (a)

$$\begin{aligned}
\lambda_1(0) &= \lambda_1^+ + \mu_1, \\
\lambda_2^- &= \lambda_2^+ + \mu_2, \\
x_1(1) &= 1, \\
x_2(1) &= 1, \\
x_1(2) &= 0, \\
x_2(2) &= 0.
\end{aligned}$$

We find $\lambda_1(0) = -6$, $\lambda_2(0) = -4$, $\lambda_1^+ = 18$, $\lambda_2^+ = 10$ and a value $J_a = 16$ for the performance index. Since the state is fully specified at a given time, the solution of the problem can be obtained by considering two completely independent subproblems with given initial and final state: the first one is defined over the interval $[0, 1]$ while the second one is defined over the interval $[1, 2]$. Actually, the above equations imply the independence of the values taken on by the vector λ before and after time 1.

Case (b)

$$\begin{aligned}
\lambda_1(0) &= \lambda_1^+ + \mu_1, \\
\lambda_2^- &= \lambda_2^+, \\
x_1(1) &= 1, \\
x_1(2) &= 0, \\
x_2(2) &= 0.
\end{aligned}$$

We find $\lambda_1(0) = -12$, $\lambda_2(0) = -6$, $\lambda_1^+ = 12$, and a value $J_b = 12$ for the performance index.

Case (c)

$$\begin{aligned} \lambda_1(0) &= \lambda_1^+, \\ \lambda_2^- &= \lambda_2^+ + \mu_2, \\ x_2(1) &= 1, \\ x_1(2) &= 0, \\ x_2(2) &= 0. \end{aligned}$$

We find $\lambda_1(0) = 6$, $\lambda_2(0) = 2$, $\lambda_2^+ = 4$ and a value $J_c = 4$ for the performance index.

Case (d)

$$\begin{aligned} \lambda_1(0) &= \lambda_1^+ + \mu_1, \\ \lambda_2^- &= \lambda_2^+ + \mu_2, \\ -\frac{(\lambda_2^-)^2}{2} + \lambda_1(0) &= -\frac{(\lambda_2^+)^2}{2} + \lambda_1^+, \\ x_1(t_1) &= 1, \\ x_2(t_1) &= 1, \\ x_1(2) &= 0, \\ x_2(2) &= 0. \end{aligned}$$

We find $\lambda_1(0) = -11.86$, $\lambda_2(0) = -6.16$, $\lambda_1^+ = 12.08$, $\lambda_2^+ = 7.88$, $t_1 = 0.84$ and a value $J_d = 14.03$ for the performance index.

Case (e)

$$\begin{aligned} \lambda_1(0) &= \lambda_1^+ + \mu_1, \\ \lambda_2^- &= \lambda_2^+, \\ -\frac{(\lambda_2^-)^2}{2} + \lambda_1(0)x_2(t_1) &= -\frac{(\lambda_2^+)^2}{2} + \lambda_1^+ x_2(t_1), \\ x_1(t_1) &= 1, \\ x_1(2) &= 0, \\ x_2(2) &= 0. \end{aligned}$$

The second and third equations imply that $\lambda_1(0) = \lambda_1^+$ or $x_2(t_1) = 0$. The first alternative should be discarded as it is not consistent with the other equations. Thus it follows that $\lambda_1(0) = -12$, $\lambda_2(0) = -6$, $\lambda_1^+ = 12$, $t_1 = 1$ and a value $J_e = 12$ for the performance index results. Due to the particular nature of the problem data this solution coincides with the solution of case (b).

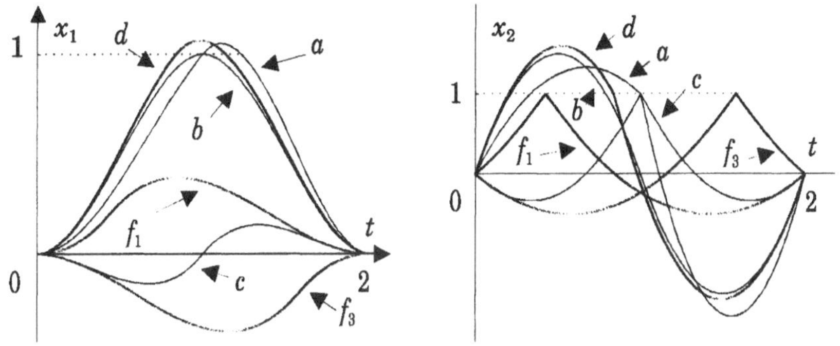

Figure 6.21: Example 6.28: state responses.

Case (f)

$$
\begin{aligned}
\lambda_1(0) &= \lambda_1^+, \\
\lambda_2^- &= \lambda_2^+ + \mu_2, \\
-\frac{(\lambda_2^-)^2}{2} + \lambda_1(0) &= -\frac{(\lambda_2^+)^2}{2} + \lambda_1^+, \\
x_2(t_1) &= 1, \\
x_1(2) &= 0, \\
x_2(2) &= 0.
\end{aligned}
$$

Three different solutions are found

$$
\begin{cases} \lambda_1(0) = 3 \\ \lambda_2(0) = -\sqrt{3} \\ \lambda_2^+ = 3 \\ t_1 = 1 - \sqrt{3}/3, \end{cases} \quad
\begin{cases} \lambda_1(0) = 6 \\ \lambda_2(0) = 2 \\ \lambda_2^+ = 4 \\ t_1 = 1, \end{cases} \quad
\begin{cases} \lambda_1(0) = 3 \\ \lambda_2(0) = \sqrt{3} \\ \lambda_2^+ = 3 \\ t_1 = 1 + \sqrt{3}/3, \end{cases}
$$

which yield the three values $J_f = 3$, $J_f = 4$, $J_f = 3$ for the performance index. Note that the second solution coincides with the solution of case (c) and that the controls relevant to the first and third solution are opposite in sign to each other.

The peculiarities of the considered problems may be appreciated by looking at Figs. 6.20, 6.21 where the state trajectories (covered in a clockwise sense) and responses are reported: the labels f_1 and f_3 denote the curves concerning the first and third solution of case (f). Finally, the values of the performance indices deserve some attention: as expected, their mutual relations are $J_a \geq J_b \geq J_c$, $J_a \geq J_d$, $J_b \geq J_e$, $J_c \geq \min\{J_f\}$, $J_d \geq J_e \geq \min\{J_f\}$.

6.3.5 Global instantaneous inequality constraints

The optimal control problems to be considered are characterized by global instantaneous inequality constraints on functions of the state and control variables. Thus, the customary problem statement is completed by expressions of the kind

$$w(x(t), u(t), t) \leq 0, \quad t_0 \leq t \leq t_f$$

where the vector valued function w and a suitable number (to be specified in the sequel) of derivatives with respect to x and t are continuous.

The way these problems are approached is similar to that presented in Subsection 6.3.3 and exploits the material in Subsection 6.3.4. First assume that each one of the r components of w explicitly depends on u, so that these constraints may be thought of as limiting only the values which can be taken on by the control while minimizing the hamiltonian function. Thus the constraints may adequately be taken into account by inserting them into the hamiltonian function itself through suitable multipliers yielding

$$H^*(x, u, t, \lambda_0, \lambda, \mu) := H(x, u, t, \lambda_0, \lambda) + \mu' w(x, u, t).$$

As is well known, the multipliers μ_i, $i = 1, 2, \ldots, r$ must satisfy the conditions

$$\mu_i \begin{cases} \geq 0, & w_i(x, u, t) = 0, \\ = 0, & w_i(x, u, t) < 0. \end{cases}$$

Then all previous results can be exploited with reference to H^*.

On the other hand, if one or more of the components of w do not explicitly depend on u, the situation above can still be recovered by noticing that if the condition $w_i(x(t), t) = 0$ holds over some (time) interval, then all total time derivatives of such a function must be identically zero over that interval. Assume that each state variable is either directly or indirectly affected by the control: then it must result, for some $q \geq 1$, that

$$\frac{d^q w_i(x(t), t)}{dt^q} := w_i^*(x(t), u(t), t)$$

since $\dot{x}$ is a function of u. By denoting with ν_i the smallest value of q corresponding to which the q-th derivative explicitly depends on u, the fact that $w_i(x(t), t) = 0$, $t_0 \leq t_{i,j} \leq t \leq t_{i,j+1} \leq t_f$ is equivalent to the $\nu_i + 1$ constraints

$$w_i^*(x(t), u(t), t) \leq 0 \qquad t_{i,j} \leq t \leq t_{i,j+1},$$

$$\left.\frac{d^h w_i(x(t), t)}{dt^h}\right|_{t=t_{i,j}} = 0 \qquad h = 0, 1, 2, \ldots \nu_i - 1.$$

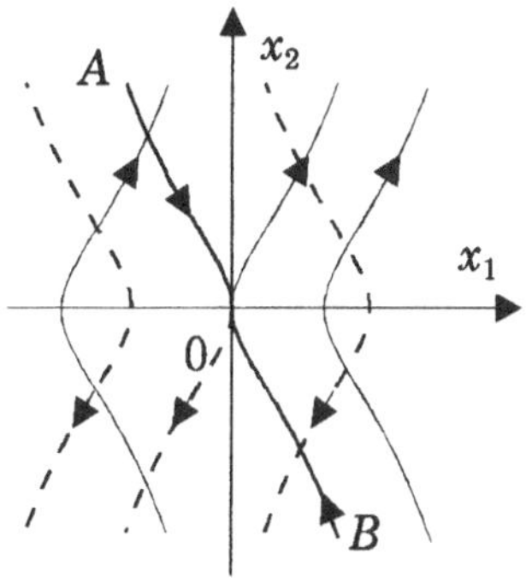

Figure 6.22: Example 6.29: state trajectories and switching curve $A-0-B$.

The first constraint allows the requirement $w_i \leq 0$ to be satisfied in a not binding way after the time $t_{i,j+1}$, while the remaining ν_i constraints may entail that λ and/or the hamiltonian function are discontinuous.

Example 6.29 Consider a system constituted by a unitary mass cart which moves without friction along a straight rail: it must be steered in the shortest time from a given (generic) initial situation $x(0) = x_0$ to $x(t_f) = 0$ while complying with the constraint $-(1 + x_2^2/2) \leq u \leq 1 + x_2^2/2$. Thus the problem refers to the system $\dot{x}_1 = x_2$, $\dot{x}_2 = u$ with given initial and final state subject to the constraint

$$w(x, u, t) = \begin{bmatrix} u - (1 + \frac{x_2^2}{2}) \\ -(u + 1 + \frac{x_2^2}{2}) \end{bmatrix} \leq 0.$$

The modified hamiltonian function is

$$H^* = 1 + \lambda_1 x_2 + \lambda_2 u + \mu_1(u - (1 + \frac{x_2^2}{2})) - \mu_2(u + 1 + \frac{x_2^2}{2})$$

while the nature of the constraints implies that at least one of the two multipliers is zero, so that

$$u_h = \begin{cases} 1 + \frac{x_2^2}{2}, & \lambda_2 < 0, \\ -(1 + \frac{x_2^2}{2}), & \lambda_2 > 0. \end{cases}$$

The state trajectories corresponding to the two ways the control is allowed to depend on x_2 are the curves described by the equation

$$x_1 = k - \text{sign}(\lambda_2) \ln(1 + \frac{x_2^2}{2}).$$

Some trajectories corresponding to various values of the constant k are reported in Fig. 6.22. It is apparent that the system can be driven from any initial state to the

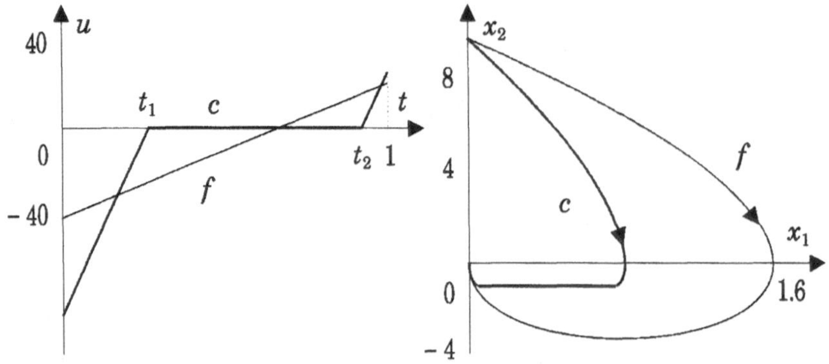

Figure 6.23: Example 6.30: control responses and state trajectories when the final time is given.

origin by switching the control at most once from $u = 1 + x_2^2/2$ to $u = -1 - x_2^2/2$, or vice versa: the switching takes place when the state reaches the curve $A - 0 - B$ which is therefore referred to as a *switching curve.* This behaviour is consistent with the choice $\mu_1 = \mu_2 = 0$ which entails $\lambda_2(t) = \lambda_2(0) - \lambda_1(0)t$.

Example 6.30 With reference to the problem described in Example 6.29, let $x_1(0) = 0$, $x_2(0) = 10$, $x(1) = 0$ and the performance index be

$$J = \int_0^1 \frac{u^2}{2} dt.$$

Furthermore the inequality constraint $x_2(t) \geq -1$ must be verified. The possibility is discussed that the constraint is binding over a time interval $\mathcal{T} := [t_1, t_2]$: since its total time derivative is $-u$, the problem must be dealt with where $x_2(t_1) = -1$ and the hamiltonian function is $H^* = u^2/2 + \lambda_1 x_2 + \lambda_2 u - \mu u$, if the choice $\lambda_0 = 1$ has been performed. Note that $u_h = -(\lambda_2 - \mu)$: thus it follows that $u_h = -\lambda_2$ outside the interval $\mathcal{T}$ where the constraint is binding, while $\lambda_2 - \mu = 0$ inside $\mathcal{T}$. By requiring that: (i) both λ_1 and the hamiltonian function be continuous at t_1 (the equality constraint does not explicitly depend on the first state component and time); (ii) the control is feasible (state constraints at t_1 and at the final time); (iii) both λ_2 and μ are continuous at t_2 (thus $\lambda_2(t_2) = 0$), the following equations are obtained:

$$\begin{aligned}
\lambda_1^+ &= \lambda_1(0), \\
0 &= \lambda_2(0) - \lambda_1(0)t_1, \\
-1 &= 10 - \lambda_2(0)t_1 + \frac{\lambda_1(0)t_1^2}{2}, \\
0 &= x_2(t_2) + \frac{\lambda_1^+(1-t_2)^2}{2},
\end{aligned}$$

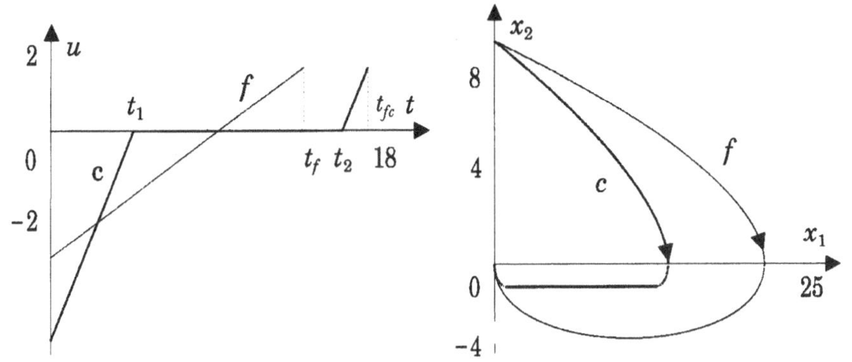

Figure 6.24: Example 6.30: control responses and state trajectories when the final time is not given.

$$0 = x_1(t_2) + x_2(t_2)(1 - t_2) + \frac{\lambda_1^+(1-t_2)^3}{6},$$
$$0 = \lambda_2^+ - \lambda_1^+(t_2 - t_1),$$

where $x(t_2)$ is evaluated according to the (known) form of the control function and λ^+ is the right limit of λ. The solution of these equations is $t_1 = 0.26$, $t_2 = 0.92$, $\lambda_1(0) = 312.21$, $\lambda_2(0) = 82.88$, $\lambda_2^+ = 204.35$. For the same problem a control which satisfies the NC when the inequality constraint is ignored is $u(t) = -40 + 60t$. For a significant comparison of the two situations the values of the performance indices are $J_c = 312.2$ and $J_f = 200$, respectively. The control responses and the state trajectories are reported in Fig.6.23 when the constraint is or is not taken into account (curves labelled with c and f, respectively).

Now assume that the control interval is free and the performance index is

$$J = \int_0^{t_f} (1 + \frac{u^2}{2})dt.$$

The procedure to be followed is truly similar to that presented relative to the previous version of the problem: thus only the transversality condition has to be added to the set of equations above, yielding

$$\lambda_1^+ = \lambda_1(0),$$
$$0 = \lambda_2(0) - \lambda_1(0)t_1,$$
$$-1 = 10 - \lambda_2(0)t_1 + \frac{\lambda_1(0)t_1^2}{2},$$
$$0 = x_2(t_2) + \frac{\lambda_1^+(t_f - t_2)^2}{2},$$
$$0 = x_1(t_2) + x_2(t_2)(t_f - t_2) + \frac{\lambda_1^+(t_f - t_2)^3}{6},$$

$$0 = \lambda_2^+ - \lambda_1^+(t_2 - t_1),$$
$$0 = 1 - \lambda_1^+.$$

Note that the last equation is the transversality condition written for a time which is interior to the interval where the constraint is binding, thus implying $u = 0$. The solution of these equations is $t_1 = 4.69$, $t_2 = 16.26$, $\lambda_1(0) = 1$, $\lambda_2(0) = 4.69$, $\lambda_2^+ = 11.56$, $t_f = t_{fc} = 17.66$. The control and final time satisfying the NC when the constraint is ignored are $u(t) = -2.83 + 0.3t$ and $t_f = t_{ff} = 14.14$: the performance indices corresponding to the two situations are $J_c = 35.34$ and $J_f = 28.28$, respectively. The control responses and the state trajectories are shown in Fig.6.24 when the constraint is or is not taken into account (curves labelled with c and f, respectively).

6.4 Singular arcs

In many optimal control problems the hamiltonian function is *linear* with respect to the control variable u (see the problems described in Examples 6.2, 6.8, 6.21, 6.22, 6.26, 6.29), i.e., it takes the form

$$H(x, u, t, \lambda_0, \lambda) = \alpha'(x, t, \lambda_0, \lambda)u + \beta(x, t, \lambda_0, \lambda). \tag{6.33}$$

In such cases it might happen that, corresponding to a given control u^*, there exist $\lambda_0^* \geq 0$ and a pair of functions (x^*, λ^*) which are solutions of the equations

$$\dot{x}(t) = f(x(t), u^*(t), t),$$
$$\dot{\lambda}(t) = -\left.\frac{\partial H(x, u^*(t), t, \lambda_0^*, \lambda(t))}{\partial x}\right|'_{x=x^*(t)}$$

and such that one or more of the components of α is zero when $t \in [t_1, t_2], t_1 < t_2$. In this time interval the state trajectory is referred to as a *singular arc* and the control components with index equal to the index of the zero components of α are called *singular components of the control.* Whenever the whole vector α is zero, the control is referred to as a *singular control.* When a singular arc exists within the interval $[t_1, t_2]$ the pair (x^*, u^*) is called a *singular solution* over that interval.

In order to be part of a solution of an optimal control problem a singular solution must satisfy the condition stated in the forthcoming theorem which requires suitable differentiability properties of the functions α and β: such properties are here assumed to hold. Moreover, $\alpha^{(i)}(x(t), t, \lambda_0, \lambda(t), u(t))$ denotes, for $i = 0, 1, 2, \ldots$, the total i-th time derivative of α where the available expression of $\dot{x}$ and $\dot{\lambda}$ has been exploited. As an example,

$$\alpha^{(1)}(x(t),t,\lambda_0,\lambda(t),u) = \left.\frac{\partial\alpha(x,t,\lambda_0,\lambda)}{\partial t}\right|_{x=x(t),\ \lambda=\lambda(t)} + \left.\frac{\partial\alpha(x,t,\lambda_0,\lambda(t))}{\partial x}\right|_{x=x(t)} f(x(t),u,t) - \left.\frac{\partial\alpha(x(t),t,\lambda_0,\lambda)}{\partial\lambda}\right|_{\lambda=\lambda(t)} \left.\frac{\partial H(x,u,t,\lambda_0,\lambda(t))}{\partial x}\right|_{x=x(t)}^{\prime}$$

Theorem 6.3 *Let the hamiltonian function corresponding to a given problem be of the form (6.33) and $(u^o(\cdot),x^o(\cdot))$ an optimal pair. Furthermore, let $\lambda_0^o \geq 0$ and $\lambda^o(\cdot)$ be a solution of the auxiliary system whose existence is guaranteed by the Maximum Principle. A necessary condition for the pair $(u^o(\cdot),x^o(\cdot))$ to include a singular solution over the interval $[t_1,t_2]$ is that*

(i) $$\frac{d^r\alpha(x^o(t),t,\lambda_0^o,\lambda^o(t))}{dt^r} = 0,\ r=0,1,2,\ldots,\ t\in[t_1,t_2],$$

(ii) $$\frac{\partial\alpha^{(p)}(x^o(t),t,\lambda_0^o,\lambda^o(t),u)}{\partial u} = 0,\ p=1,3,5,\ldots,\ t\in[t_1,t_2],$$

(iii) $$(-1)^q\frac{\partial\alpha^{(2q)}(x^o(t),t,\lambda_0^o,\lambda^o(t),u)}{\partial u} \geq 0,\ q=1,2,3,\ldots,\ t\in[t_1,t_2].$$

If $\bar{q}$ is the smallest value of q corresponding to which condition (iii) holds with the strict inequality sign, the check of the conditions above has to be stopped at $\bar{r}=2\bar{q}$, $\bar{p}=2\bar{q}-1$, $\bar{q}$.

Example 6.31 Consider the system $\dot{x}_1 = x_2$, $\dot{x}_2 = u$. The state of this system has to be driven in a short time from a given initial value x_0 to the final value $x(t_f)=0$ while keeping the system velocity small and avoiding positive positions. For these requests a convenient performance index could be

$$J = \int_0^{t_f}(1+\frac{x_2^2}{2}+x_1)dt$$

to be minimized by a suitable choice of the control which must comply with the constraint $-2 \leq u(t) \leq 2$, $0 \leq t \leq t_f$. By setting $\lambda_0 = 1$, the hamiltonian function is $H = 1 + x_2^2/2 + x_1 + \lambda_1 x_2 + \lambda_2 u$, so that $u_h = -2\text{sign}(\lambda_2)$ if $\lambda_2 \neq 0$ and $\dot{\lambda}_1 = -1$, $\dot{\lambda}_2 = -x_2 - \lambda_1$. A singular control might exist if $\lambda_2 = 0$, $\lambda_1 = -x_2$, $u = 1$ and $x_1 = x_2^2/2 - 1$, the last equation being a consequence of the transversality condition. In view of these results the variable u can take on only the values ∓ 2 yielding the trajectories $x_1 = \mp x_2^2/4 + k$ or the value 1 yielding the trajectory $x_1 = x_2^2/2 - 1$ which constitutes a singular arc (heavy curve in Fig. 6.25). Thus the origin can be reached only if $u = \pm 2$ and the consistency of the trajectories $P_3^{(1)} - P_2 - P_1 - 0$,

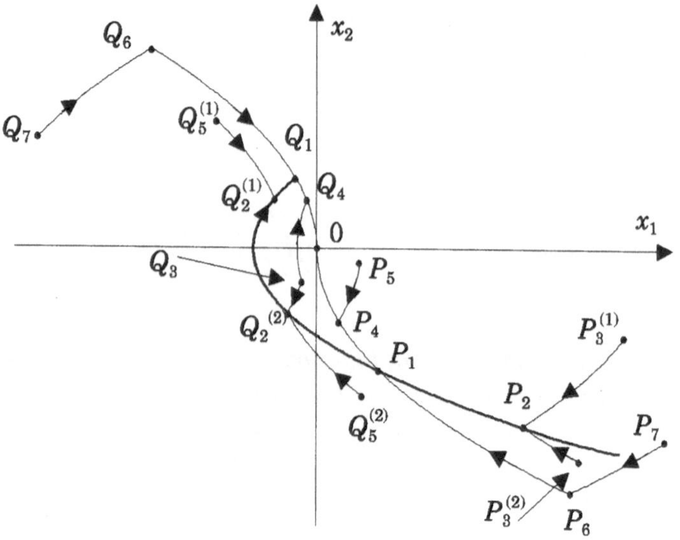

Figure 6.25: Example 6.31: state trajectories.

$P_3^{(2)} - P_2 - P_1 - 0$, $P_5 - P_4 - 0$, $Q_5^{(1)} - Q_2^{(1)} - Q_1 - 0$, $Q_5^{(2)} - Q_2^{(2)} - Q_1 - 0$, $Q_3 - Q_2^{(2)} - Q_1 - 0$, $Q_3 - Q_4 - 0$, $P_7 - P_6 - P_1 - 0$ and $Q_7 - Q_6 - Q_1 - 0$ with the NC has to be investigated (see Fig. 6.25). This analysis can easily be performed if the new time $\tau := t_f - t$ is adopted and the various involved functions of τ are marked with the superscript *.

Trajectories $0 - P_1 - P_2 - P_3^{(i)}$: The point P_1, with coordinates $(1, -2)$, is reached at time $\tau_1 = 1$ since $x_2^*(\tau) = -2\tau$. Thus the trajectory at hand satisfies the NC only if $\lambda_2^*(0) = -1/2$ (transversality condition), $0 = \lambda_2^*(\tau_1) = -1 + \lambda_1^*(0)$ so that $\lambda_1^*(0) = 1$. Let τ_2 be the time when the point P_2 is reached: it follows that $d\lambda_2^*(\tau)/d\tau = 3(\tau - \tau_2) \geq 0$, if the path continues towards $P_3^{(1)}$, while $d\lambda_2^*(\tau)/d\tau = -(\tau - \tau_2) \leq 0$, if the path continues towards $P_3^{(2)}$. These conclusions have been drawn by taking into account that $x_2^*(\tau) = x_2^*(\tau_2) \pm 2(\tau - \tau_2)$ in the two cases, with $x_2^*(\tau_2) = -(1 + \tau_2)$. Therefore the control does not further switch once the path towards either of the two points $P_3^{(i)}$ has been chosen.

Trajectory $0 - P_4 - P_5$: If τ_4 is the time when the point P_4 is reached, then it follows that $\lambda_2^*(\tau_4) = 0$ and $\lambda_1^*(0) = \tau_4/2 + 1/(2\tau_4)$, since $\lambda_2^*(0) = -1/2$. Thus $d\lambda_2^*(\tau)/d\tau > 0$, if the path continues towards the point P_5 and no further control switches can occur.

Trajectories $Q_5^{(i)} - Q_2^{(i)} - Q_1 - 0$: A discussion equal to the one carried on when considering the trajectories $0 - P_1 - P_2 - P_3^{(i)}$ leads to the conclusion that they are consistent with the NC.

Trajectories $Q_3 - Q_2^{(2)} - Q_1 - 0$ and $Q_3 - Q_4 - 0$: A discussion of the same kind as above allows us to conclude that the origin can be reached when starting from the point Q_3 by following two different trajectories which satisfy the NC.

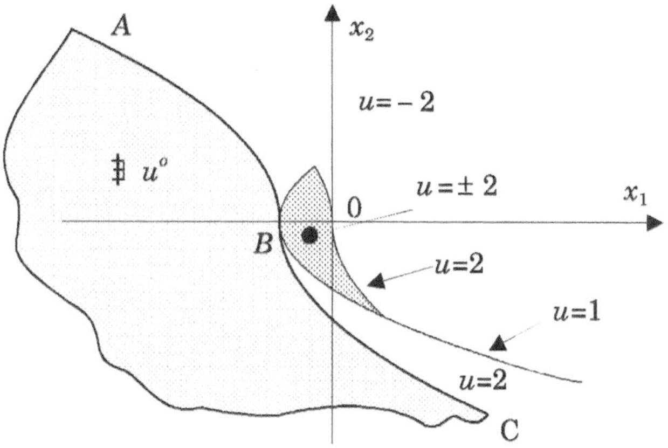

Figure 6.26: Example 6.31: values to be given to the control variable.

Trajectory $P_7 - P_6 - P_1 - 0$: This trajectory does not satisfy the NC. In fact, letting τ_6 be the time when the point P_6 is reached, it is necessary that $\lambda_2^*(\tau_6) = -1/2 + \lambda_1^*(0)\tau_6 - \tau_6^2/2 = 0$: this equation supplies values of τ_6 greater than 1 only if $\lambda_1^*(0) > 1$. However, in this case λ_2^* vanishes also for $\tau = \tau^* < 1$: at this time $d\lambda_2^*(\tau)/d\tau > 0$ so that a control switch takes place.

Trajectory $Q_7 - Q_6 - Q_1 - 0$: This case is similar to the preceding one and the same conclusion holds, namely that the trajectory does not satisfy the NC.

The analysis above is conveniently summarized in Fig. 6.26 where the state space is subdivided into regions which specify the control complying with the NC (if it exists). Note that two choices for u can be made in a region, while the NC cannot be satisfied on the left of the curve $A - B - C$ defined by the equation $x_1 = -x_2|x_2|/4 - 1$: therefore, problems where the initial state belongs to this last region do not admit an optimal solution.

Example 6.32 Consider the system shown in Fig. 6.27 which is constituted by three cylindrical tanks, the first and second ones having a unitary section, while the section of the third one is A. An outgoing flow kx_1 proportional to the liquid level originates from the first tank which is fed through an incoming flow u_1, the value of which can freely be selected between 0 and $\nu > 0$. The outgoing flow can arbitrarily be subdivided into two flows u_2 and u_3 which feed the remaining two tanks where the liquid level is denoted by x_2 and x_3, respectively. Thus the system is described by the equations $\dot{x}_1 = u_1 - kx_1$, $\dot{x}_2 = u_2$, $\dot{x}_3 = u_3/A$, $kx_1 - u_2 - u_3 = 0$, $0 \leq u_1 \leq \nu$, $u_2 \geq 0$, $u_3 \geq 0$, and the optimal control problem is to steer the system in the shortest possible time from the given initial state x_0 to the situation $x_2(t_f) = x_{2f} \geq x_{20}$, $x_3(t_f) = x_{3f} \geq x_{30}$. No requirements are set on the first tank level. By letting $\lambda_0 = 1$ and handling in the obvious way the equality constraint on the control variables,

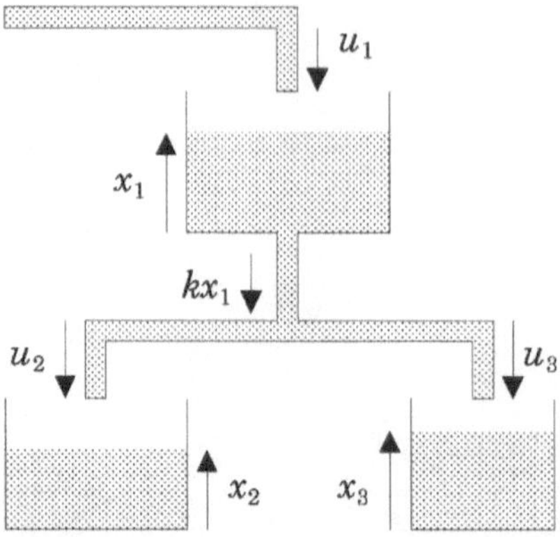

Figure 6.27: The system considered in Example 6.32.

the hamiltonian function is $H = 1 + \lambda_1(u_1 - kx_1) + \lambda_2 u_2 + \lambda_3(kx_1 - u_2)/A$. As a consequence we get

$$u_{h1} = \begin{cases} 0, & \lambda_1 > 0 \\ \nu, & \lambda_1 < 0, \end{cases}$$

$$u_{h2} = \begin{cases} 0, & \lambda_2 - \lambda_3/A > 0 \\ kx_1, & \lambda_2 - \lambda_3/A < 0, \end{cases}$$

and $\lambda_2(t) = \lambda_2(0)$, $\lambda_3(t) = \lambda_3(0)$,

$$\lambda_1(t) = \frac{\lambda_3(0)}{A}(1 - e^{k(t-t_f)})$$

where the orthogonality condition $\lambda_1(t_f) = 0$ has already been imposed. Note that $\lambda_3(0) \neq 0$, since otherwise the transversality condition ($1 + \lambda_2 u_2 = 0$, in such a case) could not be satisfied because of the form of u_2 and the constancy of λ_2. Thus the sign of λ_1 does not vary and u_1 is constant with value either 0 or ν. Since λ_2 and λ_3 are constant as well, the conclusion can be drawn that if $\lambda_2(0) - \lambda_3(0)/A$ is not zero, i.e., if no component of the control is singular, either $u_2 = 0$ or $u_3 = 0$, which is not a feasible choice whenever $\Delta_2 := x_{2f} - x_{20} > 0$ and $\Delta_3 := x_{3f} - x_{30} > 0$. By assuming that these inequalities both hold, the value of u_1 consistent with the NC is searched. Incidentally, note that conditions (*i*)–(*iii*) of Theorem 6.3 hold with the equality sign for all r, p, q if $\lambda_2(0) - \lambda_3(0)/A = 0$.

If $u_1(\cdot) = 0$ it follows that $\lambda_3(0) > 0$ (recall the expression for λ_1). Feasibility requires that

$$\Delta_2 = \int_0^{t_f} u_2 dt,$$

$$\Delta_3 = \frac{1}{A}\left[\int_0^{t_f} kx_1 dt - \Delta_2\right],$$

and

$$1 - e^{-kt_f} = \frac{A\Delta_3 + \Delta_2}{x_1(0)},$$

since $x_1(t) = x_1(0)e^{-kt}$: thus the transversality condition, set at $t = 0$, implies

$$1 + \frac{k\lambda_3(0)}{A}(x_1(0) - A\Delta_3 - \Delta_2) = 0.$$

This fact calls for $\lambda_3(0)$ to be negative since the term between brackets is positive because of the previous equation where the left term ranges between 0 and 1. Thus a contradiction results.

If $u_1(\cdot) = \nu$ it follows that $\lambda_3(0) < 0$, while feasibility calls for

$$A\Delta_3 + \Delta_2 = (x_1(0) - \frac{\nu}{k})(1 - e^{-kt_f}) + \nu t_f$$

since $x_1(t) = [kx_1(0) - \nu]e^{-kt} + \nu$ and t_f can be determined. The transversality condition set at $t = 0$, requires, in view of the expression for $\lambda_1(0)$,

$$\lambda_3(0) = -\frac{A}{kx_1(0)e^{-kt_f} + \nu(1 - e^{-kt_f})}.$$

Since the denominator is positive, it follows that $\lambda_3(0) < 0$ as required. As for the response of u_2, it is not uniquely determined by imposing the NC: among the infinite allowable choices the simplest one is setting it at the constant value $u_2 = \Delta_2/t_f$.

6.5 Time optimal control

The so-called *minimum time* problems are now considered. They deserve particular attention in view of both the significance of the available theoretical results and the number and importance of applications.

Basically, a minimum time problem consists in steering the system in the shortest time from a suitable point of a given set $\bar{S}_0$ of the allowable initial states to a suitable point of a given set $\bar{S}_f$ of the allowable final states. Of course these two sets are disjoint. In the subsequent discussion the system under control is described by linear time-invariant equations, the initial state is given and the final state belongs to a regular variety $\bar{S}_f$, namely

$$\begin{aligned} \dot{x}(t) &= Ax(t) + Bu(t), \\ x(0) &= x_0, \\ x(t_f) &\in \bar{S}_f. \end{aligned}$$

The performance index is simply the elapsed time and has to be minimized by selecting the control inside the set $\bar{\Omega}$ of piecewise continuous functions which take on values in U. The set U is a *closed, convex, bounded polyhedron,* i.e., a

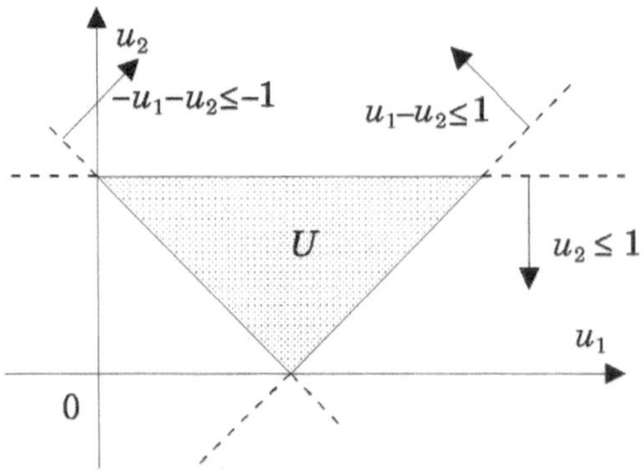

Figure 6.28: A closed, convex, bounded polyhedron U.

set resulting from the intersection of a finite number s of closed half planes: thus it can be defined by the equation (see also Fig. 6.28)

$$U = \{u |\ Su \leq \alpha,\ \alpha \in R^s\}. \tag{6.34}$$

The forthcoming results require that the matrices A, B and the set U satisfy a crucial assumption.

Assumption 6.1 *Given an arbitrary vector $v \neq 0$ aligned with any one of the edges of U, the set of vectors Bv, ABv, A^2Bv,..., $A^{n-1}Bv$ is linearly independent.*

For the considered class of problems the hamiltonian function is $H = \lambda_0 + \lambda'(Ax + Bu)$ so that the auxiliary system is $\dot{\lambda} = -A'\lambda$, the general solution of which has the form $\lambda(t) = e^{-A't}\lambda(0)$. If u^o is an optimal control there should exist a $\lambda^o(0) \neq 0$ corresponding to which the quantity $\lambda^{o\prime}(0)e^{-At}Bu$ is minimized by u^o ($\lambda^o(0)$ must not be zero since otherwise $\lambda_0^o = 0$ as well, because of the transversality condition). Assumption 6.1 guarantees that, corresponding to *any* solution $\lambda^*(\cdot) \neq 0$ of the auxiliary system, the minimization of the hamiltonian function yields a unique control $u^*(\cdot) \in \bar{\Omega}$, referred to as *extremal*, which is such that

$$\lambda^{*\prime}(t)Bu^*(t) = \min_{u \in U} \lambda^{*\prime}(t)Bu,\ 0 \leq t \leq t_f.$$

More precisely, the following result can be proved.

Theorem 6.4 *Let Assumption 6.1 hold. Then, for any nonzero solution of the auxiliary system there exists a unique control which minimizes the hamiltonian*

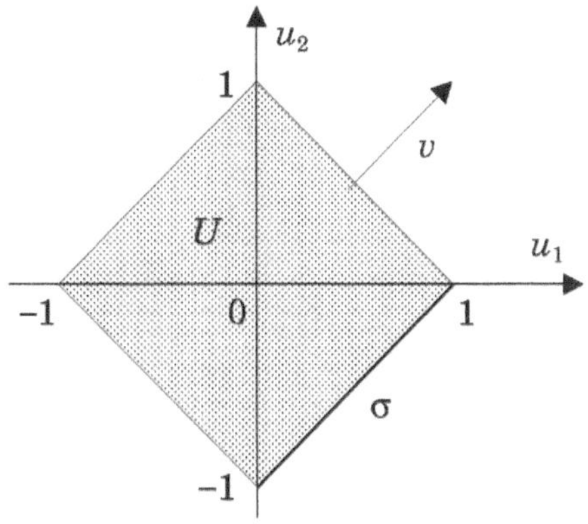

Figure 6.29: The polyhedron U of Example 6.33.

function. It is piecewise constant and takes on values corresponding to the vertices of U.

This theorem is particularly meaningful since it specifies the nature of any optimal control. As far as uniqueness of the extremal control is concerned, it has to be meant to hold almost everywhere, i.e., apart from those instants (*switching times*) where a discontinuity occurs since u *jumps* from a vertex of the polyhedron U to another.

The following simple example sheds light on the role played by Assumption 6.1.

Example 6.33 Let

$$A = \begin{bmatrix} 1 & -1 & 0 \\ 0 & 0 & 1 \\ 0 & 0 & 0 \end{bmatrix}, \quad B = \begin{bmatrix} 0 & 1 \\ 0 & 0 \\ 1 & 0 \end{bmatrix}, \quad S = \begin{bmatrix} 1 & 1 \\ -1 & 1 \\ -1 & -1 \\ 1 & -1 \end{bmatrix}, \quad \alpha = \begin{bmatrix} 1 \\ 1 \\ 1 \\ 1 \end{bmatrix}$$

and $v = \begin{bmatrix} 1 & 1 \end{bmatrix}'$, which is aligned with the edge σ of U (see Fig. 6.29). It is easy to check that Assumption 6.1 is not verified. Consistent wth this, the general solution of the auxiliary system is

$$\lambda(t) = \begin{bmatrix} \lambda_1(0)e^{-t} \\ \lambda_2(0) + \lambda_1(0)(1 - e^{-t}) \\ \lambda_3(0) + \lambda_1(0)(1 - t - e^{-t}) - \lambda_2(0)t \end{bmatrix}$$

so that

$$\lambda'(t)Bu = [\lambda_3(0) + \lambda_1(0)(1 - t - e^{-t}) - \lambda_2(0)t]u_1 + \lambda_1(0)e^{-t}u_2.$$

By choosing $\lambda_2(0) = \lambda_3(0) = -\lambda_1(0)$, $\lambda_1(0) > 0$, we get $\lambda'(t)Bu = \lambda_1(0)e^{-t}(u_2 - u_1)$ and the hamiltonian function is minimized by any pair (u_1, u_2) belonging to the edge σ.

The number of switching times of an optimal control is anyhow *finite* and generally depends on A, B, x_0, $\bar{S}_f$ and U. However, for given A, B, U, it might be *unbounded* as x_0 and/or $\bar{S}_f$ vary. Corresponding to particular subclasses of problems it is possible to set an upper bound to such a number, as stated in the following theorem.

Theorem 6.5 *Let Assumption 6.1 hold and P be a parallelepiped defined by*

$$P = \{u|\ a_i \le u_i \le b_i,\ i = 1, 2, \ldots, m\}$$

Then, if $U = P$ and all eigenvalues of A are real, each component of an extremal control commutes at most $n - 1$ times.

When the set of allowable final states shrinks to a single point x_f it is possible to claim that if a solution exists it is unique.

Theorem 6.6 *Let Assumption 6.1 hold and $\bar{S}_f = \{x_f\}$. Then the optimal control, if it exists, is unique.*

This theorem (which is *not* an *existence* theorem) states that if $\bar{u}$ and u^* are two optimal controls defined over the intervals $[0, \bar{t}_f]$ and $[0, t_f^*]$, respectively, which transfer the state of the system from x_0 to x_f, then they do coincide, i.e., $\bar{t}_f = t_f^*$ and $\bar{u}(t) = u^*(t)$ for (almost) all t.

Obviously, an optimal control has to be sought within the set of those extremal controls which are also *feasible*, that is which drive the state of the system from x_0 to $\bar{S}_f$. In some particular cases the feasible extremal controls are unique.

Theorem 6.7 *Let Assumption 6.1 hold and $\bar{S}_f = \{x_f\} = 0$. Moreover, let the origin of R^m be an interior point of U. Then there exists at most one extremal feasible control.*

This theorem (which is *not* an *existence* theorem) states that if $\bar{u}$ and u^* are two extremal controls defined over the intervals $[0, \bar{t}_f]$ and $[0, t_f^*]$, respectively, which transfer the state of the system from x_0 to 0, then they do coincide, that is $\bar{t}_f = t_f^*$ and $\bar{u}(t) = u^*(t)$, for (almost) all t. This fact implies that if an extremal feasible control has been found, then it is the optimal control, obviously provided that an optimal control exists.

As for the existence of an optimal control, two significant results are available.

Theorem 6.8 *Let Assumption 6.1 hold. If a feasible control exists, then an optimal control exists too.*

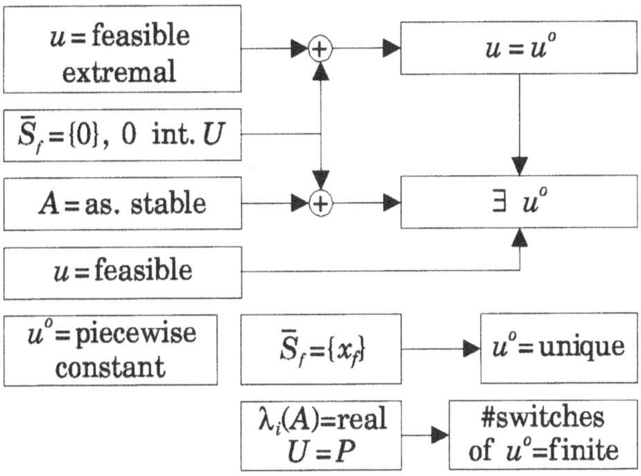

Figure 6.30: Summary of the results for time optimal control problems.

Theorem 6.9 *Let Assumption 6.1 hold, $\bar{S}_f = \{x_f\} = 0$, the origin of R^m be an interior point of U and the real parts of all the eigenvalues of A be negative. Then the optimal control exists for each initial state.*

The results concerning time optimal problems are summarized in Fig. 6.30.

Example 6.34 Consider a simple positioning problem defined by the system $\dot{x}_1 = x_2$, $\dot{x}_2 = u$ with given initial and final state, $x(0) = x_0$ and $x(t_f) = x_f$, respectively. The control variable is constrained to belong to the set $U = \{u|\ u_m \le u \le u_M,\ u_m u_M < 0\}$. Assumption 6.1 holds and Theorem 6.5 can be applied so that each extremal control (hence the optimal control, too) may commute at most once. Indeed, the nonzero solutions of the auxiliary system are $\lambda_1(t) = \lambda_1(0)$, $\lambda_2(t) = \lambda_2(0) - \lambda_1(0)t$: thus the sign of λ_2 may change at most once. Moreover λ_2 can be zero at an isolated time only since, otherwise, $\lambda_1(0) = 0$, $\lambda_2(0) = 0$ and $\lambda_0 = 0$ as well, because of the transversality condition and the NC would be violated. By noticing that

$$u_h = \begin{cases} u_m, & \lambda_2 > 0 \\ u_M, & \lambda_2 < 0 \end{cases}$$

it is easy to check that the control can switch at most once. The trajectories of the system corresponding to the two allowed values for the control are shown in Fig. 6.31: they are the parabolas defined by the equation $x_1 = k + x_2^2/\alpha$ where $\alpha = 2u_m$ or $\alpha = 2u_M$. In the first case the trajectories are covered in the sense of decreasing x_2 while in the second case they are covered in the opposite sense. It is apparent (see Fig. 6.31) that for each pair (x_0, x_f) there exists a feasible extremal control. Therefore also the optimal control exists. Now consider Fig. 6.32 and let the final state x_f^+ be such that $x_{f2}^+ > 0$. If the initial state does not belong to the region D^+ delimited by the line $A^+ - x_f^+ - P^+$ (made out of pieces of the trajectories passing

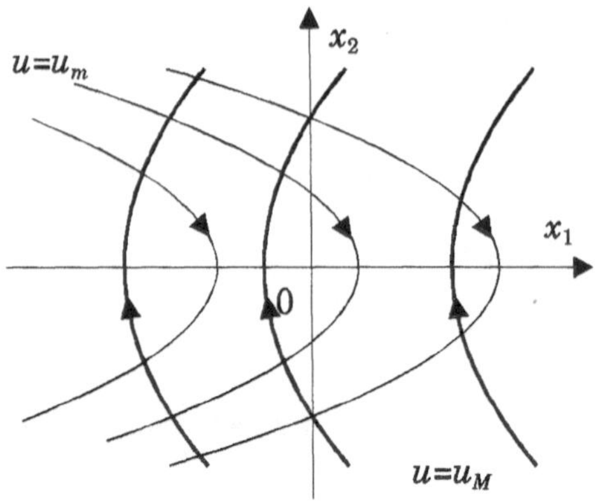

Figure 6.31: Example 6.34: the trajectories corresponding to $u = u_M$ (heavy line) and to $u = u_m$ (light line).

through x_f^+ and corresponding to $u = u_m$ and $u = u_M$) and the line σ (defined by the equation $x_1 = x_{M1} - x_2^2/(2u_m)$, $x_{M1} = x_{f1}^+ - (x_{f2}^+)^2/(2u_M)$), then there exists only one feasible extremal control. Otherwise, two extremal controls exist (in the quoted figure refer to the initial state d): the extremal control which is also optimal is the one which requires passing through the point i_1 (the trajectory $d - i_1 - x_f^+$ is covered in a shorter time since in each point the controlled object possesses a greater velocity than while covering the trajectory $d_i - 2 - x_f^+$). The line $A^+ - x_f^+ - P^+ - B^+$ is the locus where the optimal control switches from u_M to u_m or vice versa. It is appropriately referred to as a *switching curve* as it allows one to implement the solution in a closed loop form: the value to be given to the control depends only on the system state ($u = u_m$ to the right of the curve, $u = u_M$ to the left of the curve). A similar discussion applies when the final state x_f^- is such that $x_{f2}^- < 0$ or its second component is zero (see Fig. 6.32).

Finally, consider the particular case when $x(t_f) = 0$. Thanks to Theorem 6.7, there exists a unique feasible extremal control: this fact can easily be checked by looking at the trajectories corresponding to the two extreme values the control can take on. The form of the switching curve is the same as the previous case and its equation is $x_1 = x_2^2/(2u_m)$, $x_2 \geq 0$, $x_1 = x_2^2/(2u_M)$, $x_2 \leq 0$. As before, the switching curve partitions the state space into two subregions where the control is $u = u_m$ (region to the right of the curve) or $u = u_M$ (region to the left of the curve).

Example 6.35 Consider a problem similar to the one presented in Example 6.34. The initial state is zero ($x(0) = 0$) while the final state must belong to the set $\bar{S}_f = \{x | \; x_1 = 0, \; x_2^2 = 1\}$: thus the system has to be drawn back to the initial position but with a specified kinetic energy. The control variable must comply with the constraint $|u(t)| \leq u_M \neq 0$. Assumption 6.1 obviously holds so that, thanks to Theorems 6.4 and 6.5, it can be concluded that an optimal control, if it exists,

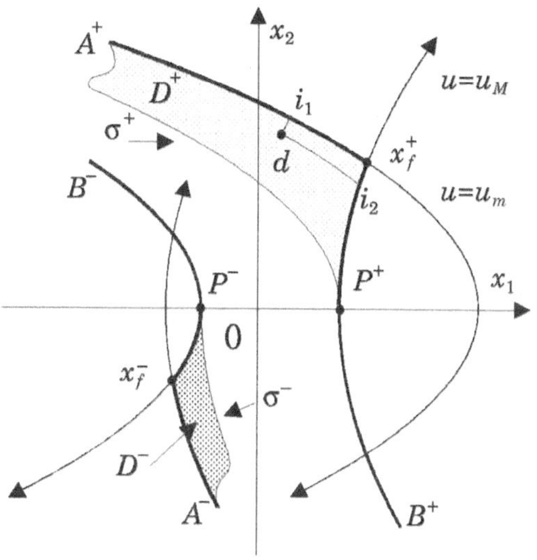

Figure 6.32: Example 6.34: the switching curves $A^+ - x_f^+ - P^+ - B^+$, $B^- - P^- - x_f^- - A^-$ and the sets of initial states (shadowed regions) where two feasible extremal controls exist when $x_2(t_f) > 0$ (D^+) or $x_2(t_f) < 0$ (D^-).

switches at most once from $-u_M$ to u_M or vice versa. In view of the form of the system trajectories corresponding to $u = \pm u_M$, which are parabolas defined by the equations $x_1 = \pm x_2^2/(2u_M)$ (see Fig. 6.33 where $u_M = 1$), and the fact that two final states are admissible, the conclusion can be drawn that two feasible extremal controls exist which are also optimal. Note that, consistent with this, Theorem 6.6 can not be applied.

Example 6.36 Again consider a positioning problem concerning a pointwise object with unitary mass. It moves in a plane subject to two forces u_1 and u_2 acting along the axis. By denoting with x_1 and x_3 the coordinates of the object, with x_2 and x_4 the corresponding time derivatives, the system equations are

$$\begin{aligned}\dot{x}_1 &= x_2,\\ \dot{x}_2 &= u_1,\\ \dot{x}_3 &= x_4,\\ \dot{x}_4 &= u_2,\end{aligned}$$

which can legitimately be viewed as describing two second order independent subsystems. The initial state is the origin ($x(0) = 0$), while, as for the final state, the object has to be steered, at rest and in the shortest time, to any point of the set

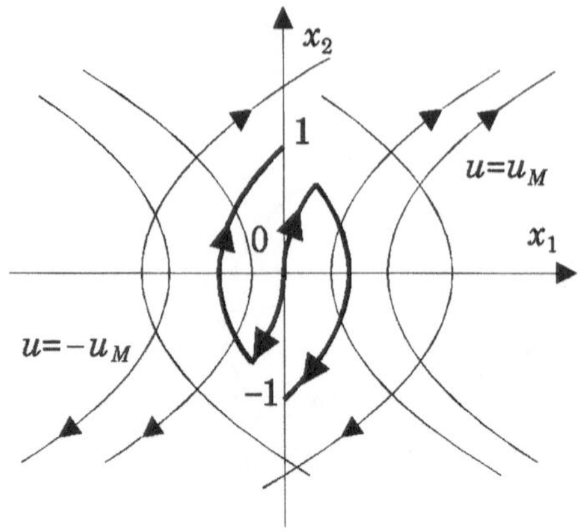

Figure 6.33: Example 6.35: two optimal trajectories (heavy curves) and generic trajectories corresponding to $u = \pm u_M$.

$\bar{S}_f = \{x | \ x_3 = x_1^2/2 - \vartheta, \ x_2 = x_4 = 0\}$, $\vartheta > 0$ (see Fig. 6.34 where the orthogonal projection Π of $\bar{S}_f$ on the $x_1 - x_3$ plane is shown when $\vartheta = 2$). The constraints $u_{mi} \leq u_i \leq u_{Mi}$, $u_{mi}u_{Mi} < 0$, $i = 1, 2$, are set on the control actions. Assumption 6.1 does not hold: actually nonzero solutions of the auxiliary system

$$\begin{aligned}
\dot{\lambda}_1 &= 0, \\
\dot{\lambda}_2 &= -\lambda_1, \\
\dot{\lambda}_3 &= 0, \\
\dot{\lambda}_4 &= -\lambda_3
\end{aligned}$$

exist which do not uniquely determine the control minimizing the hamiltonian function $H = 1 + \lambda_1 x_2 + \lambda_2 u_1 + \lambda_3 x_4 + \lambda_4 u_2$ (for instance, $\lambda_2 = 1$, $\lambda_1 = \lambda_3 = \lambda_4 = 0$). However the extremal controls satisfying the NC can be selected out of the infinite ones. The orthogonality condition requires (recall that $\lambda_i(t) = \lambda_i(0)$, $i = 1, 3$) $\lambda_1(0) = -\lambda_3(0) x_1(t_f)$, so that the following cases can occur:

(a) $\lambda_1(0) = \lambda_3(0) = 0$,

(b) $\lambda_1(0) = 0$, $\lambda_3(0) \neq 0$, $x_1(t_f) = 0$,

(c) $\lambda_1(0) \neq 0$, $\lambda_3(0) \neq 0$.

The first case must be discarded since it implies that λ_2 and/or λ_4 always have the same sign, thus entailing that u_1 and/or u_2 are constant: this fact prevents them from being feasible because $x_2(t_f) = 0$ and $x_4(t_f) = 0$ can not simultaneously occur.

The second case is consistent with $\lambda_2 \doteq 0$, $u_1 = 0$ (singular control component) so that $x_1(t_f) = 0$ and the final position is P_0 (see Fig. 6.34). On the contrary, $\lambda_4(0)$

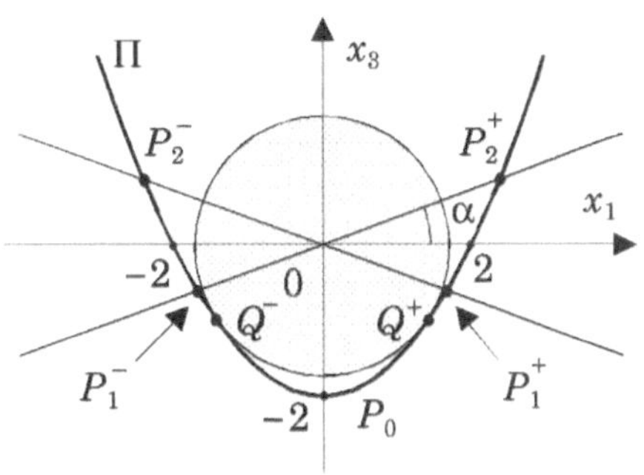

Figure 6.34: Example 6.36: the set of admissible final positions (curve Π) and final positions (P_0, P_1^+ and P_1^-) corresponding to extremal controls which satisfy the NC.

must be nonzero and such as to call for a control u_2 which steers the second subsystem from $x_3(0) = x_4(0) = 0$ to $x_4(t_f) = 0$ and $x_3(t_f) = -\vartheta$. By exploiting the material in Example 6.30 it is easy to conclude that

$$u_2(t) = \begin{cases} u_{m2}, & 0 \le t < \tau, \quad \tau = -\sqrt{2\vartheta u_{m2} u_{M2}/(u_{m2} - u_{M2})}/u_{m2}, \\ u_{M2}, & \tau < t \le t_f, \quad t_f = \sqrt{2\vartheta (u_{m2} - u_{M2})/(u_{m2} u_{M2})}. \end{cases}$$

The value for τ and t_f have been computed by taking into account that: (i) the time required to pass from $x_4(0) = 0$ to $x_4^* = x_4(\tau) < 0$, when $u = u_{m2}$, is $\tau = x_4^*/u_{m2}$, (ii) the time required to pass from $x_4^* < 0$ to $x_4 = 0$, when $u = u_{M2}$, is $t_f - \tau = -x_4^*/u_{M2}$, (iii) x_4^* is the second coordinate, in the $x_3 - x_4$ plane, of the intersection of the trajectory corresponding to $u = u_{m2}$ and starting at the origin, with the trajectory corresponding to $u = u_{M2}$ and ending at the point with coordinate $(-\vartheta, 0)$. The value for $\lambda_4(0)$ is given by the transversality condition $1 + \lambda_4(0) u_m = 0$, while the value for $\lambda_3(0)$ can be found by imposing $\lambda_4(\tau) = 0$.

Finally, the third case requires that both $\lambda_2(0)$ and $\lambda_4(0)$ be nonzero in order that $x_2(t_f) = x_4(t_f) = 0$: the two control components switch once. The (final) time when both subsystems are driven to the state $[\xi_i \ 0]'$ by a control of such a sort can be computed by exploiting the previous considerations relevant to the time required to cover a trajectory. For $i = 1, 2$, it results that

$$t_{fi} = \varphi_i \sqrt{|\xi_i|}, \quad \varphi_i := \sqrt{\frac{2(u_{mi} - u_{Mi})}{u_{mi} u_{Mi}}}.$$

By imposing that $t_{f1} = t_{f2}$ the conclusion is drawn that the final positions consistent with the case under consideration are those resulting from the intersection of the straight lines defined by the equations $x_3 = \pm(\varphi_1/\varphi_2)^2 x_1$ with the parabola Π (see Fig. 6.34, where $\alpha := \tan^{-1}((\varphi_1/\varphi_2)^2)$). However, the points P_2^+ and P_2^- are

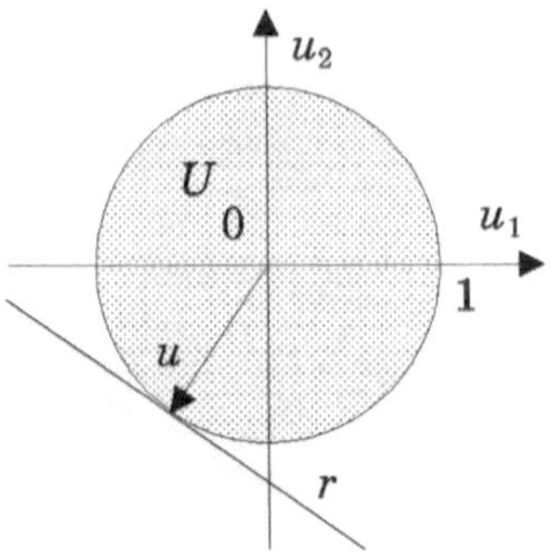

Figure 6.35: Example 6.36: the minimization of H in geometric terms when U is a circle.

reached by means of an extremal control which does not satisfy the orthogonality condition. Indeed, consider the final position P_2^+ which implies $x_i(t_f) > 0$, $i = 1, 2$: it is necessary that $u_i(0) > 0$, $i = 1, 2$ so that $\lambda_i(0) < 0$, $i = 2, 4$. In order that the two control components switch it must result that $\lambda_i(0) < 0$, $i = 1, 3$, which is inconsistent with the orthogonality condition. A similar reasoning leads to saying that also point P_2^- is reached by means of an extremal control which does not satisfy the NC, while points P_1^+ and P_1^- are reached by means of extremal controls which satisfy the NC. As a conclusion three final states have been located corresponding to feasible extremal controls. The times required to drive the system at rest to the points P_0 and P_1^+ (the time required to drive the system to the point P_1^- obviously equals the time required to reach P_1^+) are

$$t_f(P_0) = \varphi_2\sqrt{\vartheta},$$
$$t_f(P_1^+) = \varphi_1\sqrt{\sqrt{\beta^2 + 2\vartheta} - \beta},$$

where $\beta := (\varphi_1/\varphi_2)^2$. Thus, the final optimal point, if it exists, is the one among them which requires the least final time. It is not difficult to check that $t_f(P_0) > t_f(P_1^+)$ for each ϑ, u_{mi}, u_{Mi}, so that the solution of the problem cannot be unique.

Now consider a similar problem where the set U rather than being a parallelepiped is a circle centered at the origin of R^2 with unitary radius. If λ_2 and λ_4 are not simultaneously zero (this fact can occur only at isolated times, otherwise the NC are violated), the minimization of the hamiltonian function implies that the vector u_h has unitary norm and is orthogonal to the straight line r defined by the equation $\lambda_2 u_1 + \lambda_4 u_2 = k$ (see Fig. 6.35). Thus the actual problem is selecting the direction of the resulting force, its magnitude being anyway unitary. By imposing the NC, we find the final position which is closest to the initial one and belongs to the parabola Π. Since this parabola is symmetric with respect to the x_3 axis, two such points exist when $\vartheta > 1$, namely, Q^+ and Q^- with coordinates $(\sqrt{2(\vartheta - 1)}, -1)$

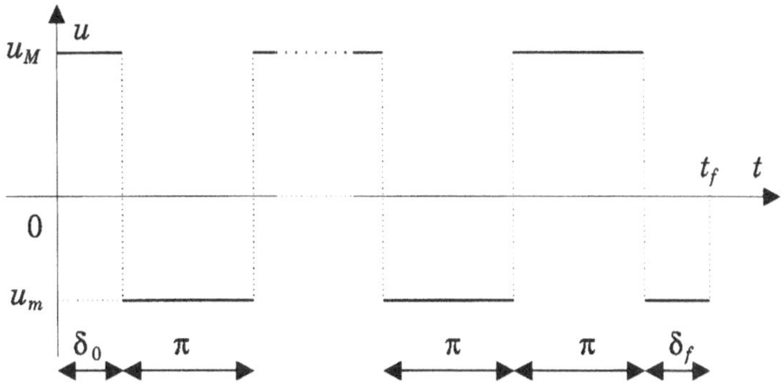

Figure 6.36: Example 6.37: shape of an extremal control.

and $(-\sqrt{2(\vartheta-1)}, -1)$, respectively, while, if $0 < \vartheta \leq 1$ the final point is unique and coincides with P_0.

Example 6.37 Consider a minimum time problem defined by a harmonic oscillator described by the equations

$$\frac{d\xi_1}{d\tau} = \xi_2,$$
$$\frac{d\xi_2}{d\tau} = -\omega_n^2\xi_1 + v$$

with given yet generic initial state ξ_0 and final state $\xi(t_f) = 0$. The control variable must belong to the set $V = \{v | \; v_m \leq v \leq v_M, \; v_m v_M < 0\}$. For the sake of convenience, define the new variables $x_1 := \omega_n\xi_1$, $x_2 := \xi_2$, $t := \omega_n\tau$, $u := v/\omega_n$, so that the system equations become $\dot{x}_1 = x_2$, $\dot{x}_2 = -x_1 + u$ and the new control variable u must comply with the constraint $u_m := v_m/\omega_n \leq u \leq v_M/\omega_n := u_M$. Assumption 6.1 holds so that the minimization of the hamiltonian function uniquely determines u_h. In fact,

$$u_h = \begin{cases} u_m, & \lambda_2 > 0 \\ u_M, & \lambda_2 < 0 \end{cases}$$

and $\lambda_2(t) = \alpha\sin(t+\varphi)$ with $\alpha > 0$. Therefore, a generic extremal control is a function like the one shown in Fig. 6.36 where $0 \leq \delta_0 \leq \pi$ and $0 \leq \delta_f \leq \pi$. Corresponding to the two limit values u_m and u_M which can be taken on by u, the system trajectories are circumferences centered in $(u_m, 0)$ and $(u_M, 0)$ respectively, which are covered in a clockwise sense. Note that π is the time required to cover half these circumferences. We now search the initial states corresponding to which there exists a feasible control of the form shown in Fig. 6.36 (or ending with $u(t_f) = u_M$ and/or beginning with $u(0) = u_M$). In view of Theorems 6.6, 6.7, 6.8 such a control is the optimal control. By making specific reference to the control shown in Fig. 6.36, it is easy to draw the corresponding trajectory which ends at the origin: it is shown in Fig. 6.37. The location of point P_1 where the last ($:= \nu$-th) switch takes place can vary, as δ_f varies,

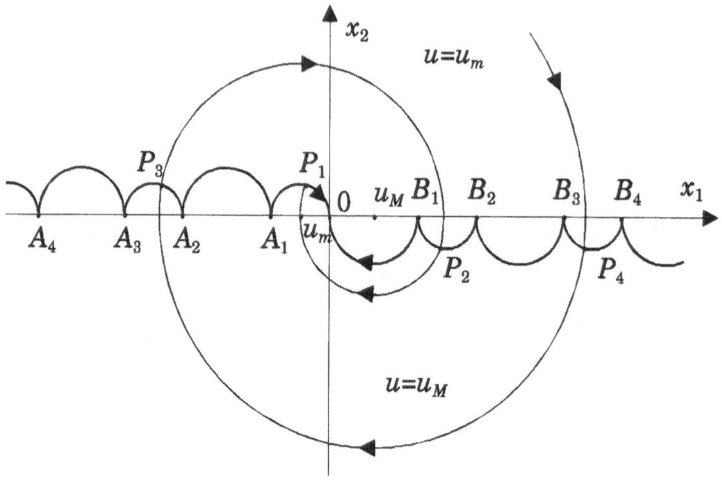

Figure 6.37: Example 6.37: optimal trajectories and switching curve (heavy line).

between 0 and A_1. Corresponding to this, the point P_2 where the $(\nu - 1)$-th switch occurs varies between B_1 and B_2, the point P_3 where the $(\nu - 2)$-th switch takes place varies between A_2 and A_3 and so on. By repeating these arguments when the value of the feasible extremal control at the end of the control interval is u_M, it is easy to conclude that: (i) a feasible extremal control exists for each initial state, (ii) the number of switching times is unbounded as the initial state varies (note that Theorem 6.5 does not apply), (iii) the curve

$$\cdots - A_4 - A_3 - A_2 - A_1 - 0 - B_1 - B_2 - B_3 - B_4 - \cdots$$

is the locus where the switches from u_m to u_M or vice versa occur. This *switching* curve partitions the state space into two subregions where the optimal control is u_m or u_M thus allowing a closed loop implementation of the solution.

6.6 Problems

Problem 6.6.1 For $\xi > 0$ and $-\infty < \varphi < \infty$ discuss the optimal control problem defined by the system

$$\dot{x}_1 = u, \ \dot{x}_2 = \varphi u, \ x(0) = 0, \ x(1) \in \bar{S}_f, \ \bar{S}_f := \left\{ x | \, (x_1 - 1)^2 + x_2^2 - \xi = 0 \right\}$$

and the performance index

$$J = \frac{1}{2}\int_0^1 u^2 dt.$$

Problem 6.6.2 Find a control which satisfies the NC for the problem defined by the system $\dot{x}_1 = x_2$, $\dot{x}_2 = u$, $x(0) = x(2) = 0$, and the performance index

$$J = \frac{1}{2}\int_0^2 u^2 dt - \alpha x_1(\tau)$$

where $\alpha > 0$ is a given constant and $0 < \tau < 2$ has to be selected.

Problem 6.6.3 Discuss the problem defined by the system $\dot{x}_1 = x_2$, $\dot{x}_2 = u$, $x_1(0) = 0$, $x_2(0) = x_{20} > 0$, $x(t_f) = 0$, and the performance index

$$J = \int_0^{t_f} (1 + \frac{u^2}{2}) dt$$

when:

(i) no other constraints are present;

(ii) $x_1(t) \leq \bar{x}_1$, $0 \leq t \leq t_f$, $1 \leq \bar{x}_1$;

(iii) $x_2(t) \geq \bar{x}_2$, $0 \leq t \leq t_f$, $-1 \geq \bar{x}_2$;

(iv) $x_1(t) \leq 4$, $x_2(t) \geq -1$, $0 \leq t \leq t_f$.

Problem 6.6.4 Consider the optimal control problem defined by the system $\dot{x} = f(x, u)$ with given initial state $x(0)$ and performance index

$$J = \int_0^T l(x, u) dt + m(x(T))$$

where T is specified, while no other constraints are present. Let all assumptions required by the Hamilton-Jacobi theory and the Maximum Principle be verified and V be a solution of the Hamilton-Jacobi equation, corresponding to which it is possible to conclude that the pair $(x^o(\cdot), u^o(\cdot))$ is an optimal solution. Show that $\lambda_0^o = 1$ and

$$\lambda^o(t) := \left. \frac{\partial V(z,t)'}{\partial z} \right|_{z=x^o(t)}$$

constitute a vector which satisfies the NC together with $(x^o(\cdot), u^o(\cdot))$.

Problem 6.6.5 Discuss the problem defined by the system $\dot{x}_1 = x_2$, $\dot{x}_2 = u$, $x_i(0) = x_{i0} \geq 0$, $i = 1, 2$, $x(t_f) = 0$, $|u(t)| \leq 1$, $x_2(t) \geq -1$, $0 \leq t \leq t_f$ and the performance index

$$J = \int_0^{t_f} dt.$$

Problem 6.6.6 Discuss the problem defined by the first order system $\dot{x} = u$, with given initial and final state and performance index

$$J = \int_{t_0}^{t_f} (t^k + \frac{u^2}{2})dt$$

where t_f and t_0 are free and $k > 0$ is given.

Problem 6.6.7 Discuss the problem defined by the system $\dot{x}_1 = x_2$, $\dot{x}_2 = u$, $x_1(0) = 2$, $x_2(0) = 0$, $x(T) \in \bar{S}_f$, $|u(t)| \leq \beta$, $0 \leq t \leq T$ and performance index

$$J = \int_0^T |u|dt$$

where $\bar{S}_f = \{x| \; x_1^2 + x_2^2 = 1\}$ and T, $\beta > 0$ are given.

Problem 6.6.8 Discuss Problem 6.6.4 when the final time is free and the performance index is

$$J = \int_0^{t_f} l(x, u)dt + m(x(t_f), t_f).$$

Problem 6.6.9 Consider the problem defined by the system $\dot{x}_1 = x_2$, $\dot{x}_2 = u$, $x_1(0) = 1$, $x_2(0) = 0$, $x(t_f) = 0$ and the performance index

$$J = \int_0^{t_f} (1 + \frac{u^2}{2})dt$$

when (i) no other constraints are present, (ii) $|u(t)| \leq 1$, $0 \leq t \leq t_f$. Find a control which satisfies the NC.

Problem 6.6.10 Discuss the problem defined by the system $\dot{x}_1 = u_1 - u_2$, $\dot{x}_2 = u_2$, $x(0) = x_0$, $x(t_f) = 0$, $|u_1(t)| \leq 2$, $|u_2(t)| \leq 1$, $0 \leq t \leq t_f$ and the performance index

$$J = \int_0^{t_f} dt.$$

Problem 6.6.11 Discuss the problem defined on the first order system $\dot{x} = u$, $x(0) = x_0 \neq 0$, $x(1) = 0$, $u_{\min} \leq u(t) \leq u_{\max}$, $u_{\min}u_{\max} < 0$, $0 \leq t \leq 1$ and the performance index

$$J = \int_0^1 (\frac{x^2}{2} + \beta x + u)dt$$

where β is a given constant.

Problem 6.6.12 For $k = 0$ and $k = 1$ consider the problem defined by the first order system $\dot{x} = u$, $x(0) = 1$, $x(t_f) = 0$ and the performance index

$$J = \int_0^{t_f} (t^k + \frac{u^2}{2})dt.$$

Find a control which satisfies the NC.

Problem 6.6.13 Consider the problem defined by the system $\dot{x}_1 = x_2$, $\dot{x}_2 = u$, $x(0) = 0$ and the performance index

$$J = \int_0^{t_f} \frac{u^2}{2} dt + t_f^\beta - \sigma x_1(t_f)$$

where $\sigma > 0$ and β is a positive integer. Whenever possible find a control which satisfies the NC and discuss its optimality.

Problem 6.6.14 Consider the problem defined by the system $\dot{x}(t) = f(x(t), u(t))$, $x(0) = x_0$, $x(t_f) = x_f$ and the performance index

$$J = \int_0^{t_f} l(x(t), u(t), t) dt$$

where $x(t_f)$, $x(0)$ are specified, t_f is free and $l(x(t), u(t), t) = p(t) + l^*(x(t), u(t))$, $p(t)$ being a given polynomial of degree $k \geq 0$. Let the problem be not pathological, $(x^o(t), u^o(t), t_f^o)$ a triple which satisfies the NC and $\lambda^o(t), \lambda_0^o = 1$ a solution of the auxiliary system corresponding to $(x^o(t), u^o(t))$. Show that $H(x^o(t), u^o(t), \lambda_0^o \lambda^o(t)) = p(t) - p(t_f^o)$.

Problem 6.6.15 Let β_1 and β_2 be given real numbers. Discuss the problem defined by the system $\dot{x}_1 = x_2$, $\dot{x}_2 = u$, $x(0) = x_0$, $x(t_f) = 0$, $|u(t)| \leq 1$, $0 \leq t \leq t_f$ and the performance index

$$J = \frac{1}{2} \int_0^{t_f} \left[1 + \beta_1 x_1 + \beta_2 x_2 + x_2^2\right] dt.$$

Problem 6.6.16 Find a control which satisfies the NC for the problem defined by the system $\dot{x}_1 = x_2$, $\dot{x}_2 = u$, $x_1(0) = 1$, $x_2(0) = 0$, $x_1(t_f) = 0$, $|x_2(t_f)| \leq 1$ and the performance index

$$J = \int_0^2 \frac{u^2}{2} dt$$

when:

(i) $x_1(1) = x_2(1) = -2$;
(ii) $x_1(1) = -2$;
(iii) $x_2(1) = -2$;
(iv) $x_1(\tau) = x_2(\tau) = -2$, $0 \leq \tau \leq 2$;
(v) $x_1(\tau) = -2$, $0 \leq \tau \leq 2$;
(vi) $x_2(\tau) = -2$, $0 \leq \tau \leq 2$.

Problem 6.6.17 Consider the minimum time problem defined by the linear system $\dot{x} = Ax + Bu$ where

$$A = \begin{bmatrix} 0 & 1 \\ 0 & -1 \end{bmatrix}, \quad B = \begin{bmatrix} 1 & 0 \\ 1 & 1 \end{bmatrix}$$

when the polyhedron U is the set $U = \{u| \ Su \leq \alpha, \ \alpha \in R^s\}$, with

$$S = \begin{bmatrix} 1 & 1 \\ -1 & -1 \\ 2 & 1 \\ -2 & -1 \end{bmatrix}, \ \alpha = \begin{bmatrix} 1 \\ 1 \\ 3 \\ 3 \end{bmatrix}.$$

Problem 6.6.18 Discuss the problem defined by the first order system $\dot{x} = x + u$, $x(0) \neq 0$, $x(T) = 0$, $|u(t)| \leq 1$, $0 \leq t \leq T$ and the performance index

$$J = \int_0^T |u|dt$$

where T is given.

Chapter 7

Second variation methods

7.1 Introduction

The necessary conditions presented in Chapter 6 originate from inspection of the perturbations caused to the performance index when the problem variables undergo a (specified) set of perturbations which comply with the constraints. As a matter of fact, these conditions state, in a quantitative way, the obvious requirement that an optimal solution must impose, namely that the corresponding performance index should not decrease when any admissible perturbation is given to such a solution. In general terms, what can actually be performed is an analysis constrained to inspection of the *first variation* of the performance index caused by *small* variations of the solution (see Remark 6.8 in Subsection 6.2.2 of Chapter 6). More specifically, *strong* control perturbations were considered, i.e., perturbations which were required to cause *small* deviations of the state motion without being themselves of small amplitude.

Here we examine the perturbations undergone by the performance index in a somehow more accurate way, as also we evaluate the *second variation* terms even if the allowed control perturbations δu are restrained to be *weak*, i.e., such that the relation

$$\sup_{t_0 \leq t \leq t_f} \sqrt{\delta u'(t)\delta u(t)} < \varepsilon$$

holds. Under suitable continuity assumptions concerning the function $f(x, u, t)$ which describes the system behaviour, the perturbation δx_u of the state motion caused by δu satisfies a relation of the same kind, provided that finite time intervals are considered.

Independently of the considered class of perturbations, the evaluation of both the first variation $[\delta J]_1$ and second variation $[\delta J]_2$ of the performance index allows us to proceed along two significant directions. The first one is pursued in Section 7.2 and leads to *local sufficient* optimality conditions which result from the requirement that $[\delta J]_1 = 0$ and $[\delta J]_2 > 0$ whatever is the (nonzero) perturbation selected inside the set of the admissible ones. Whenever these conditions are satisfied one can legitimately state that the solution at hand is locally optimal. The second direction is pursued in Section 7.3 and deals with the problem of finding the perturbations to be given to the solution at hand, optimal when the initial state is the nominal one, in order to preserve optimality (in a given sense) also when the initial state is changed. Rather surprisingly, the solvabilty conditions for this problem, referred to as the *neighbouring optimal control*, coincide with those ensuring local optimality.

We adopt the following notations. Let $\varphi(z_1, z_2, \dots, z_s, t)$ be a generic function. Then

$$\varphi^*_{\gamma\delta}(\tau) = \frac{\partial}{\partial\gamma}\frac{\partial\varphi(z_1, z_2, \dots, z_s, t)}{\partial\delta}\bigg|_{z_1=z_1^*(\tau),\dots,z_s=z_s^*(\tau),t=\tau}$$

where γ and δ are any two of the arguments of φ. In particular, the partial derivative can be performed with respect to only one of the function arguments or even with respect to none of them, this second instance simply referring to the function evaluation. As an example, for the hamiltonian function $H(x, u, t, \lambda) = l(x, u, t) + \lambda' f(x, u, t)$ we obtain

$$\begin{aligned} H^o(t) &= H(x^o(t), u^o(t), t, \lambda^o(t)), \\ H^o_x(t) &= \frac{\partial H(x, u, t, \lambda)}{\partial x}\bigg|_{x=x^o(t),u=u^o(t),\lambda=\lambda^o(t)}, \\ H^o_{xu}(t) &= \frac{\partial}{\partial x}\frac{\partial H(x, u, t, \lambda)}{\partial u}\bigg|_{x=x^o(t),u=u^o(t),\lambda=\lambda^o(t)} \end{aligned}$$

7.2 Local sufficient conditions

Weak local sufficient optimality conditions are now presented with reference to optimal control problems similar to those dealt with in Section 6.2 of Chapter 6, i.e., exhibiting simple constraints only. Here the term local is due to considering the performance index perturbation up to the second order, while the term weak reflects the nature of the admissible control perturbations.

More in detail, the problems on which the attention is focused are always assumed to be not pathological and defined by

$$\dot{x}(t) = f(x(t), u(t), t), \tag{7.1a}$$

$$x(t_0) = x_0, \tag{7.1b}$$

$$S_f = \{(x,t)|\ \alpha(x,t) = 0\}, \tag{7.1c}$$

$$J = m(x(t_f), t_f) + \int_{t_0}^{t_f} l(x(t), u(t), t)dt. \tag{7.1d}$$

In the preceding equations (7.1) the functions f (which are n-vectors), m, l, α (which are q-vectors, with $q \leq n+1$ components α_i) are continuous together with all first and second derivatives, l is not identically zero, x_0 and t_0 are given and the set S_f, which is the set of feasible final events (if not free, in this case the only requirement is that the final time be greater than the initial one) is a regular variety. No further constraints are present either on the control or state variables. A triple which satisfies the necessary conditions of the Maximum Principle is denoted by (x^o, u^o, t_f^o): it satisfies the equations

$$\dot{x}^o(t) = f(x^o(t), u^o(t), t), \tag{7.2a}$$

$$x^o(t_0) = x_0, \tag{7.2b}$$

$$0 = \alpha^o(t_f^o) \tag{7.2c}$$

and allows us to determine two vectors λ^o and ϑ^o with n and q components, respectively, such that (recall Remark 6.7 and Theorem 6.2 of Subsections 6.2.1 and 6.2.2 of Chapter 6)

$$\dot{\lambda}^o(t) = -H_x^{o\prime}(t), \tag{7.3a}$$

$$\lambda^o(t_f^o) = m_x^{o\prime}(t_f^o) + \alpha_x^{o\prime}(t_f^o)\vartheta^o, \tag{7.3b}$$

$$H^o(t_f^o) = -m_t^o(t_f^o) - \alpha_t^{o\prime}(t_f^o)\vartheta^o, \tag{7.3c}$$

$$H^o(t) \leq H(x^o(t), u, t, \lambda^o(t)),\ \forall u,\ t \in [t_0, t_f^o]. \tag{7.3d}$$

Under the assumption that $H_{uu}^o(t)$ is nonsingular for $t \in [t_0, t_f^o]$, let, for the sake of convenience in notation,

$$R(t) := H_{uu}^o(t), \tag{7.4a}$$

$$A(t) := f_x^o(t) - f_u^o(t)R^{-1}(t)H_{ux}^o(t), \tag{7.4b}$$

$$B(t) := f_u^o(t), \tag{7.4c}$$

$$Q(t) := H_{xx}^o(t) - H_{xu}^o(t)R^{-1}(t)H_{ux}^o(t), \tag{7.4d}$$

$$K(t) := B(t)R^{-1}(t)B'(t), \tag{7.4e}$$

$$\alpha_{xx}^o := \sum_{i=1}^{q} \alpha_{ixx}^o \vartheta_i^o, \tag{7.4f}$$

$$S_1 := m_{xx}^o(t_f^o) + \alpha_{xx}^o(t_f^o), \tag{7.4g}$$

$$S_2 := \alpha_x^{o\prime}(t_f^o), \tag{7.4h}$$

$$S_3 := S_1 f^o(t_f^o) + m_{xt}^{o\prime}(t_f^o) + \alpha_{xt}^{o\prime}(t_f^o)\vartheta^o + H_x^{o\prime}(t_f^o), \tag{7.4i}$$

$$S_4 := [f^{o\prime}(t_f^o)S_1 + 2(m_{xt}^o(t_f^o) + \vartheta^{o\prime}\alpha_{xt}^o(t_f^o)) + H_x^o(t_f^o)]f^o(t_f^o)$$
$$+m_{tt}^o(t_f^o) + \vartheta^{o\prime}\alpha_{tt}^o(t_f^o) + H_t^o(t_f^o), \tag{7.4j}$$
$$S_5 := \alpha_t^o(t_f^o). \tag{7.4k}$$

It is now possible to present four results which refer to the following cases:

(a) $(x(t_f), t_f) \in S_f$,
(b) $x(t_f) \in \bar{S}_f$, $t_f = T =$ given, $\bar{S}_f =$ regular variety,
(c) $(x(t_f), t_f) =$ free,
(d) $x(t_f) =$ free, $t_f = T =$ given.

Only the proof of the fourth result is reported here, as the remaining ones (especially that pertaining to the first case) are fairly complex.

Relative to case (a) the following theorem holds, where the definitions (7.4) are exploited.

Theorem 7.1 *Let (x^o, u^o, t_f^o) be a triple which satisfies the necessary optimality conditions (7.2)-(7.3). Further assume that $R(t) > 0$, $t \in [t_0, t_f^o]$ and there exist solutions of the equations*

$$\dot{P}_1(t) = -P_1(t)A(t) - A'(t)P_1(t) + P_1(t)K(t)P_1(t) - Q(t),$$
$$\dot{P}_2(t) = -[A(t) - K(t)P_1(t)]'\,P_2(t),$$
$$\dot{P}_3(t) = -[A(t) - K(t)P_1(t)]'\,P_3(t),$$
$$\dot{P}_4(t) = P_3'(t)K(t)P_3(t),$$
$$\dot{P}_5(t) = P_2'(t)K(t)P_3(t),$$
$$\dot{P}_6(t) = P_2'(t)K(t)P_2(t),$$

with the boundary conditions

$$P_1(t_f^o) = S_1, \quad P_2(t_f^o) = S_2, \quad P_3(t_f^o) = S_3,$$
$$P_4(t_f^o) = S_4, \quad P_5(t_f^o) = S_5, \quad P_6(t_f^o) = 0,$$

such that $P_4(t) > 0$, $t \in [t_0, t_f^o]$, $P_6(t) < 0$, $t \in [t_0, t_f^o)$. Then (x^o, u^o, t_f^o) is a locally optimal triple, in a weak sense, for the problem (7.1).

Some comments on the essence of the assumptions in the above theorem are in order: (i) The sign definition of matrix R, though not required by the NC, is often satisfied as naturally matching with the minimization of the hamiltonian function; (ii) The equations for P_i can be integrated one at a time and, apart from the first one (which is a Riccati equation), are all linear, the unknowns being three matrices (P_1, $n \times n$, P_2, $n \times q$, P_6, $q \times q$), two vectors (P_3, $n \times 1$, P_5, $q \times 1$) and a scalar (P_4).

Example 7.1 Consider the optimal control problem defined by the system $\dot{x} = u$, $x(0) = 0$, the performance index

$$J = \frac{1}{2}\int_0^{t_f} u^2 dt + x(t_f)$$

and the set $S_f = \{(x,t)|\ xt + 1 = 0\}$. It is easy to check that there exists only one solution which satisfies the NC, namely

$$u(t) = -\frac{2}{3},\ x(t) = -\frac{2}{3}t,\ t_f = \sqrt{\frac{3}{2}}.$$

The possibility of exploiting Theorem 7.1 is now explored. We get $A = Q = 0 = S_1$, $B = R = K = 1$, $S_2 = \sqrt{1.5}$, $S_3 = -(3S_2)^{-1}$, $S_4 = -4S_3/3$, $S_5 = -S_2^{-1}$ and $P_1(t) = 0$, $P_2(t) = S_2$, $P_3(t) = S_3$, $P_4(t) = -S_3 + 2t/27$, $P_5(t) = -1/\sqrt{6} - t/3$, $P_6(t) = 3(t - S_2)/2$. The conditions of the theorem are fulfilled, thus the control found via NC is locally optimal in a weak sense.

Example 7.2 Consider the optimal control problem defined by the system $\dot{x}_1 = x_2$, $\dot{x}_2 = u$, $x(0) = 0$, $x_2(t_f) = -1$ and the performance index

$$J = \frac{1}{2}\int_0^{t_f} u^2 dt + 2t_f^k + \sigma x_1(t_f)$$

where $\sigma = \pm 1$, $k = 1, 3$ and t_f is free. First we determine the solutions which satisfy the NC.

For $k = 1$ they are

$$\begin{aligned} x_1(t) &= \frac{1}{6}\lambda_1(0)t^3 - \frac{1}{2}\lambda_2(0)t^2, \\ x_2(t) &= \frac{1}{2}\lambda_1(0)t^2 - \lambda_2(0)t, \\ u(t) &= \lambda_1(0)t - \lambda_2(0), \end{aligned}$$

where $\lambda_1(0) = \sigma$, and, if $\sigma = 1$, $\lambda_2(0) = 2$, $t_f = t_{f1} = 2 - \sqrt{2}$ or $t_f = t_{f2} = 2 + \sqrt{2}$, while, if $\sigma = -1$, $\lambda_2(0) = 2$, $t_f = t_{f3} = \sqrt{6} - 2$ or $\lambda_2(0) = -2$, $t_f = t_{f4} = \sqrt{6} + 2$. We now explore the possibility of exploiting Theorem 7.1 is. Note that only some of the quantities involved in the sufficient conditions depend on the particular solution selected: they are $H_x = [0\ \lambda_1(0)]'$, $S_3 = [0\ \sigma]'$, $S_4 = -\sigma\lambda_2(t_f)$, where the superscript "o" has been omitted for the sake of simplicity in notation. Subsequently we find $P_1(t) = 0$, $P_2(t) = [0\ 1]'$, $P_3(t) = S_3$. As for P_4, it results that $P_4(t) = P_4(0) + \sigma^2 t$, $P_4(0) = -\sigma(\lambda_2(t_f) + \sigma t_f)$, so that the sign requirement on P_4 is fulfilled if and only if $P_4(0) > 0$. This happens only when $\sigma = -1$, $\lambda_2(0) = 2$ and $t_f = \sqrt{6} - 2$. The equations for P_5 and P_6 admit solutions and the sign of P_6 complies with the relevant request. Thus a solution which is locally optimal in a weak sense has been found. Further light can be shed on these conclusions by applying the Hamilton-Jacobi theory to the problem with given, though generic, final time. The optimal value of the performance index (as a function of the final time) can be found in this way,

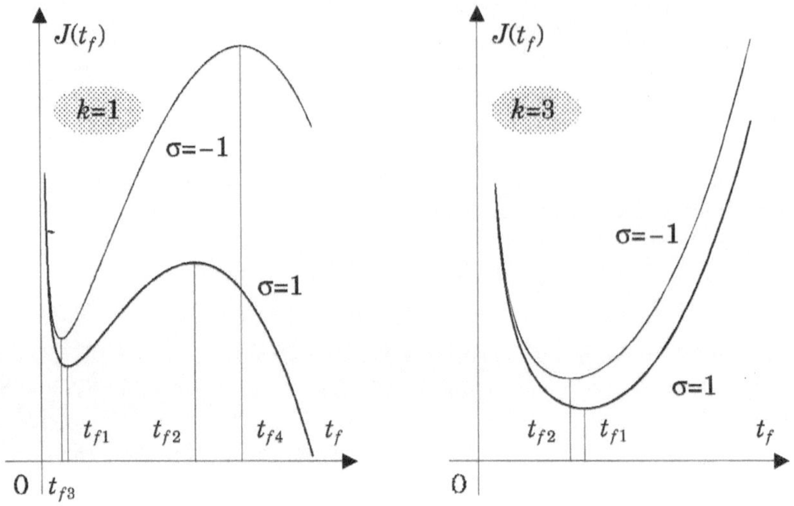

Figure 7.1: Example 7.2: the optimal value of the performance index.

yielding $J(t_f) = \beta - (\sigma + \beta^2/2)t_f + \sigma\beta t_f^2/2 - \sigma^2 t_f^3/6 + 2t_f$, where $\beta := (\sigma t_f^2 + 2)/(2t_f)$. In Fig. 7.1 the plots of this function are shown and the following facts are established: the problem does not admit an optimal solution, the local minimum is determined by the sufficient conditions only when $\sigma = -1$, the sufficient conditions cannot be verified when the final time is t_{f2} or t_{f4}.

For $k = 3$ the NC are satisfied by

$$x_1(t) = \frac{1}{6}\lambda_1(0)t^3 - \frac{1}{2}\lambda_2(0)t^2,$$
$$x_2(t) = \frac{1}{2}\lambda_1(0)t^2 - \lambda_2(0)t,$$
$$u(t) = \lambda_1(0)t - \lambda_2(0),$$

where $\lambda_1(0) = \sigma$ and $\lambda_2(0) = \sqrt{12}t_f$, with

$$t_f = \begin{cases} t_{f1} = \sqrt{\frac{2}{2\sqrt{12}-1}}, & \sigma = 1, \\ t_{f2} = \sqrt{\frac{2}{2\sqrt{12}+1}}, & \sigma = -1. \end{cases}$$

We now explore the possibility of exploiting Theorem 7.1. Note that only some of the quantities involved in the sufficient conditions depend on the particular solution selected: they are $H_x = [0 \ \lambda_1(0)]'$, $S_3 = [0 \ \sigma]'$, $S_4 = -\sigma\lambda_2(t_f) + 12t_f$, where the superscript "o" has been omitted for the sake of simplicity in notation. Subsequently we find $P_1(t) = 0$, $P_2(t) = [0 \ 1]'$, $P_3(t) = S_3$. As for P_4, it results that $P_4(t) = P_4(0) +$

$\sigma^2 t$, $P_4(0) = -\sigma\lambda_2(t_f)+(12-\sigma^2)t_f$, so that the sign requirement on P_4 is fulfilled for $\sigma = \pm 1$. The equations for P_5 and P_6 admit solutions and the sign of P_6 complies with the relevant request. Thus the solutions which satisfy the NC are locally optimal in a weak sense. Further light can be shed on this conclusions by resorting again to the Hamilton-Jacobi theory. The optimal value of the performance index as a function of the final time can be found, yielding $J(t_f) = \beta-(\sigma+\beta^2/2)t_f+\sigma\beta t_f^2/2-\sigma^2 t_f^3/6+2t_f^3$, where $\beta := (\sigma t_f^2 + 2)/(2t_f)$. The plots of this function (see Fig. 7.1) show that the solutions which have been found are globally optimal.

In the second case (constraints on the final state, given final time) the following theorem holds where reference is made only to those among the NC (7.2), (7.3) which apply to the problem at hand. Furthermore, the (obvious) modifications which must be performed in order to take care of the peculiarity of the problem have been made relative to the functions α, m and the notations defined by eqs. (7.4).

Theorem 7.2 *Let (x^o, u^o) be a pair which satisfies the necessary optimality conditions (7.2), (7.3). Moreover, assume that $R(t) > 0$, $t \in [t_0, T]$ and there exist solutions of the equations*

$$\begin{aligned}
\dot{P}_1(t) &= -P_1(t)A(t) - A'(t)P_1(t) + P_1(t)K(t)P_1(t) - Q(t),\\
\dot{P}_2(t) &= -\left[A(t) - K(t)P_1(t)\right]' P_2(t),\\
\dot{P}_6(t) &= P_2'(t)K(t)P_2(t)
\end{aligned}$$

with the boundary conditions

$$\begin{aligned}
P_1(T) &= S_1,\\
P_2(T) &= S_2,\\
P_6(T) &= 0
\end{aligned}$$

such that $P_6(t) < 0$, $t \in [t_0, T)$. Then, (x^o, u^o) is a locally optimal pair in a weak sense for problem (7.1) with $t_f = T$.

Example 7.3 Consider the optimal control problem defined by the system $\dot{x}_1 = x_2$, $\dot{x}_2 = u$, $x(0) = 0$ and the performance index

$$J = \frac{1}{2}\int_0^1 u^2 dt + 2x_1(1).$$

The final state is constrained by the equation $x_1 - x_2^2/2 + 1 = 0$. First we determine the solutions which satisfy the NC. In particular, we obtain $u = \lambda_1(0)t - \lambda_2(0)$ where the vector $\lambda(0)$ can be specified by enforcing feasibility and the orthogonality condition. It turns out that

$$\lambda_2(0) = \frac{\lambda_1(0)(4 - \lambda_1(0))}{2(3 - \lambda_1(0))},$$

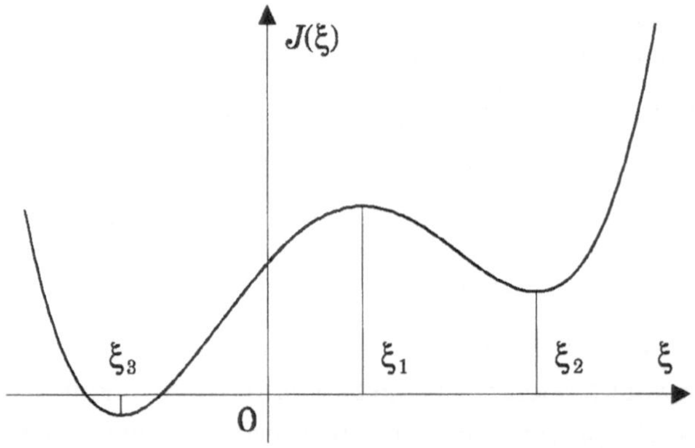

Figure 7.2: Example 7.3: the optimal value of the performance index: the final states $x_2(1) = \xi_i$ correspond to the values β_i of $\lambda_1(0)$.

$\lambda_1(0)$ being a solution of the equation $-2\lambda_1(0)^3 + 39\lambda_1(0)^2 - 180\lambda_1(0) + 216$. Three values for $\lambda_1(0)$ result, namely $\lambda_1(0) = \beta_1 := 13.3746$, $\lambda_1(0) = \beta_2 := 4.2052$, $\lambda_1(0) = \beta_3 := 1.9273$. We now explore the possibility of exploiting Theorem 7.2. Note that only some of the quantities involved in the sufficient conditions depend on the particular solution selected: they are $S_1 = \text{diag}[0, 2 - \lambda_1(0)]$, $S_2 = [1 \ \lambda_2(0) - \lambda_1(0)/2]'$, where the superscript "o" has been omitted for the sake of simplicity in notation. In a neighbourhood of $t = 1$ the solution of the equation for P_1 is $P_1 = \text{diag}[0, (2 - \lambda_1(0))/(1 + (1 - t)(2 - \lambda_1(0))]$ which can be extended up to $t = 0$ only when $\lambda_1(0) = \beta_3$. Corresponding to this, the equations for P_2 and P_6 admit solutions and the sign condition for P_6 is verified. As a conclusion, the local sufficient conditions are satisfied relative to the third value of $\lambda_1(0)$. Further light can be shed on these conclusions and a deeper understanding of the local sufficient conditions can be gained by first noticing that the final state is feasible if $x(1) = x_2(1)/2 - 1$. Then, by resorting to the Hamilton-Jacobi theory it is easy to check that, for a given final state $[\xi^2/2 - 1 \ \xi]'$, the optimal value of the performance index is $J(\xi) = a_1(1 + \xi - \xi^2/2) - a_2\xi + a_1a_2/2 - (a_1^2/6 + a_2^2/2) + 2(\xi^2/2 - 1)$, where $a_1 := 6(2 + \xi - \xi^2)$, $a_2 := 3(2 + \xi - \xi^2) - \xi$. The plot of this function is shown in Fig. 7.2 and fully motivates the reason why the local sufficient conditions could not have been verified relative to the value β_1 which entails the final state specified by ξ_1. The conditions of Theorem 7.2 hold for the, globally optimal, solution corresponding to β_3 which entails the final state specified by ξ_3.

In the third case (no constraints on the final state and time) the following theorem holds where reference is made only to those among the NC (7.2), (7.3) which apply to the problem at hand. Furthermore, the (obvious) modifications which must be performed in order to take care of the peculiarity of the problem, have been made relative to the functions α, m and the notations defined by eqs. (7.4).

Theorem 7.3 *Let (x^o, u^o, t_f^o) be a triple which satisfies the necessary optimality conditions (7.2), (7.3). Moreover, let $R(t) > 0$, $t \in [t_0, t_f^o]$ and assume that there exist solutions of the equations*

$$\begin{aligned}
\dot{P}_1(t) &= -P_1(t)A(t) - A'(t)P_1(t) + P_1(t)K(t)P_1(t) - Q(t),\\
\dot{P}_3(t) &= -\left[A(t) - K(t)P_1(t)\right]' P_3(t),\\
\dot{P}_4(t) &= P_3'(t)K(t)P_3(t)
\end{aligned}$$

with the boundary conditions

$$\begin{aligned}
P_1(t_f^o) &= S_1,\\
P_3(t_f^o) &= S_3,\\
P_4(t_f^o) &= S_4,
\end{aligned}$$

and such that $P_4(t) > 0$, $t \in [t_0, t_f^o]$. Then (x^o, u^o, t_f^o) is a locally optimal triple in a weak sense for problem (7.1a), (7.1b), (7.1d).

Example 7.4 Consider the optimal control problem defined by the first order system $\dot{x} = u$, $x(0) = 0$ and the performance index

$$J = \frac{1}{2}\int_0^{t_f} (2 + u^2)dt + 2(x^2(t_f) - 1)^2.$$

First we determine the solutions which satisfy the NC. Note that the data imply the existence of an even number of such solutions: thus only half of them will be explicitly mentioned, the remaining ones being trivially deducible. We obtain that $u = \sqrt{2}$ and two values for t_f (the positive solutions of the equation $16t_f^3 - 8t_f + 1 = 0$), namely, $t_{f1} = 0.63$ and $t_{f2} = 0.13$ and the corresponding final states $x_1 = 0.90$ and $x_2 = 0.18$, respectively. We now explore the possibility of exploiting Theorem 7.3. The quantities involved in the sufficient conditions which depend on the particular solution selected are $S_1 = m_{xx}$, $S_3 = \sqrt{2}m_{xx}$, $S_4 = 2m_{xx}$, with $m_{xx} = 11.26$ when the first value for t_f is selected, otherwise $m_{xx} = -7.20$. The equations for P_1, P_3 and P_4 admit solutions corresponding to both situations, but the sign requirement on P_4 is satisfied only when $m_{xx} > 0$. Thus the sufficient conditions are met for t_{f1} and x_1. By resorting to the Hamilton-Jacobi theory it is possible to conclude that this solution is indeed optimal, since the optimal value of the performance index as a function of the final state x_f (see Fig. 7.3) is $J(x_f) = \sqrt{2x_f} + 2(x_f^2 - 1)^2$.

Finally, in the fourth case (free final state, given final time) the following theorem holds where, as done in the two preceding cases, reference is made only to those among the NC (7.2), (7.3) which apply to the problem at hand. Furthermore, the (obvious) modifications which must be performed in order to take care of the peculiarity of the problem, have been made relative to the functions α, m and the notations defined by eqs. (7.4).

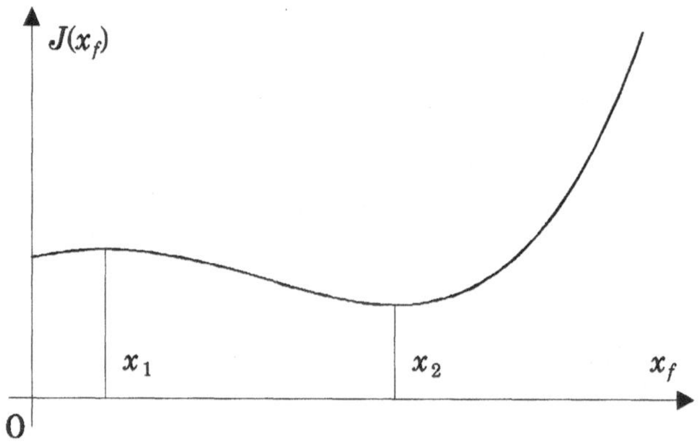

Figure 7.3: Example 7.4: the optimal value of the performance index.

Theorem 7.4 *Let (x^o, u^o) be a couple which satisfies the necessary optimality conditions (7.2), (7.3). Moreover, let $R(t) > 0$, $t \in [t_0, T]$ and assume that there exists a solution of the equation*

$$\dot{P}_1(t) = -P_1(t)A(t) - A'(t)P_1(t) + P_1(t)K(t)P_1(t) - Q(t)$$

with the boundary condition

$$P_1(T) = S_1.$$

Then (x^o, u^o) is a locally optimal pair in a weak sense for problem (7.1a), (7.1b), (7.1d) with $t_f = T$.

Proof. For the sake of simplicity in notation only the arguments which are essential for the understanding of the forthcoming discussion will be displayed in the relevant functions. Assume that the problem is not pathological, then, for any function λ it follows that $l(x, u, t) = H(x, u, t, \lambda) - \lambda'\dot{x}$ so that the variation $[\delta J]_{1,2}$ of the performance index which accounts for the terms up to the second order and is induced by the perturbation $(\delta x, \delta u)$ of the couple (x^o, u^o) is given by

$$\begin{aligned}
[\delta J]_{1,2} &= [J(x^o + \delta x, u^o + \delta u) - J(x^o, u^o)]_{1,2} \\
&= \int_0^T \left[H(x^o + \delta x, u^o + \delta u, t, \lambda^o) - H^o - \lambda^{o\prime}\delta\dot{x}\right]_{1,2} dt \\
&\quad + m(x^o(T) + \delta x(T)) - m(x^o(T)) \\
&= \int_0^T \left[H_x^o \delta x + H_u^o \delta u - \lambda^{o\prime}\delta\dot{x}\right] dt + m_x^o(T)\delta x(T) \\
&\quad + \frac{1}{2}\int_0^T \begin{bmatrix} \delta x' & \delta u' \end{bmatrix} \begin{bmatrix} H_{xx}^o & H_{xu}^o \\ H_{ux}^o & H_{uu}^o \end{bmatrix} \begin{bmatrix} \delta x \\ \delta u \end{bmatrix} dt \\
&\quad + \frac{1}{2}\delta x'(T) m_{xx}^o(T)\delta x(T),
\end{aligned}$$

provided that $\lambda = \lambda^o$. In the relation above the first two terms in the right side account for the first variation $[\delta J]_1$ of the performance index, while the remaining two terms account for the second variation $[\delta J]_2$. By performing an integration by parts of the term $-\lambda^{o\prime}\delta\dot{x}$ and taking into account that $\delta x(0) = 0$, the conclusion can be drawn that $[\delta J]_1 = 0$, if the triple (x^o, u^o, λ^o) verifies the NC. As for the term $[\delta J]_2$, if the assumptions in the theorem are satisfied we get

$$[\delta J]_2 = \frac{1}{2}\int_0^T \begin{bmatrix} \delta x' & \delta u' \end{bmatrix} \begin{bmatrix} H^o_{xx} & H^o_{xu} \\ H^o_{ux} & H^o_{uu} \end{bmatrix} \begin{bmatrix} \delta x \\ \delta u \end{bmatrix} dt$$
$$+\frac{1}{2}\delta x'(T) m^o_{xx}(T)\delta x(T) + \int_0^T \delta x' P_1 \left[f^o_x \delta x + f^o_u \delta u - \delta\dot{x}\right] dt$$

since, by disregarding terms which would entail variations of the performance index of order higher than two, the dependence of δx on δu is given by

$$\delta\dot{x} = f^o_x \delta x + f^o_u \delta u, \ \ \delta x(0) = 0.$$

By integrating by parts the term $\delta x' P_1 \delta\dot{x}$ (recall that P_1 is a symmetric matrix), noticing that $2w'Sz = w'Sz + z'S'w$ for any matrix S and couple of vectors (w, z) and exploiting the nonsingularity of H^o_{uu}, we obtain

$$\begin{aligned}
[\delta J]_2 &= \frac{1}{2}\int_0^T \Big\{\delta x' \left[H^o_{xx} + P_1 f^o_x + f^{o\prime}_x P_1 + \dot{P}_1\right] \delta x + \delta u' H^o_{uu} \delta u \\
&\quad + \delta x' \left[H^o_{xu} + P_1 f^o_u\right] \delta u + \delta u' \left[H^o_{ux} + f^{o\prime}_u P_1\right] \delta x \Big\} dt \\
&= \frac{1}{2}\int_0^T \Big\{\delta x' \left[\dot{P}_1 + P_1 A + A' P_1 - P_1 K P_1 + Q\right] \delta x \\
&\quad + \left[\delta u + \Psi^o \delta x\right]' H^o_{uu} \left[\delta u + \Psi^o \delta x\right]\Big\} dt \\
&= \frac{1}{2}\int_0^T \left[\delta u + \Psi^o \delta x\right]' H^o_{uu} \left[\delta u + \Psi^o \delta x\right] dt
\end{aligned}$$

where $\Psi^o := (H^o_{uu})^{-1}(H^o_{ux} + f^{o\prime}_u P_1)$. Therefore it follows that $[\delta J]_{1,2} > 0$ unless $\delta u = -\Psi^o \delta x$. In view of the above mentioned relation between δx and δu, this fact implies that

$$\delta\dot{x} = (A - K P_1)\delta x, \ \ \delta x(0) = 0.$$

Hence it follows that $\delta x(\cdot) = 0$ and, consequently, $\delta u(\cdot) = 0$ as well.

Example 7.5 Consider the optimal control problem defined by the system $\dot{x}_1 = x_2$, $\dot{x}_2 = u$, $x(0) = 0$ and the performance index

$$J = \frac{1}{2}\left\{\int_0^1 u^2 dt + 3x_1^3(1) + \beta x_1^4(1)\right\}$$

where β is either equal to 0 or 0.6. First we determine the solutions which satisfy the NC. In particular, we obtain that $u = \lambda_1(0)t - \lambda_2(0)$ where the constants $\lambda_i(0)$ are

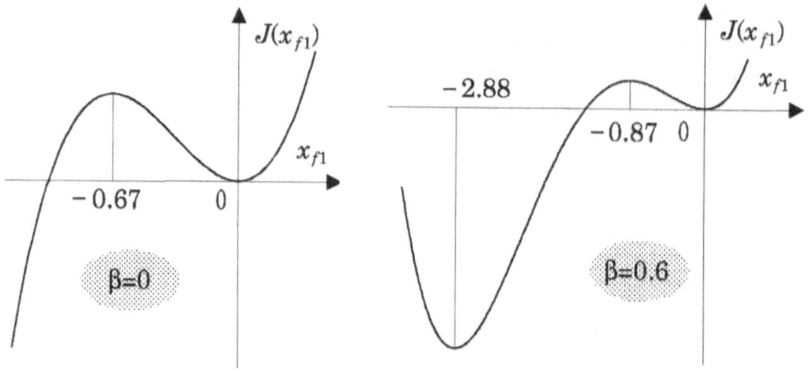

Figure 7.4: Example 7.5: the optimal value of the performance index.

determined by enforcing the orthogonality condition, yielding $\lambda_2(0) = \lambda_1(0)$ where $\lambda_1(0)$ solves the equation $2\beta\lambda_1^3(0) - 13.5\lambda_1^2(0) + 27\lambda_1(0) = 0$. Thus it follows that

$$\beta = 0 \Rightarrow \begin{cases} \lambda_1(0) = 0, & x_1(1) = 0 \\ \lambda_1(0) = 2, & x_1(1) = -0.67, \end{cases}$$

$$\beta = 0.6 \Rightarrow \begin{cases} \lambda_1(0) = 0, & x_1(1) = 0 \\ \lambda_1(0) = 8.65, & x_1(1) = -2.88 \\ \lambda_1(0) = 2.60, & x_1(1) = -0.87. \end{cases}$$

We now explore the possibility of exploiting Theorem 7.4. The equation for P_1 admits a solution corresponding to the first value of $\lambda_1(0)$ when $\beta = 0$ and to the first and second values of $\lambda_1(0)$ when $\beta = 0.6$: moreover the required assumptions are verified. By resorting to the Hamilton-Jacobi theory the optimal value of the performance index can easily be found as a function of the first component x_{f1} of the final state, namely $J(x_{f1}) = 1.5x_{f1}^2 + 1.5x_{f1}^3 + \beta x_{f1}^4/2$. This function is plotted in Fig. 7.4 in the two cases: all aspects of the problem are fairly clear.

Remark 7.1 *(A computational algorithm)* A particularly simple and easily understandable algorithm is now presented with the aim of showing how the knowledge of second order terms can fruitfully be exploited. Similarly to the algorithm described in Remark 6.11 of Subsection 6.2.2 of Chapter 6, the idea is to identify a solution which satisfies the local sufficient conditions. The perturbation of the solution at hand (which does not comply with the NC) is determined by evaluating the induced effects up to second order terms. In essence, the actual objective functional is approximated by a quadratic functional, while the algorithms which only consider first order terms performs a linear approximation. Therefore an increase in the rate of convergence should be expected whenever the information carried by the second order terms are exploited, especially in the vicinity of a (local) minimum where, in general, the methods based on the first variations are not very efficient. For the sake of simplicity the algorithm to be presented refers to the particularly simple class of optimal control problems defined by the system

$$\dot{x}(t) = f(x(t), u(t), t), \ x(t_0) = x_0 \tag{7.5}$$

and the performance index

$$J = \int_{t_0}^{T} l(x(t), u(t), t)dt + m(x(T)). \tag{7.6}$$

In these equations t_0, x_0 and T are given and the functions f, l, m, possess the continuity properties which have often been mentioned. No further constraints are present on either the state and control variables and the problem is assumed not to be pathological. Then it is possible to describe the i-th iteration of the algorithm by denoting with the superscript $^{(i)}$ the variable or parameter available at that iteration.

Algorithm 7.1

1. *Compute $x^{(i)}$ by integrating eqs. (7.5) with $u = u^{(i)}$ over the interval $[t_0, T]$;*
2. *Compute $\lambda^{(i)}$ by integrating the equations*

$$\dot{\lambda}(t) = -H_x^{(i)\prime}(t), \ \lambda(T) = m_x^{(i)\prime}(T)$$

over the interval $[t_0, T]$;

3. *If*

$$R^{(i)}(t) := H_{uu}^{(i)}(t) > 0, \ t \in [t_0, T]$$

find the solutions $P^{(i)}$ and $z^{(i)}$ (if they exist) of the equations

$$\begin{aligned}\dot{P}(t) &= -P(t)A^{(i)}(t) - A^{(i)\prime}P(t) - Q^{(i)}(t) + P(t)K^{(i)}(t)P(t), \\ \dot{z}(t) &= -\left[A^{(i)}(t) - K^{(i)}(t)P^{(i)}(t) - B^{(i)}(t)(R^{(i)})^{-1}(t)H_{ux}^{(i)}(t)\right]' z(t) \\ &\quad + v^{(i)}(t) + P^{(i)}(t)w^{(i)}(t)\end{aligned}$$

with

$$P(T) = m_{xx}^{(i)}(T), \ z(T) = 0$$

where

$$\begin{aligned}A^{(i)}(t) &:= f_x^{(i)}(t) - f_u^{(i)}(t)(R^{(i)})^{-1}(t)H_{ux}^{(i)}(t), \\ Q^{(i)}(t) &:= H_{xx}^{(i)}(t) - H_{xu}^{(i)}(t)(R^{(i)})^{-1}(t)H_{ux}^{(i)}(t), \\ K^{(i)}(t) &:= B^{(i)}(t)(R^{(i)})^{-1}(t)B^{(i)\prime}(t), \\ B^{(i)}(t) &:= f_u^{(i)}(t), \\ v^{(i)}(t) &:= H_{xu}^{(i)}(t)(R^{(i)})^{-1}(t)H_u^{(i)\prime}(t), \\ w^{(i)}(t) &:= B^{(i)}(t)(R^{(i)})^{-1}(t)H_u^{(i)\prime}(t).\end{aligned}$$

4. *If*

$$\| H_u^{(i)}(t) \| < \varepsilon, \; t \in [t_0, T]$$

where $\varepsilon > 0$ *is a suitable scalar, then the couple* $(x^{(i)}, u^{(i)})$ *satisfies the local sufficient conditions with an approximation the more accurate the smaller the scalar* ε *is. Otherwise, let*

$$u^{(i+1)}(t) := u^{(i)}(t) + \delta u^{(i)}(t)$$

where

$$\delta u^{(i)}(t) := -(R^{(i)})^{-1}(t) \left\{ \left[B^{(i)\prime}(t) P^{(i)}(t) + H_{ux}^{(i)}(t) \right] \vartheta^{(i)}(t) \right.$$
$$\left. + B^{(i)\prime}(t) z^{(i)}(t) + H_u^{(i)\prime}(t) \right\},$$

$\vartheta^{(i)}$ *being the solution of the equation*

$$\dot{\vartheta}(t) = \left[A^{(i)}(t) - K^{(i)}(t) P^{(i)}(t) \right] \vartheta^{(i)}(t) - K^{(i)}(t) z^{(i)}(t) - w^{(i)}(t)$$

with the boundary condition

$$\vartheta(t_0) = 0.$$

The i-th iteration of this algorithm can be justified as follows. First, it is easy to check that the perturbation $\delta u^{(i)}$ of the control $u^{(i)}$ induces the variation

$$[\delta J]_{1,2}^{(i)} = \frac{1}{2} \delta x^{(i)\prime}(T) m_{xx}^{(i)}(T) \delta x^{(i)}(T) + \int_{t_0}^{T} \{ H_u^{(i)}(t) \delta u^{(i)}(t)$$
$$+ \frac{1}{2} \begin{bmatrix} \delta x^{(i)\prime}(t) & \delta u^{(i)\prime}(t) \end{bmatrix} \begin{bmatrix} H_{xx}^{(i)}(t) & H_{xu}^{(i)}(t) \\ H_{ux}^{(i)}(t) & H_{uu}^{(i)}(t) \end{bmatrix} \begin{bmatrix} \delta x^{(i)}(t) \\ \delta u^{(i)}(t) \end{bmatrix} \} dt$$

in the performance index, provided that the first and second order terms are taken into account and $\lambda^{(i)}$ is selected as in step 2. In this equation $\delta x^{(i)}$ is the solution of

$$\delta \dot{x}(t) - f_x^{(i)}(t) \delta x(t) + B^{(i)}(t) \delta u^{(i)}(t), \; \delta x(t_0) - 0.$$

A significant choice for $\delta u^{(i)}$ is selecting, if possible, that function which minimizes $[\delta J]_{1,2}^{(i)}$ while complying with the constraint that the last equation imposes on $\delta x^{(i)}$ and $\delta u^{(i)}$. If what stated at step 3 holds, the Hamilton-Jacobi theory can be applied to this optimal control problem and the expression for $\delta u^{(i)}$ given at step 4 follows. If the inequality there is satisfied, the solution at hand verifies the local sufficient conditions with an approximation the more accurate the smaller the scalar ε is.

Example 7.6 Consider the problem presented in Example 6.20 which is defined by the system $\dot{x} = u$, $x(0) = 0$ and the performance index

$$J = \int_0^1 [x(t) + u^2(t)] dt.$$

As before the choice $u^{(1)}(t) = 1$, $t \in [0,1]$, is made and $\varepsilon = 0.001$ is set into the algorithm stopping condition. We find $A^{(1)}(t) = 0$, $B^{(1)}(t) = 1$, $R^{(1)}(t) = 2$, $Q^{(1)}(t) = 0$, $P^{(1)}(1) = 0$, $k^{(1)}(t) = 0$, so that the differential equations at step 3 have the solutions $P^{(1)}(t) = 0$ and $z^{(1)}(t) = 0$. The stopping condition is not verified (step 4), thus we set

$$\delta u^{(1)}(t) = \frac{1}{2} + \frac{1}{2}t, \; u^{(2)}(t) = -\frac{1}{2} + \frac{1}{2}t.$$

Corresponding to this new control we find $H_u^{(2)}(t) = 0$ and the local sufficient conditions are met.

7.3 Neighbouring optimal control

The problem to be considered in the present section deserves particular attention both from the applicative and theoretical point of view. Indeed, its solution leads, from one hand, to designing a control system which preserves the power of the optimal synthesis methods also in the face of uncertainties, thus allowing their actual exploitation, while, on the other hand, unexpected deep connections are established with the material of the preceding section which refers to (seemingly) unrelated topics. More in detail, consider the same optimal control problem discussed in Section 7.2, i.e., the problem defined by the dynamical system

$$\dot{x}(t) = f(x(t), u(t), t), \tag{7.7a}$$
$$x(t_0) = x_0, \tag{7.7b}$$
$$S_f = \{(x,t) | \; \alpha(x,t) = 0\}, \tag{7.7c}$$
$$J = m(x(t_f), t_f) + \int_{t_0}^{t_f} l(x(t), u(t), t) dt. \tag{7.7d}$$

In the preceding equations (7.7) the functions f (which are n-vectors), m, l, α (which are q-vectors, with $q \leq n+1$ components α_i) are continuous together with all first and second derivatives, l is not identically zero, x_0 and t_0 are given and the set S_f, which is the set of feasible final events (if not free, in this case the only request is that the final time be greater than the initial one) is a regular variety. No further constraints are present either on the control or state variables. A triple which satisfies the necessary conditions of the Maximum Principle is denoted by (x^o, u^o, t_f^o): it verifies the equations

$$\dot{x}^o(t) = f(x^o(t), u^o(t), t), \tag{7.8a}$$
$$x^o(t_0) = x_0, \tag{7.8b}$$
$$0 = \alpha^o(t_f^o) \tag{7.8c}$$

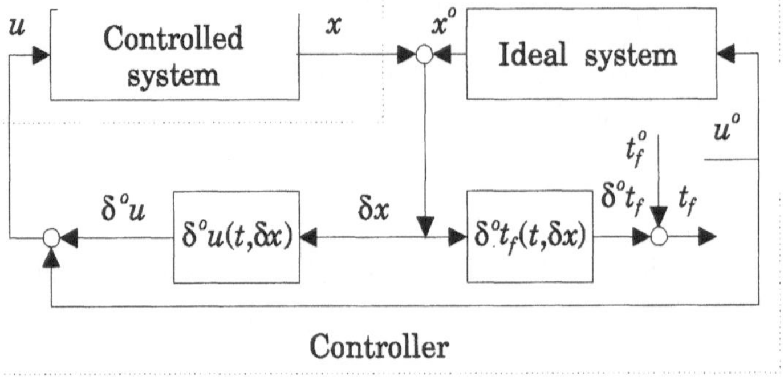

Figure 7.5: The control scheme which can be adopted when δu^o and δt_f^o are known.

and allows us to determine two vectors λ^o and ϑ^o with n and q components, respectively, such that (recall Remark 6.7 and Theorem 6.2 of Subsections 6.2.1 and 6.2.2 of Chapter 6)

$$\dot{\lambda}^o(t) = -H_x^{o\prime}(t), \tag{7.9a}$$
$$\lambda^o(t_f^o) = m_x^{o\prime}(t_f^o) + \alpha_x^{o\prime}(t_f^o)\vartheta^o, \tag{7.9b}$$
$$H^o(t_f^o) = -m_t^o(t_f^o) - \alpha_t^{o\prime}(t_f^o)\vartheta^o, \tag{7.9c}$$
$$H^o(t) \leq H(x^o(t), u, t, \lambda^o(t)), \ \forall u, \ t \in [t_0, t_f^o]. \tag{7.9d}$$

In general, the control u^o and the final time t_f^o, which are optimal for a given problem, depend on the initial state $x_0 = x(t_0)$ so that their computation should seemingly be again performed when it undergoes a change. Here we discuss the possibility of determining only the *optimal corrections* $\delta^o u(\cdot, t_0, \delta x_0)$ and $\delta^o t_f(t_0, \delta x_0)$ to be given to u^o and t_f^o which preserve optimality when the quantity δx_0 is added to the system initial state at the (initial) time t_0. Should this be possible also when the state perturbation occurs at any generic time, the control scheme of Fig. 7.5 could be implemented and the computation of the functions $\delta^o u$ and $\delta^o t_f$ done only once and for all. What can actually be computed in a fairly easy way are the perturbations $\delta_2^o u$ and $\delta_2^o t_f$ to be given to u^o and t_f^o in order that, within the set of weak perturbations, $[\delta J]_{1,2}$ is minimized, $[\delta J]_{1,2}$ being the variation of the performance index evaluated up to second order terms. When the initial state, the final time and the control are perturbed (the latter in a weak sense), it is easy to check that such a variation is given by

$$[\delta J]_{1,2} = \lambda^{o\prime}(t_0)\delta x_0 + \frac{1}{2}\begin{bmatrix} \delta x'(t_f^o) & \delta t_f \end{bmatrix}\begin{bmatrix} S_1 & S_3 \\ S_3' & S_4 \end{bmatrix}\begin{bmatrix} \delta x(t_f^o) \\ \delta t_f \end{bmatrix}$$

$$+\frac{1}{2}\int_{t_0}^{t_f^o} \left[\begin{array}{cc} \delta x'(t) & \delta u'(t) \end{array}\right] \left[\begin{array}{cc} H_{xx}^o(t) & H_{xu}^o(t) \\ H_{ux}^o(t) & H_{uu}^o(t) \end{array}\right] \left[\begin{array}{c} \delta x(t) \\ \delta u(t) \end{array}\right] dt,$$

provided that the unperturbed situation refers to a solution which satisfies the NC. In the above equations reference has been made to some of the definitions (7.4) given in Section 7.2. Note that the quantity $\lambda^{o\prime}(t_0)\delta x_0$ must be considered given and the two remaining terms in the right side of the equation constitute the second variation $[\delta J]_2$ of the performance index. Under suitable assumptions, $[\delta J]_2$ is minimized by a couple $(\delta_2^o u, \delta_2^o t_f)$ which can easily be specified as a function of the state perturbation detected at any time $\tau \in [t_0, t_f^o)$ since the initial state is arbitrary. In such a case the computed couple minimizes $[\delta J]_2$ over the time interval beginning at τ. Finally, it can be proved that the difference between the value of the performance index corresponding to the couple $(\delta^o u, \delta^o t_f)$ and the value corresponding to the couple $(\delta_2^o u, \delta_2^o t_f)$ is an infinitesimal of higher order with respect to $\delta x(\tau)$: thus the *neighbouring optimal* control which is supplied by the controller in Fig. 7.5 when the second rather than the first couple of variations is exploited, satisfactorily solves the problem.

It is now possible to present four results which refer to the following cases:

(a) $(x(t_f), t_f) \in S_f$,

(b) $x(t_f) \in \bar{S}_f$, $t_f = T =$ given, $\bar{S}_f =$ regular variety,

(c) $(x(t_f), t_f) =$ free,

(d) $x(t_f) =$ free, $t_f = T =$ given.

These are the same cases discussed in Section 7.2: rather surprisingly, they require that identical assumptions be verified, thus establishing a tight connection between the two problems. All the relevant proofs are omitted since they are rather involved (especially the one pertaining to the first case) and reference is made to the definitions (7.4). As for the first case, the following result holds.

Theorem 7.5 *Let (x^o, u^o, t_f^o) be a triple which satisfies the necessary optimality conditions (7.8), (7.9). Moreover, let $R(t) > 0$, $t \in [t_0, t_f^o]$ and assume that there exist solutions of the equations*

$$\dot{P}_1(t) = -P_1(t)A(t) - A'(t)P_1(t) + P_1(t)K(t)P_1(t) - Q(t),$$
$$\dot{P}_2(t) = -\left[A(t) - K(t)P_1(t)\right]' P_2(t),$$
$$\dot{P}_3(t) = -\left[A(t) - K(t)P_1(t)\right]' P_3(t),$$
$$\dot{P}_4(t) = P_3'(t)K(t)P_3(t),$$
$$\dot{P}_5(t) = P_2'(t)K(t)P_3(t),$$
$$\dot{P}_6(t) = P_2'(t)K(t)P_2(t)$$

with the boundary conditions

$$\begin{aligned}P_1(t_f^o) &= S_1,\\ P_2(t_f^o) &= S_2,\\ P_3(t_f^o) &= S_3,\\ P_4(t_f^o) &= S_4,\\ P_5(t_f^o) &= S_5,\\ P_6(t_f^o) &= 0\end{aligned}$$

such that $P_4(t) > 0$, $t \in [t_0, t_f^o]$, $P_6(t) < 0$, $t \in [t_0, t_f^o)$. *Then, the variations* $\delta_2^o u$ *of the control and* $\delta_2^o t_f$ *of the final time which minimize* $[\delta J]_2$ *when* $x(\tau) = x^o(\tau) + \delta x_\tau$, $\tau \in [t_0, t_f^o)$ *are given by*

$$\begin{aligned}\delta_2^o u(t) &= -R^{-1}(t)\left\{B'(t)\left[\bar{P}_1(t) - \bar{P}_2(t)\bar{P}_6^{-1}(t)\bar{P}_2'(t)\right]\right.\\ &\quad \left. + H_{ux}^o(t)\right\}\delta_2^o x(t),\ t \in [\tau, t_f^o),\\ \delta_2^o t_f &= -\left[P_3'(\tau) - P_5'(\tau)\bar{P}_6^{-1}(\tau)\bar{P}_2'(\tau)\right]P_4^{-1}(\tau)\delta x_\tau\end{aligned}$$

where $\delta_2^o x$ *is the solution of the equation*

$$\delta\dot{x}(t) = \left\{A(t) - K(t)\left[\bar{P}_1(t) - \bar{P}_2(t)\bar{P}_6^{-1}(t)\bar{P}_2'(t)\right]\right\}\delta x(t)$$

with the boundary condition $\delta x(\tau) = \delta x_\tau$ *and, for* $t \in [\tau, t_f^o)$,

$$\begin{aligned}\bar{P}_1(t) &:= P_1(t) - P_3(t)P_4^{-1}(t)P_3'(t),\\ \bar{P}_2(t) &:= P_2(t) - P_3(t)P_4^{-1}(t)P_5'(t),\\ \bar{P}_6(t) &:= P_6(t) - P_5(t)P_4^{-1}(t)P_5'(t).\end{aligned}$$

Example 7.7 Consider the first order system $\dot{x} = (x-1)u$ with $x(0) = 3$ and $x(t_f) = 2 := x_f$. The optimal control problem is defined by this system and the performance index

$$J = \frac{1}{2}\int_0^{t_f}(2 + u^2)dt$$

where t_f is free. The NC are satisfied only by the triple (the superscript "o" has been omitted)

$$\begin{aligned}x(t) &= 2e^{-\sqrt{2}t} + 1,\\ u(t) &= -\sqrt{2},\\ t_f &= \ln(2)/\sqrt{2}.\end{aligned}$$

Corresponding to this, we find that $A(t) = -2\sqrt{2}$, $B(t) = 2e^{-\sqrt{2}t}$, $Q(t) = -e^{2\sqrt{2}t}/2$, $R(t) = 1$, $S_1 = 0$, $S_2 = 1$, $S_3 = -2$, $S_4 = 2\sqrt{2}$, $S_5 = 0$ and the check can be performed whether the assumptions of Theorem 7.5 are satisfied. The equation for P_1 with the relevant boundary condition admits the solution

$$P_1(t) = \frac{t_f - t}{2\beta(t)}e^{2\sqrt{2}t},\ \beta(t) := \sqrt{2}(t - t_f) - 1.$$

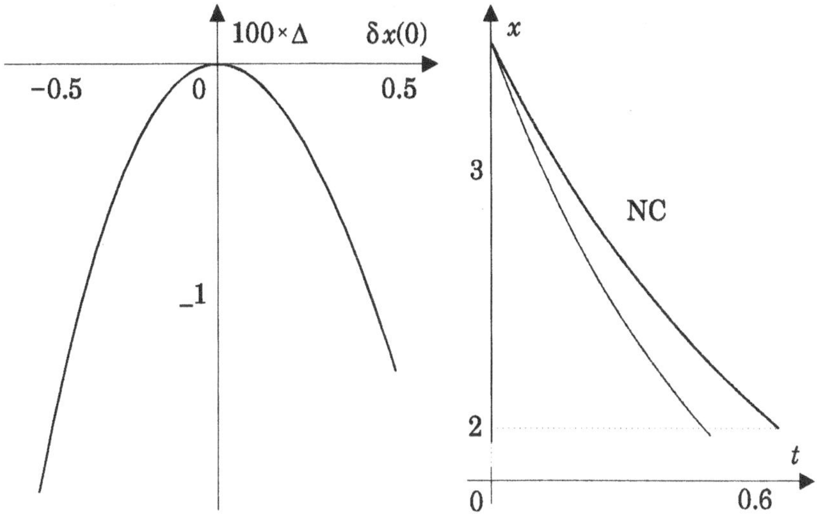

Figure 7.6: Example 7.7: violation of the constraint when $u + \delta_2 u$ is applied and state responses when $\delta x(0) = 0.5$.

All the equations for the remaining matrices P_i, with the relevant boundary conditions, admit solutions, namely,

$$P_2(t) = -\frac{e^{\sqrt{2}t}}{2\beta(t)},\ P_3(t) = \frac{e^{\sqrt{2}t}}{\beta(t)},\ P_4(t) = -\frac{2\sqrt{2}}{\beta(t)},$$

$$P_5(t) = \frac{2(t-t_f)}{\beta(t)},\ P_6(t) = \frac{t_f - t}{\beta(t)}.$$

Note that the sign conditions on P_4 and P_6 are satisfied: thus Theorem 7.5 can be exploited yielding

$$\bar{P}_1(t) = -\frac{\sqrt{2}e^{2\sqrt{2}t}}{4},\ \bar{P}_2(t) = \frac{e^{\sqrt{2}t}}{2},\ \bar{P}_6(t) = (t - t_f)$$

and

$$\delta_2 x(t, \delta x(0)) = \frac{t_f - t}{t_f} e^{-\sqrt{2}t} \delta x(0),$$

$$\delta_2 u(t, \delta x(0)) = -\frac{1}{2t_f} \delta x(0),$$

$$\delta_2 t_f(\delta x(0)) = 0.$$

As a conclusion, the control scheme of Fig. 7.5 can be implemented. It is interesting to evaluate to what extent the constraint on the final state is violated when the initial state has undergone a perturbation and the neighbouring optimal control $u + \delta_2 u$ is exploited, yielding $x(t) = (2 + \delta x(0))e^{-(\sqrt{2}+\delta x(0)/(2t_f))t} + 1$, and

$$\Delta := \frac{x(t_f) - 2}{2} = \frac{(1 + \delta x(0)/2)e^{-\delta x(0)/2} - 1}{2}$$

This function is plotted in Fig. 7.6 together with the state response when the NC are enforced or the control $u + \delta_2 u$ is applied and $\delta x(0) = 0.5$.

In the second case (constraints on the final state, given final time) the following theorem holds where reference is made only to those among the NC (7.8), (7.9) which apply to the problem at hand. Furthermore, the (obvious) modifications which must be performed in order to take care of the peculiarity of the problem, have been made relative to the functions α, m and the notation defined by eqs. (7.4).

Theorem 7.6 *Let (x^o, u^o) be a pair which satisfies the necessary optimality conditions (7.8), (7.9). Moreover, assume that $R(t) > 0$, $t \in [t_0, T]$ and there exist solutions of the equations*

$$\begin{aligned}
\dot{P}_1(t) &= -P_1(t)A(t) - A'(t)P_1(t) + P_1(t)K(t)P_1(t) - Q(t), \\
\dot{P}_2(t) &= -[A(t) - K(t)P_1(t)]' P_2(t), \\
\dot{P}_6(t) &= P_2'(t)K(t)P_2(t)
\end{aligned}$$

with the boundary conditions

$$\begin{aligned}
P_1(T) &= S_1, \\
P_2(T) &= S_2, \\
P_6(T) &= 0
\end{aligned}$$

such that $P_6(t) < 0$, $t \in [t_0, T)$. Then, the variation $\delta_2^o u$ of the control which minimizes $[\delta J]_2$ when $x(\tau) = x^o(\tau) + \delta x_\tau$, $\tau \in [t_0, T)$ is given by

$$\begin{aligned}
\delta_2^o u(t) = -R^{-1}(t) \{ B'(t) [P_1(t) - P_2(t)P_6^{-1}(t)P_2'(t)] \\
+ H_{ux}^o(t) \} \delta_2^o x(t), \ t \in [\tau, T)
\end{aligned}$$

where $\delta_2^o x$ is the solution of the equation

$$\delta \dot{x}(t) = \{ A(t) - K(t) [P_1(t) - P_2(t)P_6^{-1}(t)P_2'(t)] \} \delta x(t)$$

with the boundary condition $\delta x(\tau) = \delta x_\tau$.

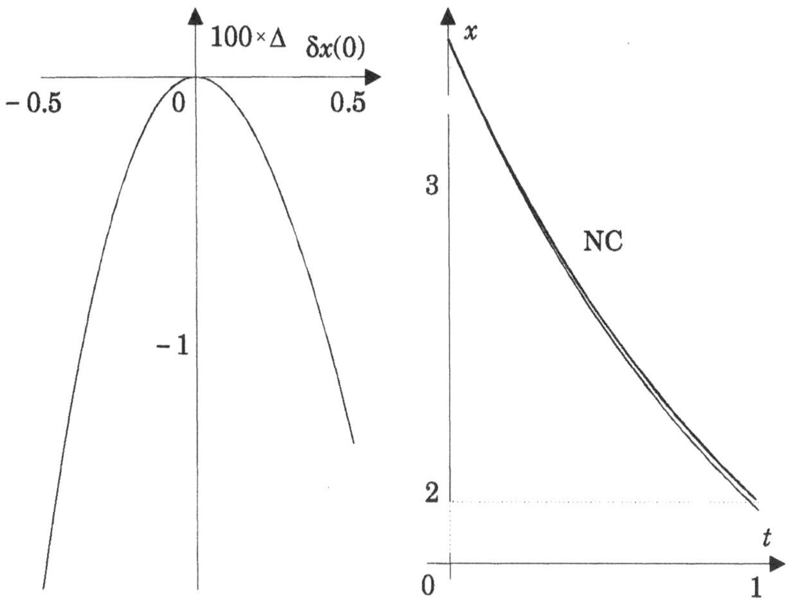

Figure 7.7: Example 7.8: violation of the constraint when $u + \delta_2 u$ is applied and state responses when $\delta x(0) = 0.5$.

Example 7.8 Consider a modified version of the problem presented in Example 7.7 where $t_f = 1$ and, consistent with this, the performance index is simply the integral of the square of the control action. The only solution which satisfies the NC is

$$\begin{aligned} x(t) &= 2e^{\gamma t} + 1, \\ u(t) &= \gamma, \\ \gamma &= -\ln(2) \end{aligned}$$

where the superscript "o" has been omitted. Corresponding to this, we find $A(t) = 2\gamma$, $B(t) = 2e^{\gamma t}$, $Q(t) = -\gamma^2 e^{-2\gamma t}/4$, $R(t) = 1$, $S_1 = 0$, $S_2 = 1$ and the check can be performed whether the assumptions of Theorem 7.6 are satisfied. The equations for P_1, P_2, P_6 with the relevant boundary condition admit the solutions

$$P_1(t) = \frac{\gamma^2(1-t)}{4\beta(t)} e^{-2\gamma t}, \quad \beta(t) := \gamma(1-t) - 1,$$

$$P_2(t) = -\frac{e^{-\gamma t}}{\beta(t)}, \quad P_6(t) = \frac{4(1-t)}{\beta(t)}.$$

Note that $P_6(t) < 0$, $t \in [0, 1)$ so that Theorem 7.6 can be applied, yielding

$$\delta_2 x(t, \delta x(0)) = (1-t)e^{\gamma t} \delta x(0) \qquad \delta_2 u(t, \delta x(0)) = -\frac{\delta x(0)}{2}$$

As a conclusion, the control scheme of Fig. 7.5 can be implemented and it is interesting to evaluate to what extent the constraint on the final state is violated when the initial state has undergone a perturbation and the neighbouring optimal control $u + \delta_2 u$ is exploited, yielding $x(1) = (2 + \delta x(0))e^{\gamma - \delta x(0)/2} + 1$ and

$$\Delta := \frac{x(1) - 2}{2} = \frac{(1 + \delta x(0)/2)e^{-\delta x(0)/2} - 1}{2}.$$

This function is plotted in Fig. 7.7 together with the state response when the NC are enforced or the control $u + \delta_2 u$ is applied and $\delta x(0) = 0.5$.

In the third case (no constraints on the final state and time) the following theorem holds where reference is made only to those among the NC (7.8), (7.9) which apply to the problem at hand. Furthermore, the (obvious) modifications which must be performed in order to take care of the peculiarity of the problem, have been made relative to the functions α, m and the notation defined by eqs. (7.4).

Theorem 7.7 *Let (x^o, u^o, t_f^o) be a triple which satisfies the necessary optimality conditions (7.8), (7.9). Moreover, let $R(t) > 0$, $t \in [t_0, t_f^o]$ and assume that there exist solutions of the equations*

$$\begin{aligned}
\dot{P}_1(t) &= -P_1(t)A(t) - A'(t)P_1(t) + P_1(t)K(t)P_1(t) - Q(t),\\
\dot{P}_3(t) &= -\left[A(t) - K(t)P_1(t)\right]' P_3(t),\\
\dot{P}_4(t) &= P_3'(t)K(t)P_3(t)
\end{aligned}$$

with the boundary conditions

$$\begin{aligned}
P_1(t_f^o) &= S_1,\\
P_3(t_f^o) &= S_3,\\
P_4(t_f^o) &= S_4
\end{aligned}$$

and such that $P_4(t) > 0$, $t \in [t_0, t_f^o]$. Then, the variations $\delta_2^o u$ of the control and $\delta_2^o t_f$ of the final time which minimize $[\delta J]_2$ when $x(\tau) = x^o(\tau) + \delta x_\tau$, $\tau \in [t_0, t_f^o)$ are given by

$$\begin{aligned}
\delta_2^o u(t) &= -R^{-1}(t)\left[B'(t)\bar{P}_1(t) + H_{ux}^o(t)\right]\delta_2^o x(t),\ t \in [\tau, t_f^o),\\
\delta_2^o t_f &= -P_3'(\tau)P_4^{-1}(\tau)\delta x_\tau
\end{aligned}$$

where $\delta_2^o x$ is the solution of the equation

$$\delta\dot{x}(t) = \left[A(t) - K(t)\bar{P}_1(t))\right]\delta x(t)$$

with the boundary condition $\delta x(\tau) = \delta x_\tau$ and, for $t \in [\tau, t_f^o)$, $\bar{P}_1(t) := P_1(t) - P_3(t)P_4^{-1}(t)P_3'(t)$.

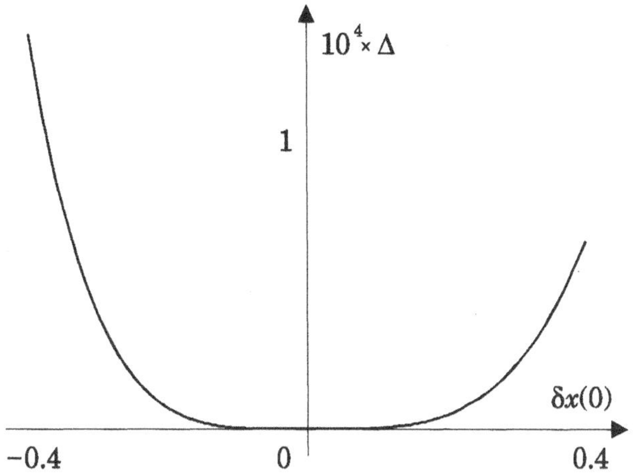

Figure 7.8: Example 7.9: performance loss when $u + \delta_2 u$ is applied.

Example 7.9 Consider a modified version of the problem described in Example 7.7 where the final state is free and a function of it is added to the performance index, namely,

$$J = \frac{1}{2}\int_0^{t_f} (2 + u^2)dt + x(t_f).$$

The NC are satisfied only by (the superscript "o" has been omitted)

$$x(t) = 2e^{-\sqrt{2}t} + 1,$$
$$u(t) = -\sqrt{2},$$
$$t_f = \frac{\ln(2/\sqrt{2})}{\sqrt{2}}.$$

Corresponding to this, we find $A(t) = -2\sqrt{2}$, $B(t) = 2e^{-\sqrt{2}t}$, $Q(t) = -e^{2\sqrt{2}t}/2$, $R(t) = 1$, $S_1 = 0$, $S_3 = -\sqrt{2}$, $S_4 = 2\sqrt{2}$ and the check can be performed whether the assumptions of Theorem 7.7 are verified. The equations for P_1, P_3, P_4 and the relevant boundary condition admit the solutions

$$P_1(t) = \frac{(t_f - t)}{2\beta(t)} e^{2\sqrt{2}t}, \; \beta(t) := \sqrt{2}(t - t_f) - 1,$$

$$P_3(t) = \frac{e^{\sqrt{2}t}}{\beta(t)}, \; P_4(t) = -\frac{2\sqrt{2}}{\beta(t)}.$$

Note that $P_4(t) > 0$, $t \in [0, t_f]$ so that Theorem 7.7 can be exploited. It follows that

$$\bar{P}_1 = -\frac{1}{2\sqrt{2}} e^{2\sqrt{2}t}$$

and

$$\delta_2 t_f(\delta x(0)) = \frac{\delta x(0)}{2\sqrt{2}},$$
$$\delta_2 u(t, \delta x(0)) = 0,$$
$$\delta_2 x(t, \delta x(0)) = e^{-\sqrt{2}t}\delta x(0).$$

As a conclusion, the control scheme of Fig. 7.5 can be implemented and it is interesting to evaluate the quantity $\Delta := (J_2 - J^o)/J^o$ (see Fig. 7.8), where

$$J_2(x_0 + \delta x_0) = \frac{2}{\sqrt{2}}(\ln(\sqrt{2}) + \delta x(0)/2) + (1 + \delta x(0)/2)\sqrt{2}e^{-\delta x(0)/2} + 1$$

is the value of the performance index relative to the neighbouring optimal control $u + \delta_2 u$ and

$$J^o(x_0 + \delta x_0) = -\sqrt{2}\ln(\frac{\sqrt{2}}{2 + \delta x(0)}) + \sqrt{2} + 1$$

is the value of the performance index relative to the control which satisfies the NC.

Finally, in the fourth case (free final state, given final time) the following theorem holds where, as done in the two preceding cases, reference is made only to those among the NC (7.8), (7.9) which apply to the problem at hand. Furthermore, the (obvious) modifications which must be performed in order to take care of the peculiarity of the problem, have been made relative to the functions α, m and the notation defined by eqs. (7.4).

Theorem 7.8 *Let (x^o, u^o) be a couple which satisfies the necessary optimality conditions (7.8), (7.9). Moreover, let $R(t) > 0$, $t \in [t_0, T]$ and assume that there exists a solution of the equation*

$$\dot{P}_1(t) = -P_1(t)A(t) - A'(t)P_1(t) + P_1(t)K(t)P_1(t) - Q(t)$$

with the boundary condition

$$P_1(T) = S_1.$$

Then, the variation $\delta_2^o u$ of the control which minimizes $[\delta J]_2$ when $x(\tau) = x^o(\tau) + \delta x_\tau$, $\tau \in [t_0, T)$ is given by

$$\delta_2^o u(t) = -R^{-1}(t)\left[B'(t)P_1(t) + H_{ux}^o(t)\right]\delta_2^o x(t), \ t \in [\tau, T)$$

where $\delta_2^o x$ is the solution of the equation

$$\delta\dot{x}(t) = \left[A(t) - K(t)\bar{P}_1(t))\right]\delta x(t)$$

with the boundary condition $\delta x(\tau) = \delta x_\tau$.

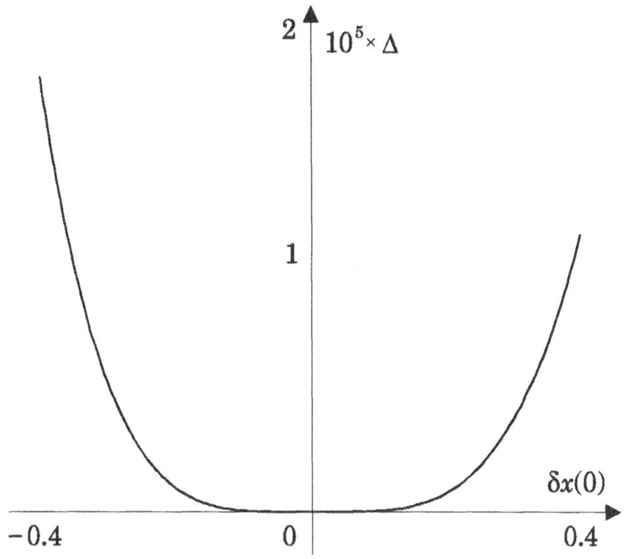

Figure 7.9: Example 7.10: performance loss when $u + \delta_2 u$ is applied.

Example 7.10 Consider a modified version of the problems described in Examples 7.7, 7.8 where the final state is free, the final time is $T := \ln 2$, and the performance index is

$$J = \frac{1}{2}\int_0^T u^2 dt + x(T).$$

The NC can be satisfied only if (the superscript "o" has been omitted)

$$\begin{aligned} x(t) &= 2e^{-t} + 1, \\ u(t) &= -1. \end{aligned}$$

Corresponding to this, we find $A(t) = -2$, $B(t) = 2e^{-t}$, $Q(t) = -e^{2t}/4$, $R(t) = 1$, $S_1 = 0$ and the check can be performed whether the assumptions of Theorem 7.8 are verified. The equation for P_1 and the relevant boundary condition admits the solution

$$P_1(t) = \frac{T-t}{4\beta(t)}e^{2t}, \ \beta(t) := t - T - 1.$$

Thus Theorem 7.8 can be exploited, yielding

$$\begin{aligned} \delta_2 u(t, \delta x(0)) &= -\frac{\delta x(0)}{2(T+1)}, \\ \delta_2 x(t, \delta x(0)) &= -\frac{\beta(t)\delta x(0)}{T+1}e^{-t}. \end{aligned}$$

As a conclusion, the control scheme of Fig. 7.5 can be implemented and it is interesting to evaluate the quantity $\Delta := (J_2 - J^o)/J^o$ (see Fig. 7.9), where

$$J_2(x_0 + \delta x_0) = \frac{(2(T+1) + \delta x(0))^2 T}{8(T+1)^2} + x_2(T),$$
$$x_2(T) = (2 + \delta x(0))e^{-(1+\delta x(0)/(2(T+1)))T} + 1$$

is the value of the performance index relative to the neighbouring optimal control $u + \delta_2 u$ and

$$J^o(x_0 + \delta x_0) = \frac{(\lambda^o(0)(2 + \delta x(0))^2 T}{2} + x^o(T),$$
$$x^o(T) = (2 + \delta x(0))\lambda^o(0) + 1$$

is the value of the performance index relative to the control which satisfies the NC, $\lambda^o(0)$ being the solution of the equation

$$1 = \lambda^o(0)e^{\lambda^o(0)(2+\delta x(0))T}.$$

7.4 Problems

Problem 7.4.1 Consider the optimal control problem defined by the system $\dot{x}_1 = x_2$, $\dot{x}_2 = u$, $x(0) = 0$, the performance index

$$J = \frac{1}{2}\left\{\int_0^{t_f} u^2 dt + t_f^2 + \gamma x_1(t_f)\right\}$$

where t_f is free and $\gamma = \pm 2$. The final state is constrained by the equation $x_1 - x_2 - 1 = 0$. Discuss the optimality of the solutions (if any) which satisfy the NC.

Problem 7.4.2 Consider the optimal control problem defined by the system $\dot{x}_1 = x_2$, $\dot{x}_2 = u$, $x(0) = 0$ and the performance index

$$J = \frac{1}{2}\int_0^1 (u^2 + 2x_1)dt.$$

The final state must comply with the constraint $x_1^2 + x_2^2 - 1 = 0$. Discuss the optimality of the solutions (if any) which satisfy the NC.

Problem 7.4.3 Consider the optimal control problem defined by the first order system $\dot{x} = u$, $x(0) = 1$ and the performance index

$$J = \frac{1}{2}\int_0^{t_f} (1 + u^2 + x)dt - 2x(t_f)$$

where both the final state and time are free. Discuss the optimality of the solutions (if any) which satisfy the NC.

Problem 7.4.4 Consider the optimal control problem defined by the system $\dot{x}_1 = x_2$, $\dot{x}_2 = u$, $x(0) = 0$ and the performance index

$$J = \frac{1}{2}\left\{\int_0^1 (u^2 + 2x_1)dt + x_2^2(1)\right\}$$

where the final state is free. Discuss the optimality of the solutions (if any) which satisfy the NC.

Problem 7.4.5 Consider the optimal control problem defined by the first order system $\dot{x} = xu$ with given initial and final state, x_0 and x_f, respectively, with $x_0 \neq x_f$, $x_0 x_f > 0$ and the performance index

$$J = \frac{1}{2}\int_0^{t_f} (2 + u^2)dt$$

where the final time is free. Find (if any) the triples (x, u, t_f) which satisfy the NC and such that Theorem 7.5 can be applied.

Problem 7.4.6 Consider the optimal control problem defined by the first order system $\dot{x} = xu$ with given initial and final state, x_0 and x_f, respectively, with $x_0 \neq x_f$, $x_0 x_f > 0$ and the performance index

$$J = \frac{1}{2}\int_0^1 u^2 dt.$$

Find (if any) the couples (x, u) which satisfy the NC and such that Theorem 7.6 can be applied.

Problem 7.4.7 Consider the optimal control problem defined by the first order system $\dot{x} = xu$ with given initial $x_0 \neq 0$ and the performance index

$$J = \frac{1}{2}\int_0^{t_f} (2 + u^2)dt + x(t_f)$$

where the final state and time are free. Find (if any) the triples (x, u, t_f) which satisfy the NC and such that Theorem 7.7 can be applied.

Problem 7.4.8 Consider the optimal control problem defined by the first order system $\dot{x} = xu$ with given initial state, x_0 with $x_0 > 0$, $x_0 x_f > 0$ and the performance index

$$J = \frac{1}{2}\int_0^1 u^2 dt + x(1)$$

where the final state is free. Find (if any) the couples (x, u) which satisfy the NC and such that Theorem 7.8 can be applied.

Problem 7.4.9 Consider the optimal control problem defined by a linear system and a linear quadratic performance index which has been discussed in Section 3.2 of Chapter 3.1 (Remark 3.3). Show that the solution given there satisfies all the assumptions of Theorem 7.4.

Problem 7.4.10 Check whether Theorem 7.2 can be applied to the optimal control problem defined by the system $\dot{x} = Ax + Bu$, $x(0) = x_0$, $x(1) = 0$ and the performance index

$$J = \frac{1}{2} \int_0^1 u^2 dt$$

where the pair (A, B) is reachable and x_0 is given.

Problem 7.4.11 Find the neighbouring optimal control for the optimal control problem defined by the system $\dot{x}_1 = x_2$, $\dot{x}_2 = u$, $x(0) = x_0$, $x(t_f) = 0$ and the performance index

$$J = \frac{1}{2} \int_0^{t_f} (2 + u^2) dt$$

where x_0 is given.

Appendix A

Basic background

A.1 Canonical decomposition

Consider an n-th order linear dynamical system $\Sigma(A, B, C, D)$. It is always possible to perform a change of variables on Σ, which is defined by a nonsingular matrix T and is such as to put into evidence the parts of the system which are

(a) *Reachable but not observable;*

(b) *Reachable and observable;*

(c) *Not reachable and not observable;*

(d) *Not reachable but observable.*

If $x_{r,no}$, $x_{r,o}$, $x_{nr,no}$, $x_{nr,o}$ denote, in the same order, the state variables of these four parts, the matrix T can be chosen in such a way that $Tx = \begin{bmatrix} x'_{r,no} & x'_{r,o} & x'_{nr,no} & x'_{nr,o} \end{bmatrix}'$ and

$$\hat{A} := TAT^{-1} = \begin{bmatrix} A_1 & A_2 & A_3 & A_4 \\ 0 & A_5 & 0 & A_6 \\ 0 & 0 & A_7 & A_8 \\ 0 & 0 & 0 & A_9 \end{bmatrix}, \quad \hat{B} := TB = \begin{bmatrix} B_1 \\ B_2 \\ 0 \\ 0 \end{bmatrix},$$

$$\hat{C} := CT^{-1} = \begin{bmatrix} 0 & C_1 & 0 & C_2 \end{bmatrix}$$

with

$$\left(\begin{bmatrix} A_1 & A_2 \\ 0 & A_5 \end{bmatrix}, \begin{bmatrix} B_1 \\ B_2 \end{bmatrix} \right) = \text{reachable}$$

$$\left(\begin{bmatrix} A_5 & A_6 \\ 0 & A_9 \end{bmatrix}, \begin{bmatrix} C_1 & C_2 \end{bmatrix} \right) = \text{observable.}$$

The matrices $\hat{A}$, $\hat{B}$, $\hat{C}$ and D specify the canonical decomposition of Σ into the above mentioned parts. A way of computing T is given in the following algorithm where

$$K_r := \begin{bmatrix} B & AB & A^2B & \cdots & A^{n-1}B \end{bmatrix},$$
$$K_o := \begin{bmatrix} C' & A'C' & (A^2)'C' & \cdots & (A^{n-1})'C' \end{bmatrix}.$$

Algorithm A.1

(i) *Compute* $n_r := \text{rank}(K_r)$ *and* $n_o := \text{rank}(K_o)$.

(ii) *Find two matrices* X_r *and* X_o *of dimensions* $n \times n_r$ *and* $n \times n_o$, *respectively, such that*

$$\text{Im}(X_r) = \text{Im}(K_r),\ \text{Im}(X_o) = \text{Im}(K_o).$$

(iii) *Find two matrices* $X_r^{\perp}$ *and* $X_o^{\perp}$ *of maximum rank and dimensions* $n \times (n - n_r)$ *and* $n \times (n - n_o)$, *respectively, such that*

$$(X_r^{\perp})'X_r = 0,\ (X_o^{\perp})'X_o = 0.$$

(iv) *Find two matrices* $X_{r,no}$ *and* $X_{nr,o}$ *of maximum rank and dimensions* $n \times (n\text{-rank}(\begin{bmatrix} X_r^{\perp} & X_o \end{bmatrix}))$ *and* $n \times (n\text{-rank}(\begin{bmatrix} X_r & X_o^{\perp} \end{bmatrix}))$, *respectively, such that*

$$X'_{r,no}\begin{bmatrix} X_r^{\perp} & X_o \end{bmatrix} = 0,\ X'_{nr,o}\begin{bmatrix} X_r & X_o^{\perp} \end{bmatrix} = 0.$$

(v) *Find two matrices* $X_{r,o}$ *and* $X_{nr,no}$ *of maximum rank and dimensions* $n \times (n\text{-rank}(\begin{bmatrix} X_r^{\perp} & X_{r,no} \end{bmatrix}))$ *and* $n \times (n\text{-rank}(\begin{bmatrix} X_{r,no} & X_o \end{bmatrix}))$, *respectively, such that*

$$X'_{r,o}\begin{bmatrix} X_r^{\perp} & X_{r,no} \end{bmatrix} = 0,\ X'_{nr,no}\begin{bmatrix} X_{r,no} & X_o \end{bmatrix} = 0.$$

(vi)
$$T^{-1} - \begin{bmatrix} X_{r,no} & X_{r,o} & X_{nr,no} & X_{nr,o} \end{bmatrix}.$$

Example A.1 Consider the system Σ defined by

$$A = \begin{bmatrix} 6 & -3 & 4 & -6 \\ 3 & 0 & 4 & -6 \\ 2 & -1 & 5 & -5 \\ 1 & 0 & 1 & -1 \end{bmatrix},\ B = \begin{bmatrix} 1 \\ 2 \\ 2 \\ 1 \end{bmatrix},\ C = \begin{bmatrix} 1 & -1 & 2 & -2 \end{bmatrix}.$$

By choosing

$$X_r = \begin{bmatrix} 1 & 2 \\ 2 & 5 \\ 2 & 5 \\ 1 & 2 \end{bmatrix},\ X_o = \begin{bmatrix} 1 & 5 \\ -1 & -5 \\ 2 & 8 \\ -2 & -8 \end{bmatrix}$$

it is possible to select

$$X_r^\perp = \begin{bmatrix} 1 & 0 \\ 0 & 1 \\ 0 & -1 \\ -1 & 0 \end{bmatrix}, \quad X_o^\perp = \begin{bmatrix} 1 & 0 \\ 1 & 0 \\ 0 & 1 \\ 0 & 1 \end{bmatrix},$$

from which it follows that

$$X_{r,no} = \begin{bmatrix} 1 \\ 1 \\ 1 \\ 1 \end{bmatrix}, \quad X_{r,o} = \begin{bmatrix} 1 \\ -1 \\ -1 \\ 1 \end{bmatrix}, \quad X_{nr,no} = \begin{bmatrix} 1 \\ 1 \\ -1 \\ -1 \end{bmatrix}, \quad X_{nr,o} = \begin{bmatrix} 1 \\ -1 \\ 1 \\ -1 \end{bmatrix}.$$

Thus we obtain

$$\hat{A} = \begin{bmatrix} 1 & -4 & 3 & 12 \\ 0 & 3 & 0 & -1 \\ 0 & 0 & 2 & 4 \\ 0 & 0 & 0 & 4 \end{bmatrix}, \quad \hat{B} = \begin{bmatrix} 1.5 \\ -0.5 \\ 0 \\ 0 \end{bmatrix}, \quad \hat{C} = \begin{bmatrix} 0 & -2 & 0 & 6 \end{bmatrix}.$$

With reference to the above defined matrix $\hat{A}$, the system Σ (the pair (A, B)) is said to be stabilizable if the two submatrices A_7 and A_9 are stable, while the system Σ (the pair (A, C)) is said to be detectable if the two submatrices A_1 and A_7 are stable. Obviously, if the system Σ is reachable (observable), it is also stabilizable (detectable).

Reachability (stabilizability) and observability (detectability) of a system can easily be checked by resorting to the following theorem, referred to as the *PBH* test, where

$$P_B(\lambda) := \begin{bmatrix} \lambda I - A & -B \end{bmatrix}, \qquad P_C(\lambda) := \begin{bmatrix} \lambda I - A \\ C \end{bmatrix}.$$

Theorem A.1 *The system $\Sigma(A, B, C, D)$ is:*

(i) *Reachable if and only if*

$$\text{rank}\,(P_B(\lambda)) = n, \ \forall \lambda.$$

(ii) *Observable if and only if*

$$\text{rank}\,(P_C(\lambda)) = n, \ \forall \lambda.$$

(iii) *Stabilizable if and only if*

$$\text{rank}\,(P_B(\lambda)) = n, \ \forall \lambda, \ \text{Re}[\lambda] \geq 0.$$

(iv) *Detectable if and only if*

$$\text{rank}\,(P_C(\lambda)) = n, \ \forall \lambda, \ \text{Re}[\lambda] \geq 0.$$

Furthermore, the values of λ corresponding to which the rank of matrix $P_B(\lambda)$ is less than n are the eigenvalues of the unreachable part of Σ, and the values of λ corresponding to which the rank of matrix $P_C(\lambda)$ is less than n are the eigenvalues of the unobservable part of Σ.

A.2 Transition matrix

Consider the linear system

$$\begin{aligned}\dot{x}(t) &= A(t)x(t) + B(t)u(t),\\ x(t_0) &= x_0\end{aligned}$$

where $A(t)$ are $B(t)$ are matrices of continuous functions. The state motion of this system is given by

$$x(t) = \Phi(t,t_0)x_0 + \int_{t_0}^{t} \Phi(t,\tau)B(\tau)u(\tau)d\tau$$

where $\Phi(t,\tau)$ is the *transition matrix* associated to $A(t)$, i.e., is the unique solution of the matrix differential equation

$$\begin{aligned}\frac{d\Phi(t,\tau)}{dt} &= A(t)\Phi(t,\tau),\\ \Phi(\tau,\tau) &= I.\end{aligned}$$

The transition matrix enjoys the following properties.

(a) $\Phi(t,\tau)$ is nonsingular $\forall t,\ \forall \tau$;
(b) $\Phi(t,\tau)^{-1} = \Phi(\tau,t),\ \forall t,\ \forall \tau$;
(c) $\Phi(t_1,t_2)\Phi(t_2,t_3) = \Phi(t_1,t_3),\ \forall t_1,\ \forall t_2,\ \forall t_3$;
(d) $\Phi'(\tau,t) = \Psi(t,\tau)$, where $\Psi(t,\tau)$ is the transition matrix associated to $-A'(t)$;
(e) If $A(t) =$ const., $\Phi(t,\tau) = e^{A(t-\tau)}$.

Example A.2 Let

$$A(t) = \begin{bmatrix} 1 & e^{-t^2} \\ 0 & 2t \end{bmatrix}.$$

The relevant transition matrices are

$$\Phi(t,\tau) = \begin{bmatrix} e^{t-\tau} & e^{-\tau^2}(e^{t-\tau}-1) \\ 0 & e^{t^2-\tau^2} \end{bmatrix}$$

and

$$\Psi(t,\tau) = \begin{bmatrix} e^{\tau-t} & 0 \\ e^{-t^2}(e^{\tau-t}-1) & e^{\tau^2-t^2} \end{bmatrix}.$$

Properties (a)–(d) can easily be checked.

A.3 Poles and zeros

Consider the n-th order linear dynamical system $\Sigma(A, B, C, D)$ with p outputs, m inputs and let $G(s) = C(sI - A)^{-1}B + D$ be the relevant transfer function with $\mathrm{rank}(G(s)) = r$, $r \leq \min(p, m)$. Furthermore, consider the (canonical) McMillan-Smith form of $G(s)$, i.e., the matrix $M(s)$ such that $G(s) = L(s)M(s)R(s)$, L and R being unimodular polynomial matrices, and

$$M(s) = \left| \begin{array}{ccccc} f_1(s) & 0 & \cdots & 0 & 0 \\ 0 & f_2(s) & \cdots & 0 & 0 \\ \vdots & \vdots & \ddots & \vdots & \vdots \\ 0 & 0 & \cdots & f_r(s) & 0 \\ 0 & 0 & \cdots & 0 & 0 \end{array} \right| \begin{array}{l} \\ \\ \\ \\ p-r \text{ rows} \end{array}$$

$$\qquad\qquad\qquad\qquad m-r \text{ columns}$$

with

$$f_i(s) = \frac{\varepsilon_i(s)}{\psi_i(s)}, \; i = 1, 2, \ldots, r$$

where

- $\varepsilon_i(s)$ and $\psi_i(s)$ are monic polynomials, $i = 1, 2, \ldots, r$,
- $\varepsilon_i(s)$ and $\psi_i(s)$ are coprime polynomials, $i = 1, 2, \ldots, r$,
- $\varepsilon_i(s)$ divides $\varepsilon_{i+1}(s)$, $i = 1, 2, \ldots, r - 1$,
- ψ_{i+1} divides ψ_i, $i = 1, 2, \ldots, r - 1$.

Based on the polynomials ε_i and ψ define the polynomials

$$\pi_p(s) := \prod_{i=1}^{r} \psi_i(s), \; \pi_{zt}(s) := \prod_{i=1}^{r} \varepsilon_i(s).$$

The polynomial π_p is the *least common multiple* of the polynomials which are the denominators of all nonzero minors of any order of $G(s)$, while the polynomial π_{zt} is the *greatest common divisor* of the polynomials which are the numerators of all nonzero minors of order r of $G(s)$, once they have been adjusted so as to have π_p as a denominator. Finally, consider the *system matrix*

$$P(s) := \begin{bmatrix} sI - A & -B \\ C & D \end{bmatrix}.$$

The following definitions can now be given.

Definition A.1 (Poles of Σ) *The poles of Σ are the roots of the polynomial π_p.*

Definition A.2 (Eigenvalues of Σ) *The eigenvalues of Σ are the roots of the characteristic polynomial of A.*

Definition A.3 (Transmission zeros of Σ) *The transmission zeros of Σ are the roots of the polynomial π_{zt}.*

Definition A.4 (Invariant zeros of Σ) *The invariant zeros of Σ are the values of s corresponding to which* $\text{rank}(P(s)) < n + \min(m,p)$.

The set of the transmission zeros of Σ is a subset of the set of invariant zeros of Σ, so that each transmission zero is also an invariant zero, while the converse is not true in general. However if Σ is minimal the two sets coincide.

A.4 Quadratic forms

Let Q be an $n \times n$ complex matrix and x an n-dimensional complex vector. The function

$$f(x) := x^{\sim} Q x$$

is a quadratic form with respect to x with kernel Q. Notice that if f is real

$$x^{\sim} Q x = \frac{1}{2}\left(x^{\sim} Q x + x^{\sim} Q x\right) = \frac{1}{2}\left(x^{\sim} Q x + x^{\sim} Q^{\sim} x\right) = \frac{1}{2} x^{\sim}(Q + Q^{\sim})x$$

so that $f(x) = x^{\sim} H x$, where $H := \frac{1}{2}(Q + Q^{\sim})$ is an hermitian matrix, i.e., a matrix which coincides with its conjugate transpose. In conclusion, the kernel of any real quadratic form can be assumed to be an hermitian matrix without loss of generality (the kernel is symmetric when working in the field of real numbers).

A sign definition property can be attached to hermitian matrices.

Definition A.5 (Positive semidefinite matrix) *An hermitian matrix Q is positive semidefinite ($Q \geq 0$) if $x^{\sim} Q x \geq 0$, $\forall x$.*

Definition A.6 (Positive definite matrix) *An hermitian matrix Q is positive definite ($Q > 0$) if it is positive semidefinite and $x^{\sim} Q x = 0 \Leftrightarrow x = 0$.*

From these definitions it follows that for any matrix B and $Q = Q^{\sim} \geq 0$ the matrix $R := B^{\sim} Q B$ is positive semidefinite.

The check whether an hermitian matrix Q is positive semidefinite (definite) can be done as shown in the forthcoming theorems where $q_{\{i_1, i_2, \ldots, i_k\}}$ denotes the determinant of the submatrix built with the elements of Q which simultaneously belong to the $i_1, i_2, \ldots, i_k$ rows and columns.

Theorem A.2 *The $n \times n$ hermitian matrix Q is positive definite (semidefinite) if and only if all its eigenvalues are positive (nonnegative).*

Thus a positive definite matrix is nonsingular and its inverse is positive definite as well.

Theorem A.3 *The $n \times n$ hermitian matrix Q is positive definite if and only if the following inequalities all hold*

$$\begin{aligned} q_{\{1\}} &> 0, \\ q_{\{1,2\}} &> 0, \\ &\vdots \\ q_{\{1,2,\ldots,n\}} &> 0. \end{aligned}$$

Theorem A.4 *The $n \times n$ hermitian matrix Q is positive semidefinite if and only if the following inequalities all hold*

$$q_{\{i_1,i_2,\ldots,i_p\}} \geq 0, \; 1 \leq i_1 < i_2 < \cdots < i_p \leq n, \; p = 1, 2, \ldots, n.$$

Hermitian positive semidefinite matrices can be factorized, i.e., they can be expressed as the product of a suitable matrix with its conjugate transpose. Thus if $Q = Q^{\sim} \geq 0$ it is possible to find a matrix C such that

$$Q = C^{\sim}C.$$

This property follows from the fact that each hermitian matrix can be diagonalized by means of a unitary matrix (a matrix the inverse of which coincides with its conjugate transpose) so that if $Q = Q^{\sim} \geq 0$ there exists a matrix U with $U^{-1} = U^{\sim}$ such that

$$Q = UDU^{\sim}$$

where D is a diagonal matrix with nonnegative elements. If $D^{\frac{1}{2}}$ is the matrix the elements of which are the square roots of the corresponding elements of D, so that $D = D^{\frac{1}{2}}D^{\frac{1}{2}}$, then the matrix

$$C := UD^{\frac{1}{2}}U^{\sim}$$

is a factorization of Q, since $(D^{\frac{1}{2}})^{\sim} = D^{\frac{1}{2}}$. An $n \times n$ matrix $Q = Q^{\sim} \geq 0$ admits many factorizations: among them are those with dimensions $\text{rank}(Q) \times n$. When Q is real, U and C can be selected as real.

A.5 Expected value and covariance

Consider the linear dynamic system

$$\dot{x}(t) = A(t)x(t) + B(t)w(t), \tag{A.1a}$$

$$x(t_0) = x_0 \tag{A.1b}$$

where A and B are matrices of continuous functions, w is a zero-mean white noise with intensity W, x_0 is a random variable independent from w with expected value $\bar{x}_0$ and variance Π_0. For this system the following results hold.

Theorem A.5

(i) *The expected value of x at time $t \geq t_0$ is*

$$E[x(t)] = \Phi(t, t_0)\bar{x}_0$$

where Φ is the transition matrix associated to A.

(ii) *The variance Π of x at time $t \geq t_0$ solves the differential equation*

$$\begin{aligned}\dot{\Pi}(t) &= A(t)\Pi(t) + \Pi(t)A'(t) + B(t)WB'(t),\\ \Pi(t_0) &= \Pi_0.\end{aligned}$$

With reference to the system (A.1), where it is assumed that $B(t) = I$, consider the functional

$$J_1 := E\left[\int_{t_0}^{t_f} x'(t)Q(t)x(t)dt + x'(t_f)Sx(t_f)\right] \tag{A.2}$$

with $Q = Q' \geq 0$ and $S = S' \geq 0$. The value of this functional can easily be computed according to the following theorem.

Theorem A.6 *The value of the functional (A.2) with eqs. (A.1) taken into account is*

$$J_1 = \mathrm{tr}\left[P(t_0)(\Pi_0 + \bar{x}_0\bar{x}_0') + W\int_{t_0}^{t_f} P(t)dt\right]$$

where P is the solution of the Lyapunov differential equation

$$\dot{P}(t) = -P(t)A(t) - A'(t)P(t) - Q(t), \; P(t_f) = S.$$

Particular yet meaningful is the case where the input noise w is not present and the initial state is known (thus $W = 0$ and $\Pi_0 = 0$) and the case where the initial state is zero (thus $\bar{x}_0 = 0$ and $\Pi_0 = 0$).

Somehow different is the situation when the functional is defined over an unbounded interval. By restraining our attention to the case where $S = 0$, both the system and the functional are time-invariant (the matrices A and Q are constant in eqs. (A.1), (A.2)), $t_0 = 0$ and $t_f = \infty$, it is easy to conclude that J_1 is no longer bounded so that it makes sense to assume either that $w(\cdot) = 0$ when the functional is as in eq. (A.2), or to set $x(0) = 0$ when the functional becomes

$$J_2 := E\left[\lim_{T\to\infty} \frac{1}{T}\int_0^T x'(t)Qx(t)dt\right].$$

Letting C be any factorization of Q and T a nonsingular matrix such that

$$\hat{A} := TAT^{-1} := \begin{bmatrix} A_1 & 0 \\ A_2 & A_3 \end{bmatrix}, \quad \hat{C} := CT^{-1} := \begin{bmatrix} C_1 & 0 \end{bmatrix}$$

where the pair (A_1, C_1) is observable, the following result holds.

Theorem A.7 *Let the matrices A and Q be constant, $t_0 = 0$ and $t_f = \infty$. Let the matrix A_1 be stable. Then,*

i) *If $w(\cdot) = 0$,*

$$J_1 = \operatorname{tr}\left[\bar{P}(\Pi_0 + \bar{x}_0\bar{x}_0')\right].$$

ii) *If $x(0) = 0$,*

$$J_2 = \operatorname{tr}\left[\bar{P}W\right]$$

where

$$\bar{P} = T'\begin{bmatrix} \bar{P}_1 & 0 \\ 0 & 0 \end{bmatrix}T.$$

$\bar{P}_1$ being the solution of the algebraic Lyapunov equation $0 = PA_1 + A_1'P + C_1'C_1$.

Appendix B

Eigenvalues assignment

B.1 Introduction

The pole placement problem consists in designing, if possible, a feedback controller R (linear, finite dimensional, time-invariant) for a given system P (linear, finite dimensional and time-invariant as well) in such a way that the eigenvalues of the resulting closed loop system are located in arbitrarily preassigned positions of the complex plane. It is obvious that only the eigenvalues of P belonging to its jointly reachable and observable part (namely the *poles* of P) can be affected by a control law which is implemented on the system output (see Fig. B.1, where this part is drawn with heavier lines). For this reason the problem at hand is referred to as the eigenvalue placement problem and P is assumed to be minimal.

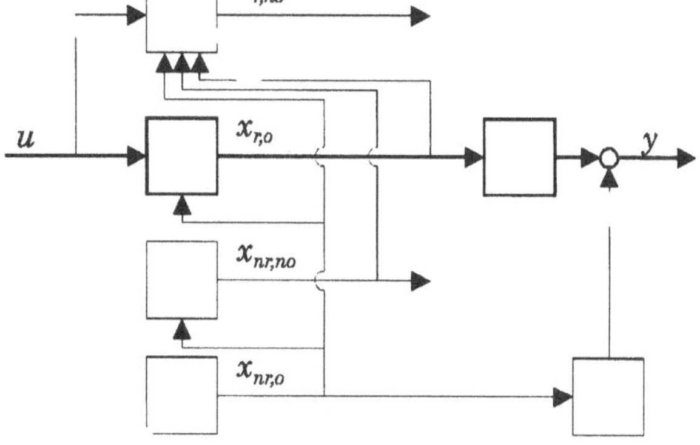

Figure B.1: The canonical decomposition of P.

The eigenvalues of the closed loop system constitute a *symmetric* set of complex numbers since the controller is described by equations with real coefficients: thus, if p is a complex number which appears in such a set n_p times, also the number $p^{\sim}$, the conjugate of p, will be present n_p times.

The next sections deal with the *eigenvalue assignment* when the state is *accessible* (the controller input is the state of P) or the state is *not accessible* (the controller input is the output of P) or asymptotic *zero error regulation* is required.

B.2 Assignment with accessible state

The eigenvalue assignment problem when the controller input is the state of the system

$$\dot{x}(t) = Ax(t) + Bu(t) \tag{B.1}$$

(where $x \in R^n$ and $u \in R^m$, $m \leq n$) can be stated in the following way.

Problem B.1 (Eigenvalue assignment with accessible state) *Find a control law*

$$u(x) = Kx \tag{B.2}$$

such that the eigenvalues of the closed loop system (B.1), (B.2) constitute a preassigned symmetric set Λ *of* n *complex numbers.*

In other words, given the set Λ, this problem amounts to selecting, if possible, a matrix K in such a way that the set of the eigenvalues of $A + BK$ coincides with Λ.

A very nice result can be proved relative to problem B.1: however, when the control variable is not a scalar it requires a preliminary fact which is stated in the following lemma where β_i denotes the i-th column of B and $[I_m]^i$ is the i-th column of the $m \times m$ identity matrix. Furthermore, the rank of matrix B is equal to the number of its columns, without a true loss of generality: thus $\beta_i \neq 0$, $\forall i$.

Lemma B.1 *Let the pair* (A, B) *be reachable. Then there exists a matrix* K_i *such that the pair* $(A + BK_i, B[I_m]^i)$ *is reachable.*

Proof. The proof of this lemma can easily be found in the literature, thus only the form of one of the infinite many matrices K_1 is here given. Consider the matrix

$$Q := \begin{bmatrix} Q_1 & Q_2 & \cdots & Q_m \end{bmatrix} \tag{B.3}$$

with

$$Q_i := \begin{bmatrix} \beta_i & A\beta_i & \cdots & A^{h_i-1}\beta_i \end{bmatrix}$$

where h_i, $i = 1, 2, \ldots, m$ are suitable nonnegative integers (if $h_i = 0$, no terms containing β_i appear in eq. (B.3)) which satisfy the relation $\sum_{i=1}^{m} h_i = n$ and are such that $A^{h_i}\beta_i$ is a linear combination of the previous columns already appearing in Q. The reachability of the pair (A, B) implies that Q is square and nonsingular. Let further: *a)* $\bar{\imath}$ be the maximal index i corresponding to which $h_i \neq 0$; *b)* $\mathcal{I}$ be the set of indexes i corresponding to which $h_i \neq 0$, deprived of the index $\bar{\imath}$; *c)* k_i, $i \in \mathcal{I}$, be the minimal index $j > i$ corresponding to which $h_j \neq 0$; *d)* $\nu_i := \sum_{j=1}^{i} h_j$, $i \in \mathcal{I}$. According to these definitions construct an $m \times n$ matrix V with the ν_i-th column, $i \in \mathcal{I}$ equal to $[I_m]^{k_i}$, while the remaining columns are zero. The matrix $K_1 := VQ^{-1}$ solves the problem.

A simple example clarifies Lemma B.1.

Example B.1 Consider the matrices

$$A = \begin{bmatrix} 0 & 1 & 0 & 0 \\ 0 & 0 & 0 & 0 \\ 0 & 0 & 0 & 1 \\ 0 & 0 & 0 & 0 \end{bmatrix}, \quad B = \begin{bmatrix} 0 & 0 \\ 1 & 0 \\ 0 & 0 \\ 0 & 1 \end{bmatrix}.$$

One finds

$$Q = \begin{bmatrix} \beta_1 & A\beta_1 & \beta_2 & A\beta_2 \end{bmatrix} = \begin{bmatrix} 0 & 1 & 0 & 0 \\ 1 & 0 & 0 & 0 \\ 0 & 0 & 0 & 1 \\ 0 & 0 & 1 & 0 \end{bmatrix}$$

and the parameters of interest are $h_1 = h_2 = 2$, $\bar{\imath} = 2$, $\mathcal{I} = \{1\}$, $k_1 = 2$, $\nu_1 = 2$: thus it follows that

$$V = \begin{bmatrix} 0 & 0 & 0 & 0 \\ 0 & 1 & 0 & 0 \end{bmatrix}, \quad K_1 = VQ^{-1} = \begin{bmatrix} 0 & 0 & 0 & 0 \\ 1 & 0 & 0 & 0 \end{bmatrix}.$$

The pair $(A + BK_1, B[I_2]^1)$ is easily checked to be reachable.

Theorem B.1 *Problem B.1 admits a solution if and only if the system is reachable.*

Proof. Necessity. In view of the discussion above, the pair (A, B) must be reachable, since, otherwise, the eigenvalues of the unreachable part could not be modified and the elements of the set Λ could not be arbitrarily fixed. *Sufficiency.* Thanks to Lemma B.1 there is no loss in generality if the system is supposed to have a single control variable, since this situation can be recovered by means of a first control law $u = K_i x + [I_m]^i v$. Then reachability of the pair (A, B) together with $m = 1$ implies the existence of a nonsingular matrix T such that

$$\dot{z} = Fz + Gu, \quad z := Tx$$

with

$$F := \begin{bmatrix} 0 & 1 & 0 & \cdots & 0 \\ 0 & 0 & 1 & \cdots & 0 \\ \vdots & \vdots & \vdots & & \vdots \\ 0 & 0 & 0 & \cdots & 1 \\ -a_1 & -a_2 & -a_3 & \cdots & -a_n \end{bmatrix}, \; G := \begin{bmatrix} 0 \\ 0 \\ \vdots \\ 0 \\ 1 \end{bmatrix}$$

where the entries a_i, $i = 1, 2, \ldots, a_n$ of F are the coefficients of the characteristic polynomial of the matrix A. If the control law is $u = K_c z$, $K_c := [k_1 \, k_2 \, \cdots \, k_n]$, the (closed loop dynamic) matrix $F + GK_c$ has the same (canonical) structure as F, because of the form of matrix G. The entries in the last row are $f_i := -a_i + k_i$, $i = 1, 2, \ldots, n$: thus the characteristic polynomial of the matrix $F+GK_c$ is $\psi_{F+GK_c}(\lambda) = \lambda^n + f_n\lambda^{n-1} + \cdots + f_2\lambda + f_1$. If $\Lambda = \{\lambda_1, \lambda_2, \ldots, \lambda_n\}$, the desired characteristic polynomial for $F+GK_c$ is $\psi_\Lambda(\lambda) = \prod_{i=1}^{n}(\lambda - \lambda_i) := \lambda^n + \varphi_n\lambda^{n-1} + \cdots + \varphi_2\lambda + \varphi_1$. By setting $\psi_\Lambda(\lambda) = \psi_{F+GK_c}(\lambda)$, the set of equations $f_i = -a_i + k_i = \varphi_i$, $i = 1, 2, \ldots, n$ is obtained and the matrix K_c is uniquely determined. In terms of the given state variables x the desired control law is $u = Kx$, with $K := K_c T$.

Remark B.1 *(Uniqueness of the control law when m=1)* The proof of Theorem B.1 shows that when the control u is scalar the matrix K is *uniquely* determined by the set Λ.

Remark B.2 *(Selection of the set Λ)* The possibility of (arbitrarily) assigning the eigenvalues of the closed loop system in principle allows us to *arbitrarily increase* the speed of response of the system itself. However this usually entails weighty consequences on the control variables effort, which might attain unacceptable limits. The forthcoming Example B.2 sheds light on these considerations either in terms of the time response of u or by evaluating the quantity

$$J_u := E[\int_0^\infty u'(t)u(t)dt] \tag{B.4}$$

where the expected value operation is performed with respect to the set of allowable initial conditions. In general, eq. (B.4) makes sense only for stable systems and stands for the expected value of the *control energy* . By assuming that the initial state is a random variable with zero mean value and unitary variance, it results that (see Theorem A.7 of Section A.5 of Appendix A)

$$J_u = \text{tr}[P] \tag{B.5}$$

where P is the solution of the Lyapunov equation

$$0 = P(A + BK) + (A + BK)'P + K'K. \tag{B.6}$$

Example B.2 Consider system (B.1) with

$$A = \begin{bmatrix} 0 & 1 & 0 \\ 0 & 0 & 1 \\ 0 & 0 & 0 \end{bmatrix}, \; B = \begin{bmatrix} 0 \\ 0 \\ 1 \end{bmatrix}.$$

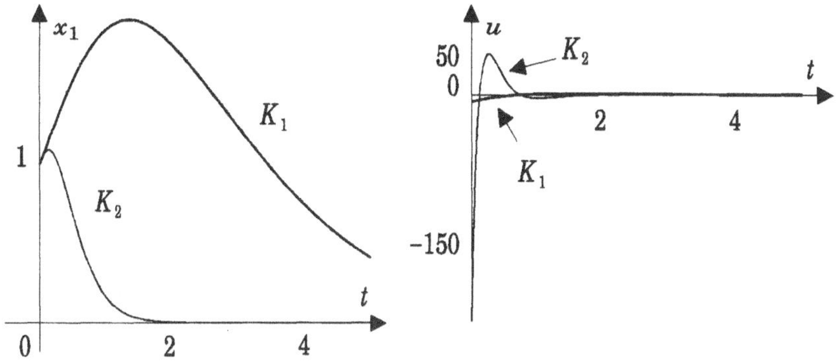

Figure B.2: Example B.2: state and control responses.

By choosing $\Lambda_1 = \{-1,-1,-1\}$ and $\Lambda_2 = \{-5,-5,-5\}$, we get

$$K_1 = -\begin{bmatrix} 1 & 3 & 3 \end{bmatrix}, \; K_2 = -\begin{bmatrix} 125 & 75 & 15 \end{bmatrix}.$$

The time responses of the first state variable x_1 and the control variable u are reported in Fig. B.2 corresponding to the initial state $x(0) = \begin{bmatrix} 1 & 1 & 1 \end{bmatrix}$. On the other hand, if the quantity J_u is evaluated by means of the eqs. (B.4)–(B.6), it results that $J_{u1} = 4$ and $J_{u2} = 815$, in the two cases. Thus it is possible to verify that reduction in duration of the transient lasting entails a much more severe involvement of the control variable.

Remark B.3 *(Multiplicity of the control law when $m > 1$)* In general, if the control variable is not a scalar, more than one control law exists which causes the eigenvalues of the closed loop system to constitute a given set Λ. This should be clear in view of the procedure presented in the proof of Theorem B.1 and the statement of Lemma B.1. Furthermore, by exploiting the fact that the reachability property is *generic* (in other words, given two random matrices F and G of consistent dimensions, the probability that they constitute a reachable pair is 1), one can state that the pair $(A + BK_i, B[I_m]^i)$ is reachable for almost all matrices K_i, provided that $[B]^i \neq 0$.

The possibility of achieving the same closed loop eigenvalue configuration by means of different control laws naturally raises the question of selecting the best one (in some sense) among them. The forthcoming example shows the benefits entailed by a wise choice of matrix K, but the relevant problem is not discussed here.

Example B.3 Consider the system (B.1) with

$$A = \begin{bmatrix} 0 & 1 & 0 \\ 0 & 0 & 1 \\ 0 & 0 & 0 \end{bmatrix}, \; B = \begin{bmatrix} 0 & 0 \\ 1 & 0 \\ 0 & 1 \end{bmatrix}.$$

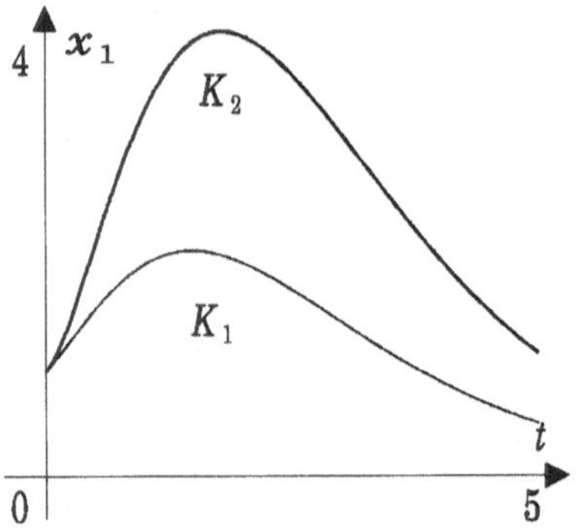

Figure B.3: Example B.3: response of x_1.

By letting $\Lambda = \{-1, -1, -1\}$, it is easy to check that the two matrices

$$K_1 = -\begin{bmatrix} 0 & 0 & -1 \\ 0.5 & 1.5 & 3 \end{bmatrix}, \; K_2 = -\begin{bmatrix} 0 & 0 & -9 \\ 0.1 & 0.3 & 3 \end{bmatrix}$$

define two different control laws which solve the placement problem. The response of the first state variable x_1 corresponding to the initial state $x(0) = \begin{bmatrix} 1 & 1 & 1 \end{bmatrix}$ is shown in Fig. B.3, while the evaluation of J_u through eqs. (B.4)–(B.6) yields $J_{u1} = 2.94$ and $J_{u2} = 17.93$, in the two cases: both outcomes suggest that the first solution has to be preferred.

Remark B.4 *(Observer of order n)* With reference to the system (B.1) consider the further output equation

$$y(t) = Cx(t) + Du(t) \tag{B.7}$$

where $y \in R^p$, $p \leq n$ and the rank of matrix C is equal to the number of its rows. Now consider the system

$$\dot{\hat{x}} = A\hat{x} + Bu + L(C\hat{x} + Du - y) \tag{B.8}$$

where L is chosen as shown in the following. By letting

$$\varepsilon := \hat{x} - x \tag{B.9}$$

from eqs. (B.1), (B.7), (B.8) it follows that

$$\dot{\varepsilon} = (A + LC)\varepsilon. \tag{B.10}$$

If the matrix $A + LC$ is stable, ε and $\hat{x}$ asymptotically tend to 0 and x, respectively, and it is possible to consider the state of the system (B.8) as a meaningful *approximation* of the state of the system (B.1). System (B.8) is referred to as an n-th order asymptotic state *observer*. If the pair (A, C) is observable, Theorem B.1, applied to

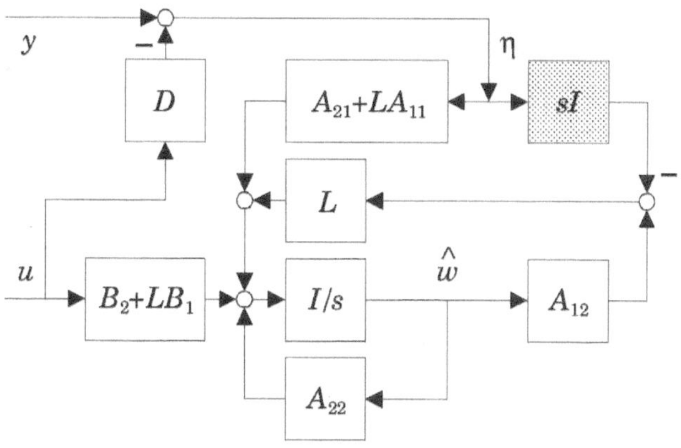

Figure B.4: The observer of order n-rank(C).

the pair (A', C') (which is reachable) guarantees the existence of a matrix L' such that the eigenvalues of $A' + C'L'$ (hence also the eigenvalues of $A + LC$) are the elements of an arbitrarily assigned symmetric set Λ_R of n complex numbers. Therefore, if the system (B.1), (B.7) is observable, system (B.8) can be designed in such a way that $\hat{x}$ asymptotically tends to x with any desired speed.

Remark B.5 *(Observer of order n-rank(C))* The information carried by the p output variables y concerning the state x is not explicitly exploited by the observer (B.8). In fact if rank$(C) = p$ the change of variables $z := Tx$ where

$$T := \begin{bmatrix} C \\ C_1 \end{bmatrix},$$

C_1 being any matrix such that T is nonsingular, implies that eq. (B.7) becomes

$$y = \begin{bmatrix} I & 0 \end{bmatrix} z + Du \tag{B.11}$$

and the first p components of z differ from the output variables because of the known term Du. Thus it is possible to assume that the state vector is $z := \begin{bmatrix} \eta' & w' \end{bmatrix}'$, where $\eta := y - Du$, i.e., that the output transformation C is as in eq. (B.11) and conclude that the asymptotic approximation of η need not be found. Then system (B.1), (B.7) becomes

$$\dot{\eta} = A_{11}\eta + A_{12}w + B_1 u, \tag{B.12a}$$
$$\dot{w} = A_{21}\eta + A_{22}w + B_2 u, \tag{B.12b}$$
$$y = \eta + Du,$$

and the problem of observing z reduces to the problem of observing the vector w which has n-rank$(C) = n - p$ components. This new problem can be stated relative

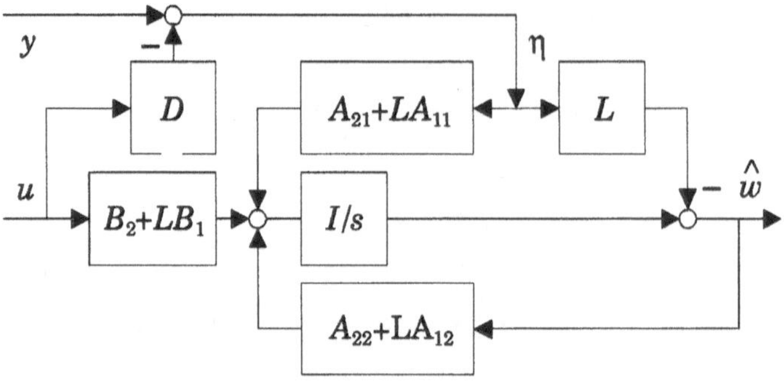

Figure B.5: The observer of order n-rank(C) without differentiator.

to the system (B.12b) with the output variable

$$\zeta := A_{12}w = \dot{\eta} - A_{11}\eta - B_1 u$$

which can be seen as known in view of eq. (B.12a), since η, u are known and, at least in principle, also $\dot{\eta}$ is known. If, as will later be shown, the pair (A_{22}, A_{12}) is observable, the observation of w can be performed by means of the $n-p$-th order device presented in Fig. B.4 where a *differentiator* is present. However the system in Fig. B.4 is readily seen to be equivalent (from the input-output point of view) to the system in Fig. B.5 which does not require a differentiator. Thus the observer of order n-rank(C) is described by

$$\dot{\psi} = (A_{22} + LA_{12})(\psi - L\eta) + (A_{21} + LA_{11})\eta + (B_2 + LB_1)u, \tag{B.13a}$$

$$\hat{w} = \psi - L, \eta \tag{B.13b}$$

$$\eta = y - Du, \tag{B.13c}$$

$$\hat{x} = T^{-1}\begin{bmatrix} \eta \\ \hat{w} \end{bmatrix},$$

where, if it is the case, also the change of variables $z = Tx$ has been taken into account. The last equation gives the observation $\hat{x}$ of x. The observability of the pair (A_{22}, A_{12}) can easily be checked if reference is made to Fig. B.6 where the system (B.12) is shown after depriving it of the input u. If the system (B.1), (B.7) is observable also, the system shown in Fig. B.6 is such. However, this last system remains observable also when $A_{21} = 0$, since an output feedback does not alter the observability properties. Finally, it is obvious that the system shown in Fig. B.6 (with $A_{21} = 0$) can not be observable if the pair (A_{22}, A_{12}) is not.

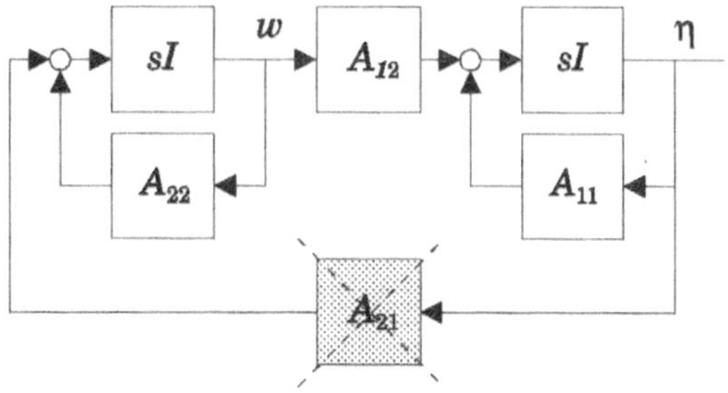

Figure B.6: Observability of the pair (A_{22}, A_{12}).

B.3 Assignment with inaccessible state

The eigenvalue assignment problem when the state of the system

$$\dot{x}(t) = Ax(t) + Bu(t), \tag{B.14a}$$
$$y(t) = Cx(t) + Du(t), \tag{B.14b}$$

(where $u \in R^m$, $y \in R^p$, $x \in R^n$, $m \leq n$ and $p \leq n$) cannot be fed to the input of the controller can be stated in the following way.

Problem B.2 (Eigenvalue assignment with inaccessible state) *Find a dynamic system of order ν and transfer function $R(s)$ such that the control law*

$$u(y, s) = R(s)y$$

causes the eigenvalues of the closed loop system to constitute a set of $n + \nu$ numbers coincident with an arbitrarily given set

$$\Lambda = \Lambda_P \cup \Lambda_R \tag{B.15}$$

where Λ_P and Λ_R are two symmetric sets of n and ν complex numbers, respectively.

By resorting to an asymptotic state observer of order n (recall Remark B.4) a very simple solution of Problem B.2 can be found, as stated in the following theorem.

Theorem B.2 *Problem B.2 admits a solution with $\nu = n$ if and only if the system (B.14) is minimal.*

Proof. Necessity. Necessity follows from the previous discussion since the eigenvalues of unreachable and/or unobservable parts of the system cannot be altered.

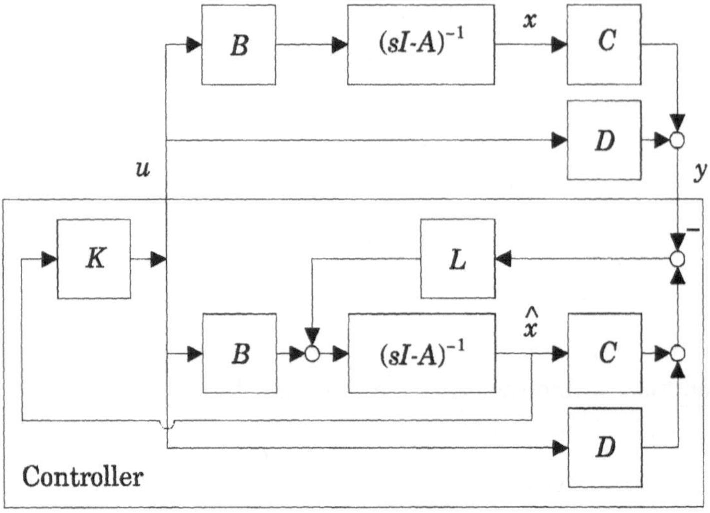

Figure B.7: The closed loop system with an n-th order controller.

Sufficiency. Let the controller be constituted by an asymptotic observer of order n (recall eq. (B.8)), namely by the system

$$\dot{\hat{x}} = A\hat{x} + Bu + L(C\hat{x} + Du - y) \tag{B.16}$$

together with the further equation

$$u = K\hat{x}. \tag{B.17}$$

By exploiting eqs. (B.9), (B.10), the dynamic matrix of the system (B.14), (B.16), (B.17) turns out to be

$$F := \begin{bmatrix} A+BK & BK \\ 0 & A+LC \end{bmatrix}.$$

The eigenvalues of the matrix F are those of the matrix $A+BK$ together with those of the matrix $A+LC$: such eigenvalues can be made to coincide with the elements of two arbitrarily given symmetric sets Λ_P and Λ_R of complex numbers, respectively, thanks to reachability of the pair (A, B) and observability of the pair (A, C).

The closed loop system resulting from the proof of Theorem B.2 is shown in Fig. B.7: it is apparent that the same control law which would have been implemented on the state of the system (B.14) in order to obtain Λ_P when the state is accessible, is implemented on the state of the observer (B.16).

Remark B.6 *(Choice of Λ_R)* The time response of the observation error ε is determined by the location in the complex plane of the elements of the set Λ_R which is

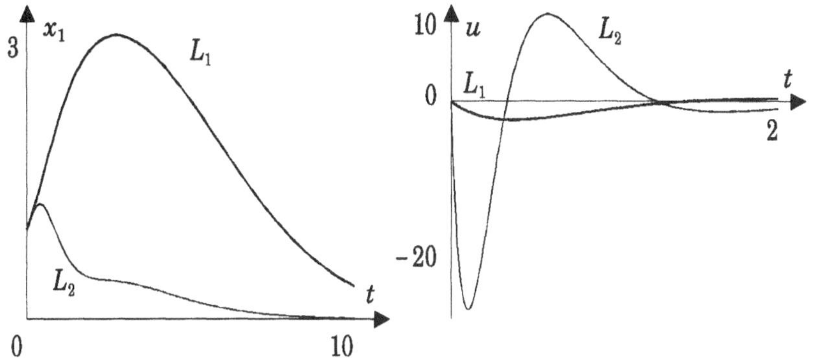

Figure B.8: Example B.4: responses of x_1 and u.

arbitrary: thus it makes sense to select them in such a way that their real parts are substantially smaller than the real parts of the elements of the set Λ_P which characterize the transient evolution of the state x. This synthesis approach can surely be adopted provided that possible unsatisfactory behaviours of the control variables are checked, as shown in the forthcoming Example B.4.

Example B.4 Consider the system (B.14), with

$$A = \begin{bmatrix} 0 & 1 & 0 \\ 0 & 0 & 1 \\ 0 & 0 & 0 \end{bmatrix}, \ B = \begin{bmatrix} 0 \\ 0 \\ 1 \end{bmatrix}, \ C = \begin{bmatrix} 1 & 0 & 0 \end{bmatrix}.$$

By choosing $\Lambda_P = \{-1, -1, -1\}$, $\Lambda_{R1} = \{-1, -1, -1\}$ and $\Lambda_{R2} = \{-5, -5, -5\}$, we obtain $K = -\begin{bmatrix} 1 & 3 & 3 \end{bmatrix}$, $L_1 = -\begin{bmatrix} 3 & 3 & 1 \end{bmatrix}'$ and $L_2 = -\begin{bmatrix} 15 & 75 & 125 \end{bmatrix}'$. The responses of the first state variable x_1 and the control variable u are reported in Fig. B.8, corresponding to the initial state $x(0) = \begin{bmatrix} 1 & 1 & 1 \end{bmatrix}$, $\varepsilon(0) = -\begin{bmatrix} 1 & 1 & 1 \end{bmatrix}$. On the other hand, if the quantity J_u is evaluated (recall eq. (B.4)), we get $J_{u1} = 12.63$ and $J_{u2} = 163.18$ in the two cases. Note the effects of the choice of a faster observer on the closed loop system performances.

Remark B.7 *(Stability of the controller)* The stability of the closed loop system which is guaranteed by the selection of stable sets Λ_P and Λ_R (that is with elements in the open left half-plane only) does not imply that the controller designed according to the preceding discussion is stable, since its dynamic matrix is $A_R := A + BK + LC$ which can be stable or not independently from stability of the matrices $A + BK$ and $A + LC$. This fact is clarified in the forthcoming Example B.5.

Example B.5 Consider the (stable) system (B.14) with

$$A = \begin{bmatrix} 0 & 1 \\ -10 & -1 \end{bmatrix}, \ B = \begin{bmatrix} 0 \\ 1 \end{bmatrix}, \ C = \begin{bmatrix} 7 & 1 \end{bmatrix}.$$

Let $\psi_{A+BK}(s) := s^2+s+1$ be the polynomial the roots of which constitute the set Λ_P and $\psi_{A+LC}(s) := s^2+4s+4$ the polynomial the roots of which constitute the set Λ_R. We obtain $K = \begin{bmatrix} 9 & 0 \end{bmatrix}$ and $L = \begin{bmatrix} -27/52 & 33/52 \end{bmatrix}'$, but the characteristic polynomial of $A+BK+LC$ is $\psi_{A+BK+LC}(s) = s^2+4s-17/52$, so that the controller is not stable.

Remark B.5 suggests that Problem B.2 could be solved by resorting to a controller of order less than n, namely, of order n-rank(C), as will be shown in the proof of Theorem B.3.

Theorem B.3 *Problem B.2 admits a solution with* $\nu = n - \text{rank}(C)$ *only if the system (B.14) is minimal. This condition is also sufficient for almost all the sets* Λ *of the form (B.15).*

Proof. Necessity. Necessity is proved as in Theorem B.2. *Sufficiency.* Assume that the system is already in the form (B.12), i.e.,

$$\begin{aligned} \dot{\eta} &= A_{11}\eta + A_{12}w + B_1 u, \\ \dot{w} &= A_{21}\eta + A_{22}w + B_2 u, \end{aligned}$$

and consider its feedback connections with a controller constituted by an observer of order n-rank(C) (recall eqs. (B.13a), (B.13b)), that is the system

$$\begin{aligned} \dot{\psi} &= (A_{22}+LA_{12})(\psi - L\eta) + (A_{21}+LA_{11})\eta + (B_2+LB_1)u, \\ \hat{w} &= \psi - L\eta \end{aligned}$$

together with the further equation

$$u = K_\eta \eta + K_w \hat{w}. \tag{B.18}$$

By choosing η, w and $\varepsilon := \hat{w} - w = \psi - L\eta - w$ as state variables, the dynamic matrix of the closed loop system is

$$F := \begin{bmatrix} A_z + B_z K_z & B_z K_w \\ 0 & A_{22} + LA_{12} \end{bmatrix}$$

where

$$A_z := \begin{bmatrix} A_{11} & A_{12} \\ A_{21} & A_{22} \end{bmatrix}, \; B_z := \begin{bmatrix} B_1 \\ B_2 \end{bmatrix}, \; K_z := \begin{bmatrix} K_\eta & K_w. \end{bmatrix}$$

The eigenvalues of the matrix F can be made to coincide with the elements of the set Λ since the pair (A_z, B_z) is reachable (this property follows from the reachability of the pair (A, B)) and the pair (A_{22}, A_{12}) is observable. The actual implementation of the control law must be performed by taking into account eq. (B.13c) and checking the solvability of the *algebraic loop* that such a relation constitutes together with eq. (B.18), i.e., by verifying that the matrix $I + (K_\eta - K_w L)D$ is nonsingular. If such a property does not hold it is sufficient to perturb the three matrices K_η, K_w and L: the amount of the perturbation is, at least in principle, arbitrarily small so that the eigenvalues of the resulting closed loop system differ from their desired values of a similarly arbitrarily small amount.

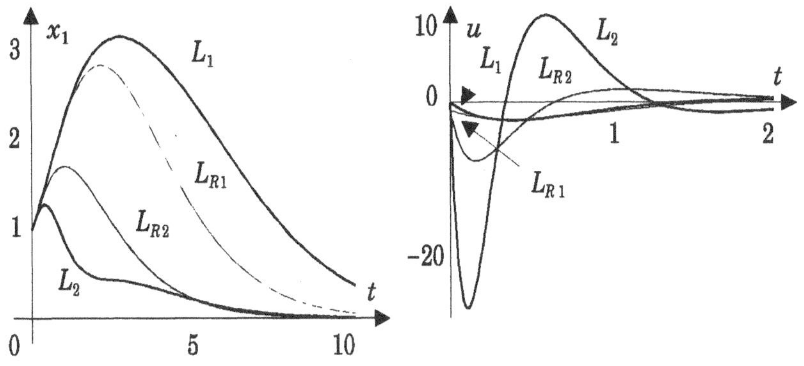

Figure B.9: Example B.4 and B.6: responses of x_1 and u.

Example B.6 Consider the system described in Example B.4 and let $\Lambda_P = \{-1, -1, -1\}$, while the set Λ_R is taken equal to either $\Lambda_{R1} = \{-1, -1\}$ or $\Lambda_{R2} = \{-5, -5\}$ since a second order controller has to be adopted. We obtain $K = -\begin{bmatrix} 1 & 3 & 3 \end{bmatrix}$, $L_{R1} = -\begin{bmatrix} 2 & 1 \end{bmatrix}'$, $L_{R2} = -\begin{bmatrix} 10 & 25 \end{bmatrix}'$. The transient responses of the first state variable x_1 and the control variable u are reported in Fig. B.9 corresponding to the initial state $x(0) = \begin{bmatrix} 1 & 1 & 1 \end{bmatrix}$, $\varepsilon(0) = -\begin{bmatrix} 1 & 1 \end{bmatrix}$. In order to allow a meaningful comparison the responses of these variables are shown in the same figure when the controllers given in the Example B.4 are exploited. On the other hand, if the quantity J_u is evaluated (recall eq. (B.4)), we obtain $J_{u1} = 8.21$ and $J_{u2} = 11.63$ in the two cases: the comparison of these values with those in Example B.4 deserves attention.

Given a system Σ resulting from the feedback connection of the systems $\Sigma_1 := (A, B, C, D)$ and $\Sigma_2 := (F, G, H, K)$, the new system $\hat{\Sigma}$ resulting from the feedback connection of the system $\Sigma_1' := (A', C', B', D')$ (which is the *transposed* system of Σ_1) with the system $\Sigma_2' := (F', H', G', K')$ (which is the *transposed* system of Σ_2) is the *transposed* system of Σ, that is $\hat{\Sigma} = \Sigma'$. Therefore the eigenvalues of Σ and Σ' are the same. This fact can suitably be exploited with reference to the problem under consideration. Indeed Theorem B.3 can be applied to the system which is the transpose of the system (B.14), yielding the following result.

Theorem B.4 *Problem B.2 admits a solution with $\nu = n$-rank(B) only if the system (B.14) is minimal. This condition is also sufficient for almost all the sets Λ of the form (B.15).*

Remark B.8 *(Stability of the reduced order controllers)* Remark B.7 applies also to the cases where the controller has been designed by exploiting a reduced order observer of the system (B.14) (controller of order n-rank(C)) or of its transpose (controller of order n-rank(B)).

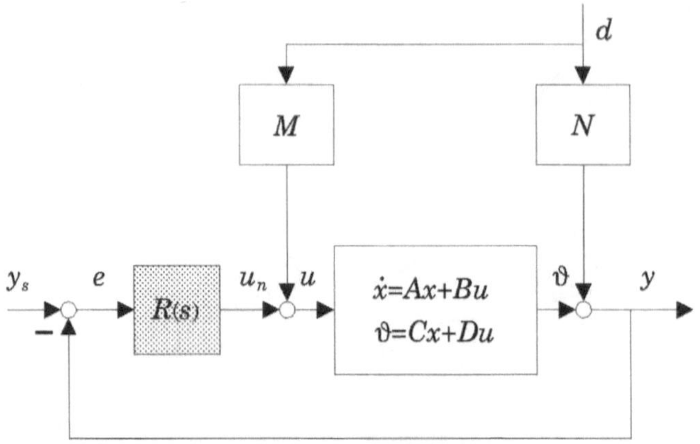

Figure B.10: The control system with disturbances.

B.4 Assignment with asymptotic errors zeroing

The problem under consideration refers to the control system shown in Fig. B.10 where the signals d and y_s are the vectors of the *disturbances* acting on the system and the *set points*, respectively. These signals are *polynomial* functions of time with degree not greater than k, i.e., they are defined by the equations

$$y_s(t) := \sum_{i=0}^{k} y_s^{(i)} t^i, \tag{B.19a}$$

$$d(t) := \sum_{i=0}^{k} d^{(i)} t^i, \tag{B.19b}$$

where, for $i = 0, 1, \ldots, k$, $y_i^{(i)}$ and $d^{(i)}$ are arbitrary vectors belonging to R^m and R^r, respectively.

The system P to be controlled is described by the equations

$$\dot{x}(t) = Ax(t) + Bu(t), \tag{B.20a}$$

$$u(t) = u_n(t) + Md(t), \tag{B.20b}$$

$$\vartheta(t) = Cx(t) + Du(t), \tag{B.20c}$$

$$y(t) = \vartheta(t) + Nd(t), \tag{B.20d}$$

where $x \in R^n$, $u \in R^m$, $\vartheta \in R^m$.

The control problem concerning the system (B.20) determines a linear controller $R(s)$ such that the eigenvalues of the resulting closed loop system are located in arbitrarily preassigned positions of the left open half-plane and the error e asymptotically vanishes corresponding to arbitrary exogenous signals y_s and d of the form (B.19). The formal statement of this problem is as follows.

Problem B.3 (Eigenvalue assignment and errors zeroing) *Given the system (B.20) and the set of exogenous signals (B.19), find a linear time-invariant controller such that:*

i) *The eigenvalues of the closed loop system are located in arbitrarily preassigned positions.*

ii) *Corresponding to any exogenous signal of the form (B.19) and any initial condition it results that*

$$\lim_{t\to\infty} e(t) = \lim_{t\to\infty} y_s(t) - y(t) = 0. \tag{B.21}$$

A solution of this problem is given in the following theorem, the proof of which supplies a way of actually designing the required controller.

Theorem B.5 *Problem B.3 admits a solution if and only if*

i) *The pair (A, B) is reachable;*

ii) *The pair (A, C) is observable;*

iii) *No invariant zeros of the system P are in the origin.*

Proof. Necessity. The necessity of conditions i) and ii) follows from the discussion in Section B.3, since it is required that the eigenvalues of the closed loop system could be arbitrarily located in the complex plane. As for condition iii), recall that it is equivalent to the request that the matrix

$$P(s) := \begin{bmatrix} sI - A & -B \\ C & D \end{bmatrix}$$

be full rank for $s = 0$, or, in other words, that

$$\det(P(0)) = \det(\begin{bmatrix} -A & -B \\ C & D \end{bmatrix}) \neq 0 \tag{B.22}$$

since the controlled system is *square*.

The condition (B.22) is violated if and only if at least one of the following facts is established:

(a) $\text{rank}(\begin{bmatrix} A & B \end{bmatrix}) < n$;

(b) $\text{rank}(\begin{bmatrix} C & D \end{bmatrix}) < m$;

(c) $\text{rank}(P(0)) < n + m$ and none of the two preceding conditions holds.

If the condition (a) is verified there exists a vector $v \neq 0$ such that $A'v = 0$ and $B'v = 0$ so that (recall the PBH test) the pair is not reachable and an already proved necessary condition is violated. If the condition (b) is verified and the disturbance d is not present, at least for one $i \in \{1, 2, \ldots, m\}$ it results that

$$y_i(\cdot) = \sum_{i \neq j=1}^{m} \alpha_j y_j(\cdot)$$

against the requirement ii) in Problem B.3. Finally, if the condition (c) is verified and the disturbance d is not present, it follows that

$$\sum_{i=1}^{m} \alpha_i y_i(\cdot) = \sum_{j=1}^{n} \beta_j \dot{x}_j(\cdot)$$

with at least one of the coefficients $\alpha_i \neq 0$. This implies that the error can not asymptotically be zero for constant, yet arbitrary, set points since $\lim_{t\to\infty} \dot{x}(t) = 0$.
Sufficiency. Consider the system described by the equations (B.20) and

$$\dot{\xi}^{(0)} = e = y_s - y, \tag{B.23a}$$
$$\dot{\xi}^{(1)} = \xi^{(0)}, \tag{B.23b}$$
$$\vdots$$
$$\dot{\xi}^{(k-1)} = \xi^{(k-2)}, \tag{B.23c}$$
$$\dot{\xi}^{(k)} = \xi^{(k-1)}, \tag{B.23d}$$

where $\xi^{(i)} \in R^m$, $i = 0, 1, \ldots, k$. If this system is reachable from the input u_n and observable from the output $\xi := \begin{bmatrix} \xi^{(0)\prime} & \xi^{(1)\prime} & \cdots & \xi^{(k-1)\prime} & \xi^{(k)\prime} \end{bmatrix}'$, then (see Fig. B.10), there exists a controller $R(s)$ constituted by the system (B.23) and the system

$$\dot{z} = Fz + G\xi, \tag{B.24}$$
$$u_n = Hz + K\xi \tag{B.25}$$

such as to guarantee the eigenvalue assignment for the resulting closed loop system. Thus, the system shown in Fig. B.10 can without loss of generality be thought of as stable. This fact implies that the steady state condition which will be shown to be consistent and characterized by $e = 0$ is the one to which the system asymptotically tends.

As a preliminary it is proved that the assumptions *i)–iii)* imply both observability from the output ξ and reachability from the input u_n for the system (B.20), (B.23). The dynamic and input matrices for the system under consideration are

$$A_A = \begin{bmatrix} A & 0_{n\times mk} & 0_{n\times m} \\ -C & 0_{m\times mk} & 0_{m\times m} \\ 0_{mk\times n} & I_{mk} & 0_{mk\times m} \end{bmatrix}, \quad B_A = \begin{bmatrix} B \\ -D \\ 0_{mk\times m} \end{bmatrix}.$$

Thus the matrix K_r for the Kalman reachability test is

$$K_r = \begin{bmatrix} B & AB & A^2B & A^3B & \cdots \\ -D & -CB & -CAB & -CA^2B & \cdots \\ 0_{m\times m} & -D & -CB & -CAB & \cdots \\ 0_{m\times m} & 0_{m\times m} & -D & -CB & \cdots \\ \vdots & \vdots & \vdots & \vdots & \end{bmatrix},$$

which can be written as $K_r = TS$, where

$$T := \begin{bmatrix} -A & -B & 0_{n\times mk} \\ C & D & 0_{m\times mk} \\ 0_{mk\times n} & 0_{mk\times m} & I_{mk} \end{bmatrix},$$

$$S := \begin{bmatrix} 0_{n\times m} & -B & -AB & -A^2B & -A^3B & \cdots \\ -I_m & 0_{m\times m} & 0_{m\times m} & 0_{m\times m} & 0_{m\times m} & \cdots \\ 0_{m\times m} & -D & -CB & -CAB & -CA^2B & \cdots \\ 0_{m\times m} & 0_{m\times m} & -D & -CB & -CAB & \cdots \\ \vdots & \vdots & \vdots & \vdots & \vdots & \end{bmatrix}.$$

The matrix T is nonsingular because of assumption *iii*), hence the rank of K_r is maximum if and only if the rank of S is such. In view of the structure of S (in particular, the position of the submatrix I_m) the rank of S is maximum if and only if the rank of the matrix obtained from S after the rows and columns which are concerned with the submatrix I_m have been deleted is such. The resulting matrix possesses the very same structure as K_r (apart from the sign of the first row of submatrices, which is of no relevance in the rank context). What was previously done with reference to K_r can also be applied to this new matrix yielding, after $k-1$ iterations, a type "S" matrix of the form

$$\begin{bmatrix} 0_{n\times m} & -B & -AB & -A^2B & \cdots \\ -I_m & 0_{m\times m} & 0_{m\times m} & 0_{m\times m} & \cdots \end{bmatrix}$$

which has maximum rank thanks to the assumption *i*). Thus the pair (A_A, B_A) is reachable. As for observability, notice that the output matrix C_A is

$$C_A = \begin{bmatrix} 0_{m(k+1)\times n} & I_{m(k+1)} \end{bmatrix}.$$

Thus the matrix K_o for the Kalman observability test is, after a suitable rearrangement of the columns,

$$K_o = \begin{bmatrix} 0_{n\times m(k+1)} & -C' & -A'C' & -(A')^2C' & \cdots \\ I_{m(k+1)} & 0_{m(k+1)\times m} & 0_{m(k+1)\times m} & 0_{m(k+1)\times m} & \cdots \end{bmatrix}.$$

In view of the assumption *ii*) this matrix has maximum rank.

In order to prove that eq. (B.21) holds at the steady state, consider the *stable* system $\dot{w} = A_c w + B_c v$ where $v = \sum_{i=0}^{k} v^{(i)} t^i$, $v^{(i)} = \text{const.}$, $i = 0, 1, \ldots, k$. At the steady state the state of this system is a polynomial function of degree not greater than k, i.e., $w = \sum_{i=0}^{k} w^{(i)} t^i$, $w^{(i)} = \text{const.}$, $i = 0, 1, \ldots, k$. In fact, the unknown vectors $w^{(i)}$, $i = 0, 1, \ldots, k$ are uniquely determined by the set of equations (which enforce the above steady state condition)

$$\begin{aligned}
0 &= A_c w^{(k)} + B v^{(k)}, \\
k w^{(k)} &= A_c w^{(k-1)} + B v^{(k-1)}, \\
&\vdots \\
2 w^{(2)} &= A_c w^{(1)} + B v^{(1)}, \\
w^{(1)} &= A_c w^{(0)} + B v^{(0)},
\end{aligned}$$

no matter how the vectors $v^{(i)}$, $i = 0, 1, \ldots, k$ (which specify the input v) have been selected, since the matrix A_c is nonsingular (the system is stable). When applied to the system in Fig. B.10, this discussion implies that eq. (B.21) is satisfied since otherwise $\xi^{(k)}$ would be a polynomial function of time of degree $k+1$ (recall eqs. (B.23)).

Remark B.9 *(Internal model principle)* The most significant part of the controller defined in the proof of Theorem B.5 (eqs. (B.23)) is constituted by a set of m non-interacting subsystems (as many as the number of controlled variables) each one of them simply consisting of a bunch of $k+1$ cascaded integrators (thus their number is just equal to the maximum degree of the polynomials exogenous inputs plus 1). Therefore each subsystem possesses the very same structure of an autonomous system capable of generating the considered family of exogenous signals. Including into the controller this block of integrators is, within the framework under consideration, only a sufficient condition for zero errors regulation, which becomes also necessary whenever the controlled plant is allowed to undergo arbitrary parametric perturbations, the magnitude of which has to be so small as to preserve the stability of the closed loop system. Indeed the integrators which could be present in the plant (nominal) description can not be exploited in achieving asymptotic zero errors regulation in case of arbitrary parameters perturbations. In the literature, the need for inserting into the controller a duplicate of the system which generates the exogenous signals whenever some kind of *robustness* is required for the closed loop system (designed so as to achieve zero errors regulation) is referred to as the *Internal model principle.* These facts are well known to anybody who is familiar with basic control theory for single input single output systems.

Remark B.10 *(Inputs with rational Laplace transform)* A similar discussion can be carried on if the exogenous signals have rational Laplace transform with poles in the closed right half plane, namely signals which are the sums and products of time functions of polynomial, increasing exponential, sinusoidal type. Once the structure of the autonomous linear system Σ_E of least order which is able to generate the whole set of considered exogenous signals has been identified, it is sufficient to insert into the controller a subsystem constituted by m copies of Σ_E (one for each error component) and enforce stability in order to guarantee asymptotic zero errors regulation.

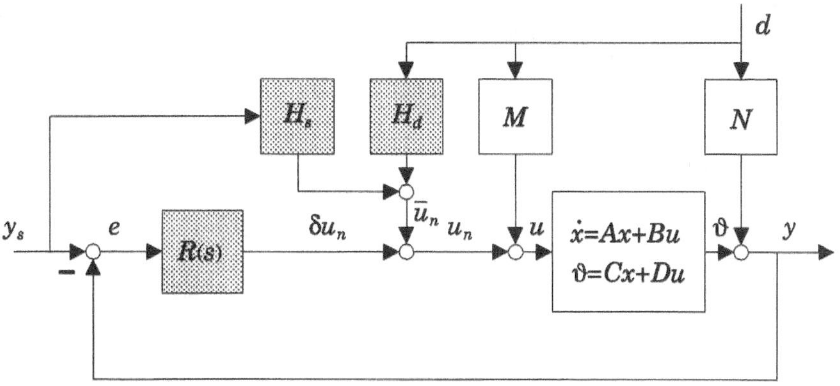

Figure B.11: The control system with constant exogenous signals.

Remark B.11 *(Controller order)* The part (B.24), (B.25) of the controller which has to assign the eigenvalues of the closed loop system (actually it must guarantee a satisfactory stability level) is connected to a system of order $n + (k + 1)m$ which possesses m control variables and $(k + 1)m$ output variables: thus its order can be bounded to n (recall Theorem B.3). However further information on the system state is carried by the controlled variables y. Thus $(k + 2)m$ variables can be fed to the input of system (B.24), (B.25) and its order lowered to $n - m$ since the relevant output matrix is readily seen to have maximum rank provided that the (reasonable) assumption is made that also the matrix C has maximum rank.

Remark B.12 *(Constant exogenous signals)* The particular case of constant exogenous signals deserves special attention from a practical point of view (the design specifications frequently call for asymptotic zero error regulation in the presence of constant set points and disturbances) and furthermore allows us to resort to a substantially different controller when the disturbance d can be measured. With reference to Fig. B.11, where H_s and H_d are nondynamical systems, it is easy to check that condition (B.21) is satisfied provided that the closed loop system is stable, $\delta u_n = 0$ and $\dot{x} = 0$ (i.e., $x = \bar{x} =$ const.) with

$$\begin{aligned} 0 &= A\bar{x} + B\bar{u}_n + BMd, \\ y_s &= C\bar{x} + D\bar{u}_n + (DM + N)d. \end{aligned}$$

These equations can be rewritten as

$$\begin{bmatrix} -A & -B \\ C & D \end{bmatrix} \begin{bmatrix} \bar{x} \\ \bar{u}_n \end{bmatrix} = \begin{bmatrix} BM & 0 \\ -(DM+N) & I \end{bmatrix} \begin{bmatrix} d \\ y_s \end{bmatrix}.$$

In view of the assumption *iii)* $\bar{u}_n$ can be chosen as

$$\bar{u}_n = H_d d + H_s y_s$$

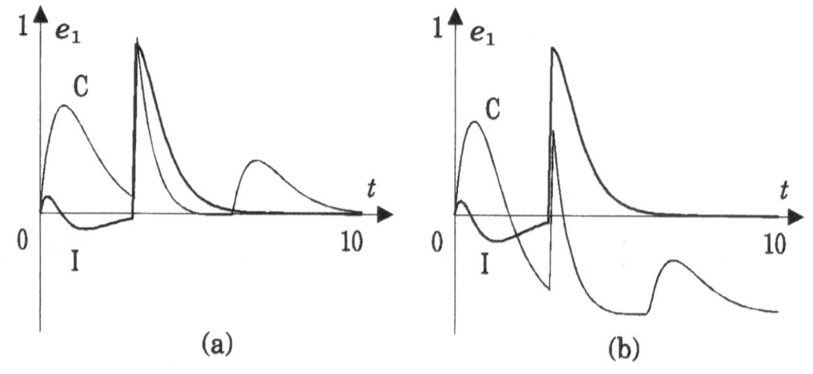

Figure B.12: Example B.7: responses of the error e_1 when a compensator (C) or an integral controller (I) is adopted: (a) nominal conditions; (b) perturbed conditions.

where

$$H_d := \begin{bmatrix} 0 & I \end{bmatrix} (P(0))^{-1} \begin{bmatrix} BM \\ -(DM+N) \end{bmatrix},$$

$$H_s := \begin{bmatrix} 0 & I \end{bmatrix} (P(0))^{-1} \begin{bmatrix} 0 \\ I \end{bmatrix}.$$

Assumptions *i)* and *ii)* obviously guarantee the existence of a controller $R(s)$ which assigns the system eigenvalues to their desired locations.

However it must be noted that, different from the controller presented in the proof of Theorem B.5, a design carried on according to these ideas does not guarantee, in general, that eq. (B.21) is verified if the system parameters undergo even small perturbations. In other words, the open loop *compensator* made out of the two subsystems H_d and H_s does not endow the design with any kind of robustness.

Remark B.13 *(Nonsquare systems)* It is quite obvious how Theorem B.5 could be applied to the case where the number of components of the control vector u_n is greater than the number of components of the output vector y, while Problem B.3 does not have a solution when the dimension of the vector y is greater than the dimension of the vector u_n.

Example B.7 Consider the system (controlled plant) (B.20) where

$$A = \begin{bmatrix} 1 & 0 & 1 \\ 0 & 1 & 0 \\ 0 & 1 & -1 \end{bmatrix}, \quad B = \begin{bmatrix} 1 & 0 \\ 0 & 1 \\ 0 & 0 \end{bmatrix},$$

$$C = \begin{bmatrix} 1 & 0 & 0 \\ 0 & 1 & 0 \end{bmatrix}, \quad D = 0, \quad M = \begin{bmatrix} 1 \\ 1 \end{bmatrix}, \quad N = \begin{bmatrix} 0 \\ 2 \end{bmatrix},$$

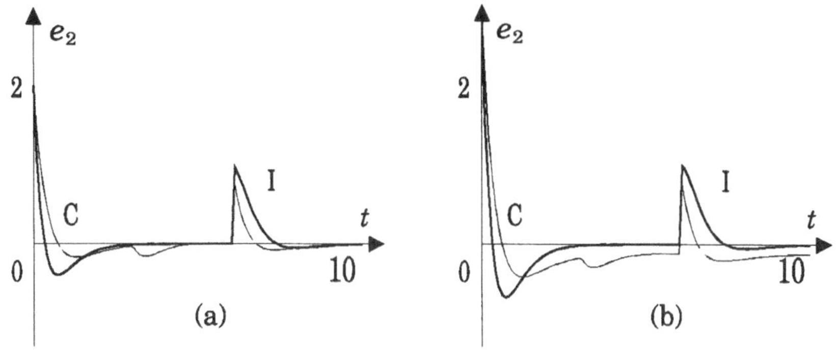

Figure B.13: Example B.7: responses of the error e_2 when a compensator (C) or an integral controller (I) is adopted: (a) nominal conditions; (b) perturbed conditions.

and let both the set point y_s and the disturbance d be constant. Problem B.3 is solved by resorting both to a controller based on an open loop compensator (see Remark B.12) and to a controller inspired by the internal model principle. In the two cases the closed loop eigenvalues are required to be equal to -2. The first design is specified by

$$H_d = \begin{bmatrix} 1 \\ 1 \end{bmatrix}, \ H_s = \begin{bmatrix} -1 & -1 \\ 0 & -1 \end{bmatrix}$$

while the system $R(s) = \Sigma(A_R, B_R, C_R, D_R)$ is defined by

$$A_R = -2, \ B_R = \begin{bmatrix} 0 & -1 \end{bmatrix}, \ C_R = \begin{bmatrix} 0 \\ -1 \end{bmatrix}, \ D_R = \begin{bmatrix} 3 & 0 \\ 1 & 4 \end{bmatrix}.$$

A controller $R(s) = \Sigma(A_R, B_R, C_R, D_R)$ which implements the second design approach and has $\begin{bmatrix} e' & y' \end{bmatrix}'$ as input is specified by

$$A_R = \begin{bmatrix} 0 & 0 & 0 \\ 0 & 0 & 0 \\ -4 & 0 & -1 \end{bmatrix}, \ B_R = \begin{bmatrix} 1 & 0 & 0 & 0 \\ 0 & 1 & 0 & 0 \\ 0 & 0 & 3 & 1 \end{bmatrix},$$

$$C_R = \begin{bmatrix} 4 & 0 & -1 \\ 0 & 8 & 1 \end{bmatrix}, \ D_R = \begin{bmatrix} 0 & 0 & -6 & 0 \\ 0 & 0 & 1 & -6 \end{bmatrix}.$$

The time responses of the error variables for the two designs are reported in Figs. B.12 and B.13 when the exogenous signals are

$$y_{s1} = \begin{cases} 0, & t < 3 \\ 1, & t > 3, \end{cases}$$

$$y_{s2} = \begin{cases} 0, & t < 6 \\ 1, & t > 6, \end{cases}$$

$$d = \ -1, \ t > 0,$$

and the initial state of the plant is zero. Now assume that in the plant description the matrix N has become 1.5 times the original one. The time responses of the error variables are reported in the quoted figures for this new situation and the same exogenous signals. It is apparent that the first design procedure does not satisfactorily face the occurred perturbation.

Appendix C

Notation

Re$[s]$ is the real part of the complex number s.

For any complex matrix function $F(s)$, $F^{\sim}(s) = F'(-s)$.

$A^{\sim}$ is the conjugate transpose of the constant matrix A.

$A^{\dagger}$ is the pseudoinverse of A. It is $AA^{\dagger}A = A$, $A^{\dagger}AA^{\dagger} = A^{\dagger}$ and, if A is square and nonsingular, $A^{\dagger} = A^{-1}$.

$\Sigma(A, B, C, D)$ is the n-th order system described by the equations

$$\begin{aligned} \dot{x} &= Ax + Bu, \\ y &= Cx + Du. \end{aligned}$$

Im(A) is the range space of A, that is the subspace spanned by the columns of A.

ker(A) is the null space of A.

rank(A) is the rank of A.

$[A]^i$ is the i-th column of matrix A.

$[A]_i$ is the i-th row of matrix A.

tr$[A]$ is the trace of the square matrix A.

I and 0 are the identity and null matrices, respectively. If their dimensions are not clear from the context, then I_m is the $m \times m$ identity matrix and $0_{r \times s}$ is the $r \times s$ null matrix.

$A =$diag$[\alpha_1, \alpha_2, \ldots, \alpha_n]$ is a square matrix with $a_{i,j} = 0$, $i \neq j$ and $a_{i,i} = \alpha_i$, $i = 1, 2, \ldots, n$.

Bibliography

It is fairly difficult to select a reasonably compact list of references out of the huge number of excellent contributions which appeared in the literature since the late 1950s. Thus we are going to mention a small number of textbooks and journal papers only, picking them out of those which are widely recognized as particularly significant for the understanding of the basic issues of optimal control theory.

With specific reference to the subjects which have been discussed in this book it is first worth mentioning the texts [2], [4], [8], [20], [22], [32] which are universally acknowledged as classical references. More recent contributions are the books [1], [3], [18], [21], [25], [29], [37]. A more detailed analysis leads to the following suggestions.

A synthetic presentation of the Hamilton-Jacobi theory is given in [20]; the Linear Quadratic problem is extensively discussed in [2], [3], [4], [9], [22], [26], [33], [41], in particular, more details about the cheap control issue can be found in [22]. The Linear Quadratic Gaussian problem is dealt with in [2], [15], [16], [22], this last book constituting a particularly suitable source for a simple yet efficient summary of basic notions of probability theory, while [12], [33] and [35] supply a deeper insight into the robustness issues concerning control systems; Riccati equations are analyzed under various view points in [7], [12], [23], [27]. A reference for stochastic control is [36]. A different approach to the Linear Quadratic Gaussian problem can be found in [10], [34]. Computational aspects are considered in [30], while interesting examples of applications can be found in [24].

The Maximum Principle, and in general the material pertaining to first order variational methods, is extensively treated in [4], [8], [21], [32]; the specific issue of singular solutions is further investigated in [5], [6], [9], while many examples of minimum time problems can be found in [4] and [32].

Second order variational methods have been presented here by following the book [8] which is a useful reference, together with [9], [14], [19], [31] and [38], for the basic aspects of computational algorithms, also for those which restrict their attention to first order effects only.

As for the observation of the state of a linear system based on the knowledge of its inputs and outputs and some specific issues concerning eigenvalues assignment see [28] and [17]. The first statement of the so called internal model principle can be found in [11] and [13].

A wide collection of examples particularly suited for didactic purposes can be found in the Italian books [26] and [27].

[1] V.M. Alekseev, V.M. Tikhomirov, S.V. Fomin, *Optimal control*, Consultants Bureau, 1987.

[2] B.D.O. Anderson, J.B. Moore, *Linear optimal control*, Prentice-Hall, 1971.

[3] B.D.O. Anderson, J.B. Moore, *Optimal control. Linear quadratic methods*, Prentice-Hall, 1989.

[4] M.Athans, P.L. Falb, *Optimal control*, Mc Graw-Hill, 1966.

[5] D.J. Bell, D.H. Jacobson, *Singular optimal control problems*, Academic Press, 1975.

[6] P. Berhnard, *Commande optimale, decentralization et jeux dynamiques*, Dunod, 1976.

[7] S. Bittanti, A.J. Laub, J.C. Willems, (ed.), *The Riccati equation*, Springer Verlag, 1991.

[8] A.E. Bryson, Y.C. Ho, Applied optimal control, Hemisphere Publ. Co., 1975.

[9] R. Burlisch, D. Kraft, (ed.), *Computational optimal control*, Birkhauser, 1994.

[10] P.Colaneri, J.C. Geromel, A. Locatelli, *Control theory and Design. A RH_2/RH_∞ viewpoint*, Academic Press, 1997

[11] E.J. Davison, *The robust control of servomechanism problem for linear time-invariant multivariable systems*, IEEE Trans. on Automatic Control, vol. AC-21, pp. 25–34, 1976.

[12] J.C. Doyle, *Guaranteed margins for LQG regulators*, IEEE Trans. on Automatic Control, vol. AC-23, pp. 756–757, 1978.

[13] B.A. Francis, W.M. Wonham, *The internal model principle of control theory*, Automatica, vol. 12, pp. 457–465, 1976.

[14] J. Gregory, *Constrained optimization in the calculus of variations and optimal control theory*, Van Nostrand Reinhold, 1992.

[15] M.S. Grewal, A.P. Andrews, *Kalman filtering: theory and practice*, Prentice Hall, 1993.

[16] M.J. Grimble, M.A. Johnson, *Optimal control and stochastic estimation*, J. Wiley, 1988.

[17] M. Heyman, *Comments on pole assignment in multi-input controllable linear systems*, IEEE Trans. on Automatic Control, vol. AC-13, pp. 748–749, 1968.

[18] L.M. Hocking, *Optimal control: an introduction to the theory and applications*, Oxford University Press, 1991.

[19] R. Kalaba, K. Spingarn, *Control, identification and input optimization*, Plenum Press, 1982.

[20] R.E. Kalman, P.L. Falb, M.A. Arbib, *Topics in mathematical system theory*, Mc Graw-Hill, 1969.

[21] G. Knowles, *An introduction to applied optimal control*, Academic Press, 1981.

[22] H. Kwakernaak, R. Sivan, *Linear optimal control systems*, J. Wiley, 1972.

[23] P. Lancaster, L. Rodman, *Algebraic Riccati equations*, Oxford Science Publications, 1995.

[24] F.L. Lewis, *Applied optimal control and estimation: digital design and inplementation*, Prentice Hall, 1993.

[25] F.L. Lewis, V.L. Syrmos *Optimal control*, J. Wiley, 1995.

[26] A. Locatelli, *Elementi di controllo ottimo*, CLUP, 1987.

[27] A. Locatelli, *Raccolta di problemi di controllo ottimo*, Pitagora, 1989.

[28] D.G. Luenberger, *An introduction to observers*, IEEE Trans. on Automatic Control, vol. AC-16, pp. 596–602, 1971.

[29] J.M. Macieiowsky, *Multivariable feedback design*, Addison Wesley, 1989.

[30] V.M. Mehrmann, *The autonomous linear quadratic control problem: theory and numerical solutions*, Springer-Verlag, 1991.

[31] A. Miele, *Gradient algorithms for the optimization of dynamic systems* in C.T. Leondes (ed.), *Control and dynamic systems*, vol. 16, pp. 1–52, Academic Press, 1980.

[32] L.S. Pontryagin, V.G. Boltyanskii, R.V. Gamkrelidze, E.F. Mishchenko, *The mathematical theory of optimal processes*, Interscience Publ., 1962.

[33] A. Saberi, B.M. Chen, P. Sannuti, *Loop transfer recovery: analysis and design*, Springer Verlag, 1993.

[34] A. Saberi, P. Sannuti, B.M. Chen, *H_2 optimal control*, Prentice Hall, 1995.

[35] M.G. Safonov, M. Athans, *Gain and phase margin for multiloop LQG regulators*, IEEE Trans. on Automatic Control, vol. AC-22, pp. 173–179, 1977.

[36] R.F. Stengal, *Stochastic optimal control: theory and application*, J. Wiley, 1986.

[37] R.F. Stengal, *Optimal control and estimation*, Dover Publications, 1994.

[38] K.L. Teo, C.J. Goh, K.H. Wong, *A unified computational approach to optimal control problems*, Longman Scientific & Technical, 1991.

List of Algorithms, Assumptions, Corollaries, Definitions, Examples, Figures, Lemmas, Problems, Remarks, Tables, Theorems

Algorithms

6 **6.1**: 175

7 **7.1**: 233

A **A.1**: 250

Assumtions

6 **6.1**: 206

Corollaries

2 **2.1**: 16

Definitions

2 **2.1**: 10, **2.2**: 11, **2.3**: 11, **2.4**: 11, **2.5**: 11, **2.6**: 12

6 **6.1**: 149, **6.2**: 149

A **A.1**: 254, **A.2**: 254, **A.3**: 254, **A.4**: 254, **A.5**: 255, **A.6**: 255

Examples

1 **1.1**: 2, **1.2**: 3, **1.3**: 4, **1.4**: 4

2 **2.1**: 11, **2.2**: 13, **2.3**: 15, **2.4**: 16, **2.5**: 16, **2.6**: 17

3 **3.1**: 24, **3.2**: 25, **3.3**: 28, **3.4**: 29, **3.5**: 30, **3.6**: 32, **3.7**: 34, **3.8**: 35, **3.9**: 37, **3.10**: 39, **3.11**: 43, **3.12**: 46, **3.13**: 47, **3.14**: 47, **3.15**: 49, **3.16**: 50, **3.17**: 52, **3.18**: 57, **3.19**: 60, **3.20**: 61, **3.21**: 62, **3.22**: 66, **3.23**: 68, **3.24**: 70, **3.25**: 73, **3.26**: 74, **3.27**: 79, **3.28**: 79, **3.29**: 80, **3.30**: 81, **3.31**: 85, **3.32**: 85, **3.33**: 86, **3.34**: 88

4 **4.1**: 96, **4.2**: 98, **4.3**: 101, **4.4**: 103, **4.5**: 104, **4.6**: 107, **4.7**: 108, **4.8**: 111, **4.9**: 114, **4.10**: 117, **4.11**: 118, **4.12**: 120

5 **5.1**: 127, **5.2**: 128, **5.3**: 130, **5.4**: 131, **5.5**: 132, **5.6**: 132, **5.7**: 132, **5.8**: 134, **5.9**: 134, **5.10**: 134, **5.11**: 135, **5.12**: 136, **5.13**: 136, **5.14**: 136, **5.15**: 137, **5.16**: 138, **5.17**: 138, **5.18**: 138, **5.19**: 139, **5.20**: 139, **5.21**: 141, **5.22**: 142, **5.23**: 143

6 **6.1**: 150, **6.2**: 152, **6.3**: 154, **6.4**: 155, **6.5**: 155, **6.6**: 156, **6.7**: 157, **6.8**: 157, **6.9**: 159, **6.10**: 160, **6.11**: 163, **6.12**: 165, **6.13**: 166, **6.14**: 169, **6.15**: 170, **6.16**: 171, **6.17**: 172, **6.18**: 173, **6.19**: 178, **6.20**: 179, **6.21**: 181, **6.22**: 183, **6.23**: 184, **6.24**: 186, **6.25**: 188, **6.26**: 189, **6.27**: 191, **6.28**: 192, **6.29**: 197, **6.30**: 198, **6.31**: 201, **6.32**: 203, **6.33**: 207, **6.34**: 209, **6.35**: 210, **6.36**: 211, **6.37**: 215

7 **7.1**: 225, **7.2**: 225, **7.3**: 227, **7.4**: 229, **7.5**: 231, **7.6**: 234, **7.7**: 238, **7.8**: 241, **7.9**: 243, **7.10**: 245

A **A.1**: 250, **A.2**: 253

B **B.1**: 261, **B.2**: 262, **B.3**: 263, **B.4**: 269, **B.5**: 269, **B.6**: 271, **B.7**: 278

Figures

1 **1.1**: 2

2 **2.1**: 14, **2.2**: 17

3 **3.1**: 22, **3.2**: 24, **3.3**: 26, **3.4**: 26, **3.5**: 28, **3.6**: 33, **3.7**: 33, **3.8**: 35, **3.9**: 35, **3.10**: 36, **3.11**: 37, **3.12**: 38, **3.13**: 40, **3.14**: 44, **3.15**: 46, **3.16**: 47, **3.17**: 48, **3.18**: 53, **3.19**: 55, **3.20**: 57, **3.21**: 58, **3.22**: 59, **3.23**: 60, **3.24**: 62, **3.25**: 63, **3.26**: 64, **3.27**: 66, **3.28**: 67, **3.29**: 70, **3.30**: 71, **3.31**: 72, **3.32**: 73, **3.33**: 74, **3.34**: 75, **3.35**: 76, **3.36**: 77, **3.37**: 79, **3.38**: 80, **3.39**: 81

4 **4.1**: 97, **4.2**: 101, **4.3**: 107, **4.4**: 108, **4.5**: 111, **4.6**: 115, **4.7**: 117, **4.8**: 119, **4.9**: 121

5 **5.1**: 142

6 **6.1**: 150, **6.2**: 155, **6.3**: 157, **6.4**: 159, **6.5**: 162, **6.6**: 164, **6.7**: 167, **6.8**: 169, **6.9**: 171, **6.10**: 172, **6.11**: 172, **6.12**: 174, **6.13**: 182, **6.14**: 183, **6.15**: 185, **6.16**: 186, **6.17**: 188, **6.18**: 190, **6.19**: 192, **6.20**: 193, **6.21**: 195, **6.22**: 197, **6.23**: 198, **6.24**: 199, **6.25**: 202, **6.26**: 203, **6.27**: 204, **6.28**: 206, **6.29**: 207, **6.30**: 209, **6.31**: 210, **6.32**: 211, **6.33**: 212, **6.34**: 213, **6.35**: 214, **6.36**: 215, **6.37**: 216

7 **7.1**: 226, **7.2**: 228, **7.3**: 230, **7.4**: 232, **7.5**: 236, **7.6**: 239, **7.7**: 241, **7.8**: 243, **7.9**: 245

B **B.1**: 259, **B.2**: 263, **B.3**: 264, **B.4**: 265, **B.5**: 266, **B.6**: 267, **B.7**: 268, **B.8**: 269, **B.9**: 271, **B.10**: 272, **B.11**: 277, **B.12**: 278, **B.13**: 279

Lemmas

3 **3.1**: 49, **3.2**: 61, **3.3**: 71, **3.4**: 77

5 **5.1**: 128, **5.2**: 139

B **B.1**: 260

Problems

2 **2.1**: 10

3 **3.1**: 21, **3.2**: 40, **3.3**: 45, **3.4**: 69

4 **4.1**: 93, **4.2**: 112

6 **6.1**: 148

B **B.1**: 260, **B.2**: 267, **B.3**: 273

Remarks

3 **3.1**: 25, **3.2**: 25, **3.3**: 27, **3.4**: 29, **3.5**: 30, **3.6**: 31, **3.7**: 34, **3.8**: 36, **3.9**: 37, **3.10**: 44, **3.11**: 46, **3.12**: 50, **3.13**: 51, **3.14**: 53, **3.15**: 55, **3.16**: 58, **3.17**: 63, **3.18**: 63, **3.19**: 66, **3.20**: 85, **3.21**: 85

4 **4.1**: 92, **4.2**: 96, **4.3**: 96, **4.4**: 98, **4.5**: 100, **4.6**: 100, **4.7**: 101, **4.8**: 113, **4.9**: 116, **4.10**: 118, **4.11**: 118, **4.12**: 120

6 **6.1**: 154, **6.2**: 154, **6.3**: 154, **6.4**: 156, **6.5**: 158, **6.6**: 161, **6.7**: 162, **6.8**: 166, **6.9**: 168, **6.10**: 168, **6.11**: 173

7 **7.1**: 232

B **B.1**: 262, **B.2**: 262, **B.3**: 263, **B.4**: 264, **B.5**: 265, **B.6**: 268, **B.7**: 269, **B.8**: 271, **B.9**: 276, **B.10**: 276, **B.11**: 277, **B.12**: 277, **B.13**: 278

Tables

4 **4.1**: 98

5 **5.1**: 131, **5.2**: 133

6 **6.1**: 179, **6.2**: 180

Theorems

2 **2.1**: 12

3 **3.1**: 22, **3.2**: 40, **3.3**: 45, **3.4**: 48, **3.5**: 49, **3.6**: 61, **3.7**: 63, **3.8**: 69, **3.9**: 72, **3.10**: 74, **3.11**: 78, **3.12**: 79, **3.13**: 82, **3.14**: 86

4 **4.1**: 94, **4.2**: 99, **4.3**: 100, **4.4**: 102, **4.5**: 103, **4.6**: 104, **4.7**: 112, **4.8**: 115, **4.9**: 116, **4.10**: 120, **4.11**: 120

5 **5.1**: 126, **5.2**: 127, **5.3**: 129, **5.4**: 132, **5.5**: 133, **5.6**: 135, **5.7**: 137, **5.8**: 137, **5.9**: 138, **5.10**: 140

6 **6.1**: 151, **6.2**: 165, **6.3**: 201, **6.4**: 206, **6.5**: 208, **6.6**: 208, **6.7**: 208, **6.8**: 208, **6.9**: 209

7 **7.1**: 224, **7.2**: 227, **7.3**: 229, **7.4**: 230, **7.5**: 237, **7.6**: 240, **7.7**: 242, **7.8**: 244

A **A.1**: 251, **A.2**: 255, **A.3**: 255, **A.4**: 255, **A.5**: 256, **A.6**: 257, **A.7**: 257

B **B.1**: 261, **B.2**: 267, **B.3**: 270, **B.4**: 271, **B.5**: 273

Index

– A –

Algebraic Riccati equation, 45, 69, 103, 117
  decomposition, 50
  feedback connection, 132
    eigenvalues, 132
  hamiltonian matrix
    Z-invariant subspaces, 129
    definition, 127
    eigenvalues, 127, 128, 132, 139
  limit of $P(\rho)$, 77
  minimal solution, 48
  sign defined solution
    existence, 137
    uniqueness, 140
  solution
    number of, 130
    sign defined, 136
    vs. Z-invariant subspaces, 129
  stabilizing solution, 133
    computation, 141
    existence, 135, 137
    properties, 133, 138
    uniqueness, 140
Auxiliary system, 149

– C –

Canonical decomposition, 249
  computation, 250
Complex numbers
  symmetric set, 260
Constraints
  complex, 148
    global instantaneous equality, 181, 187
    global instantaneous inequality, 181, 196
    integral, 180, 184
    isolated instantaneous equality, 181, 190
    nonregular variety, 180, 181
  simple, 148

– D –

Detectability
  PBH test, 251
Differential Riccati equation, 22, 40, 69, 94, 99, 100, 102, 103, 112, 116, 224, 227, 229, 230, 237, 240, 242, 244
  hamiltonian matrix
    definition, 126
  solution, 126

– E –

Eigenvalues, 254
Eigenvalues assignment
  accessible state
    definition, 260
    number of solutions, 262, 263
    selection of Λ, 262
    solution, 261
  errors zeroing
    constant exogenous signals, 277

controller order, 277
definition, 273
internal model principle, 276
nonsquare systems, 278
rational Laplace transform, 276
solution, 273
inaccessible state
choice of Λ_R, 268
controller stability, 269
definition, 267
reduced order controller stability, 271
reduced order solution, 270, 271
solution, 267

– G –

Global methods
H-minimizing control, 11
admissible control, 10
hamiltonian function, 11
H-minimizing control, 11
regularity, 11
optimal control, 11
optimal control problem
definition, 10
sufficient conditions, 12, 16

– H –

Hamilton-Jacobi equation, 12
Hamiltonian matrix, 126, 127
Hamiltonian system, 149

– I –

Internal model principle, 276

– K –

Kalman filter
normal case, 100
Du term, 100
error/filter state
incorrelation, 101
infinite horizon, 102
meaning of $Pi(t)$, 100
normal time-invariant case
infinite horizon, 103
stability, 104
singular time-invariant case
finite horizon, 106
infinite horizon, 110

– L –

Linear-quadratic problem
definition, 21
finite horizon
Du term, 34
$\frac{1}{2}$ coefficient, 25
linear-quadratic index, 27
penalty on $\dot{u}$, 36
rectangular term, 29
sign of Q and S, 30
solution, 22
stochastic problem, 37
tracking problem, 31
uniqueness of solution, 25
infinite horizon, 39
definition, 40
solution, 40
uniqueness of solution, 44
optimal regulator
cheap control, 78, 79
control of equilibrium, 46
definition, 45
existence and stability, 63
existence of solution, 49
exogenous inputs, 63, 66
frequency domain index, 55
index choice, 85
inverse problem, 82, 85, 86
minimality of $\bar{P}$, 48
penalty on $\dot{u}$, 53
positiveness of $\bar{P}$, 61

robustness, 72, 74
solution, 45
stability, 61, 63
tracking problem, 51
optimal regulator with exponential stability
definition, 69
solution, 69
Linear-quadratic-gaussian control
definition, 112
finite horizon
index value, 113
solution, 112
infinite horizon
index value, 116, 118
robustness, 120
solution, 115
stability, 118
time-invariant case, 116
Linear-quadratic-gaussian estimation
definition, 93
normal case
correlated noises, 98
error variance, 96
meaning of β^o, 96
result, 94
solution, 99
Local sufficient conditions, 222, 224, 227, 229, 230

– **M** –
Matrix
hamiltonian, 126, 127
McMillan-Smith form, 253
positive definite
check, 254, 255
definition, 254
eigenvalues, 254
positive semidefinite
check, 254, 255
definition, 254
eigenvalues, 254
factorization, 255
Maximum Principle, 148
constancy of the hamiltonian function, 168
constrained final event, 162
constrained initial event, 162
constructive use, 154
derivation, 166
free final/initial state, 156
index function of the final event, 165
integral index, 151
limits, 170, 172, 173
limits of, 154
minimum time
existence of optimal control, 208, 209
extremal control, 206
feasible control, 208
feasible extremal control, 208
optimal control uniqueness, 208
piecewise constancy of optimal control, 206
orthogonality condition, 152
pathological problems, 154, 168, 169
singular arc, 200
singular control, 200
singular control component, 200
singular solution, 200
existence, 201
time-invariant problems, 158
time-varying problems, 161
transversality condition, 152
McMillan-Smith form, 253

– **N** –
Neighbouring optimal control, 222, 237, 240, 242, 244

– O –

Observability
  PBH test, 251
  of the pair (A, Q), 49
Optimal control problem
  attitude, 4
  characterization, 4
  linear-quadratic, 21
    finite horizon, 22
    infinite horizon, 39
  positioning, 3
  regulator, 45
  rendezvous, 2
  stabilization, 4
  tracking, 31
Orthogonality condition, 152

– P –

Parallelepiped, 208
PBH test, 251
Poles, 253
Polyhedron, 205

– Q –

Quadratic form, 254
  kernel, 254

– R –

Reachability
  PBH test, 251
  from a single input, 260
Regular variety, 149

– S –

Stabilizability
  PBH test, 251
State observer
  of order n, 264
  of order n-rank(C), 265
Stochastic system, 255
  expected value of x, 256
  quadratic functional
    finite horizon, 256
    infinite horizon, 257
  variance of x, 256
Switching times, 207
  number, 208

– T –

Transition matrix, 252
  properties, 252
Transposed system, 271
Transversality condition, 152

– V –

Variational methods
  auxiliary system, 149
  first order analysis, 147, 221
    necessary conditions, 151, 165
    strong perturbations, 221
  first order conditions
    computational algorithm, 173
  hamiltonian function, 149
  hamiltonian system, 149
  optimal control problem
    definition, 148
  second order analysis, 221
    computational algorithm, 232
    local sufficient conditions, 222, 224, 227, 229, 230
    neighbouring optimal control, 222, 237, 240, 242, 244
    weak perturbations, 221

– Z –

Zeros
  invariant, 254
  transmission, 254